Thus Spoke Galileo

Cover image: Galileo and Milton, modern marble statue by Cesare Aureli, located in the entrance hall of the Marconi Building, Physics Department, University of Rome "La Sapienza"

Thus Spoke Galileo

The Great Scientist's Ideas and Their Relevance to the Present Day

Andrea Frova

AND

Mariapiera Marenzana

Translated by

Jim McManus

in collaboration with the authors

OXFORD
UNIVERSITY PRESS

OXFORD
UNIVERSITY PRESS

Great Clarendon Street, Oxford OX2 6DP
United Kingdom
Oxford University Press is a department of the University of Oxford.
It furthers the University's objective of excellence in research, scholarship,
and education by publishing worldwide. Oxford is a registered trade mark of
Oxford University Press in the UK and in certain other countries

Translation of Parola di Galileo by Andrea Frova and Mariapiera Marenzana
originally published in Italian by R.C. S Libri SpA-M ilano 1998

First published 2006
First published in paperback 2011
Reprinted 2014

Published in the United States of America by Oxford University Press
198 Madison Avenue, New York, NY 10016, United States of America

British Library Cataloguing in Publication Data
Data available

Library of Congress Cataloging in Publication Data
Data available

ISBN 978-0-19-960682-5

To Eli and Lupi

Acknowledgements

The excerpts from translations of original Galilean texts used in this book were taken from the sources listed below. For copyrighted material the authors are indebted to Publishers as indicated:

- S. Drake, *Dialogue Concerning the Two Chief World Systems*, The Modern Library, New York, 2001 (by kind permission of the University of California Press).
- M.A. Finocchiaro, *The Galileo Affair: A Documentary History*, University of California Press, Berkeley, CA, 1989 (by kind permission of the University of California Press).
- M.A. Finocchiaro, *Galileo on the World Systems*, University of California Press, Berkeley, CA, 1997 (by kind permission of the University of California Press).
- S. Drake and O'Malley, *The Assayer*, in *Controversy on the Comets of 1618*, Philadelphia, PA, 1960 (out of print, in public domain).
- S. Drake, *Galileo at Work: His Scientific Biography*, University of Chicago Press, Chicago, IL and London, 1978 (by kind permission of the University of Chicago Press).
- G. de Santillana, *The Crime of Galileo*, University of Chicago Press, Chicago, IL, 1955 (by kind permission of the University of Chicago Press).
- A. Van Helden, *Sidereus Nuncius* or *The Sidereal Messenger*, University of Chicago Press, Chicago, IL, 1989 (by kind permission of the University of Chicago Press).
- H. Crew and A. de Salvio, *Two New Sciences*, Dover Publications, New York, 1954 (in public domain).

Contents

PART VI THE INVASION OF THE INFINITESIMALS

PART VII BODIES AND THE ENVIRONMENT

PART VIII FROM MATTER TO LIGHT

PART IX ALL THAT GLITTERS . . .

EPILOGUE

Foreword

The first Italian edition of the book *Parola di Galileo* appeared in 1998. Its authors, Andrea Frova and Mariapiera Marenzana, are a gifted husband and wife team who complement each other in their various study and writing projects. Andrea, with whom I have enjoyed a long professional and personal friendship, is professor of General Physics at the "La Sapienza" University of Rome ("Sapienza" may be translated into English as "Knowledge" or "Wisdom").

Andrea has published nearly 150 scientific articles in prominent international journals, mostly dealing with optical spectroscopy of semiconductors. Over the past two decades he has combined his scientific research with the writing of books, both textbooks and books for the general public. Among the latter I mention: *Perché accade ciò che accade*, 1995 (*Why things happen the way they happen*), *Fisica nella Musica*, 2001 (*Physics in Music*) and *Bravo Sebastian*, 1989, a delightful novelized biography of Bach. He has also authored a large number of non-technical articles and essays in newspapers and magazines. These often deal with political, scientific and cultural issues, and take a stand against unethical government policies at national and international level.

Mariapiera, too, has written essays and articles in journals and cultural magazines, frequently criticizing positions manifesting bigotry and superstition, which are quite pervasive in Italy. The reader of *Thus Spoke Galileo* will recognize her expert hand and welcome the interplay of Andrea's scientific expertise and Mariapiera's linguistic and humanistic skills–in particular regarding the use of archaic Italian in creating Galileo's apocryphal autobiography (the reader of the English version will find it more difficult to appreciate this point in spite of the commendable efforts made in translation). I personally, however, would have used the better-known Zarathustran term "spake" instead of "spoke" in the title of the book, in order to give a flavor of Galilean archaic language. As I interpret it, the title of the original version—*Parola di Galileo*—may be an allusion to Catholic church services, in which readings of the Gospel end with the words "parola di Dio" (which in English services would be "This is the word of God").

Thus Spoke Galileo starts with a challenging quiz composed of seventeen questions and subtitled "persisting misconceptions". These basically cover the field of classical mechanics without any use of mathematics. The answers are given at the end of the book. This quiz should prove useful to high school and college teachers who, rather than testing their students' common sense or mathematical ability, may wish to test their powers of logical reasoning.

After the quiz, and by way of introduction to the book, comes a "posthumous" self-portrait of Galileo. The term "posthumous", of course, is a semiotic indication that this self-portrait is a "forgery" (albeit one based largely on Galileo's private writings and letters). It conveys Galileo's gentle, though firm and purposeful, nature, his abhorrence of teachings based on authority and coercion. And also his bitterness at the accusations levelled against him, at his trial and at his forced abjuration.

After the "self-portrait" we have twenty-one chapters which discuss the central issues of each of Galileo's major works. Most of these chapters end with historical and mathematical notes. They contain a rich selection of Galilean original writings on Mechanics, Optics, Materials, Astronomy (and related theological questions). Chapters 4 and 5, e.g., deal with inertia and relativity, Chapter 6 with pendulums and musical harmony, Chapter 12 deals with the concept of infinitesimal, which was to be fully developed a few decades later by his English counterpart Isaac Newton (and also by the German Gottfried Leibniz). Chapter 17 marks the transition from Mechanics to Optics. In Chapter 20 dark clouds begin to amass above Galileo following his rejection of higher (i.e., ecclesiastic) authority, culminating in his dramatic trial, condemnation and abjuration. This is amply reported and analyzed in Chapter 22.

I believe *Thus Spoke Galileo* will make readers aware that Galileo's teaching and message are no less valid today than they were in his time[1]. And also that several of the many obstacles he had to battle against in order to make his voice heard have still not disappeared.

Manuel Cardona
Director Emeritus
Max Planck Institut für Festkörperforschung

Stuttgart, December 2005

[1] For instance, the Web of Science database (WoS) contains so-called "source journals", going back to 1900. It excludes, therefore, as source journals all of Galileo's publications except reprinted ones. Nonetheless, it lists over 2300 citations to Galileo, divided among *Hard Sciences*, *Social Sciences* and *Humanities*. Looking at the authors of these citations and the titles of the articles in which they appeared, one gets a sense of the hugely broad range of Galileo's impact.

Preface to the English Edition

Galileo is well known and highly esteemed in the English-speaking world, where many of his most eminent scholars are to be found and his major works have been admirably translated and commented upon. We think, however, that despite this extensive range of studies, this book can still fill a gap. To our knowledge, no anthological text exists in the English language which enables a reader to get an overview of Galileo's all-embracing interests and extremely wide spectrum of achievements—as they are described in the words of the scientist himself. The effort required to read some of his passages in his own words will without doubt be rewarded by satisfaction at being able to ascertain at first hand that Galileo was the true inventor of science, and that his approach to knowledge retains its validity today in a variety of fields, both within scientific research and without.

The educational value of the Galilean lesson makes us hope this book will be particularly welcomed by teachers and students. After all, it was mostly the young that Galileo had in mind when he decided to write in Italian rather than Latin—the language of science of his time—as he was concerned that, owing to a lack of information and poor self-knowledge, they might make wrong choices in their studies.

Galileo's original language may present problems on a first reading—it does so even for Italians today. This is a consequence not only of the centuries that have elapsed, but also of the extraordinary richness of his vocabulary, and still more of the complex structure of his sentences, which is perfectly suited to the variegated forms his reasoning assumes as it develops. It is this last point which makes the task of translating Galileo into English particularly demanding.

For these reasons we have not embarked here, with minor exceptions, upon new translations of the original Galilean texts themselves—indeed, we consider those already available to be of excellent quality. If they inevitably tend to result in some simplification of the Galilean discourse, they nonetheless often succeed in rendering the flavour of the original writing. We have relied, by permission of the publishers, on translations by Crew and de Salvio, Drake, Finocchiaro, and Van Helden—each of whom can be considered an authority in the field. At times the choice to adopt a given translation was dictated by whether or not it included passages which were present in the original Italian anthology (this was the case, for instance, with Finocchiaro's abridged translation of the *Dialogue*). Whatever, we trust that the reader will find some interest in comparing the different styles of translation, as each contributes to giving a sense of the richness and elegance of Galileo's writing.

We would like to thank Michael Fitzpatrick for his careful reading of the translated manuscript and for his helpful suggestions.

The translators
Andrea Frova, Mariapiera Marenzana, and Jim McManus

Rome, June 30, 2005

Preface

Infelice questo nostro clima, nel quale regna una fissa resoluzione di voler esterminare tutte le novità, in particulare nelle scienze, quasi che già si sia saputo ogni scibile

How wretched is this present climate of ours, in which there reigns a fixed resolution to exterminate all novelty, especially in the the sciences, as if all knowledge had already been acquired.

(Galileo, Letter to Diodati, December 18, 1635)

This book aims to provide a representative selection of Galileo's original writings, for the attention mainly of the young, but also of anyone to whom Galileo may be known only indirectly, and by way of stereotyped ideas and images that have been passed down through the centuries.

We make no claim to originality or deepening of the analysis regarding the historical and biographical facts of Galileo's life, for which we have relied on the works of authoritative scholars both from Italy and abroad. Rather we hope we have been able to shed light on Galileo's scientific discoveries and their subsequent development, and also on the logical mechanism by which his reasoning advanced—analyses, associations, extensions, returns, further clarifications—in a gradual, but relentless, building to the conclusion.

The excerpts from Galileo's writings offered to the reader, though few in comparison to his work as a whole, have been selected to give a comprehensive view of his manifold interests in both the scientific and cultural fields, of his prodigious curiosity for all natural phenomena, of his inexhaustible capacity for posing questions to himself and searching for answers by reasoning. And also of his exultation at making discoveries.

The anthological nature of the book foresees a selective reading of individual chapters according to interest, meaning that each argument had to be 'self-contained'. This has made repetition of certain concepts unavoidable, and we apologize for this to those who may wish to read the entire book from beginning to end.

The excerpts are accompanied by historical and scientific notes, and introduced by summaries outlining the main points of each argument. Footnotes are generally limited to those indispensable to the understanding of the text, as we believe that Galileo's elegant language, though extremely rich in vocabulary and characterized by long complex sentences, will be accessible to readers once they become familiar with it.[1]

Indeed, Galileo's language, when compared to the stiff and codified language of modern science, surprises us with its wealth of equivalent terms—whether verbs, adverbs, or nouns. Despite this, however, any seeking after ornamentation was alien to his style, which De Sanctis described as "built on things and thought alone, stripped of any pretentiousness or mannerism". The stylistic choices which Galileo continually made, though they were not influential as to the clarity of the message, appear to have been prompted by his musical and prosodic sensitivity, as well as by his extraordinary gift for moulding language to the demands of fluidity and naturalness. "*Il discorrere è come il correre*",[2] wrote Galileo—a statement that sums up his stylistic approach. As Italo Calvino pointed out in his *Lezioni americane*:[3] "style as a method of thinking and as literary taste: rapidity, agility of reasoning, economy of arguments, but also imaginative use of examples, are, for Galileo, essential requirements for thinking properly".

The literary value of Galileo's writings has been well expressed by Natalino Sapegno: "His work, in terms of richness of human

[1] This statement does not apply to the same extent to the English translation, where most of the difficulties of Galileo's archaic language have been removed.

[2] *Il discorrere è come il correre*—discussing is like running.

[3] English translation by Patrick Creagh, *Six Memos for the Next Millennium/the Charles Eliot Norton Lectures 1985–86* (Vintage International).

content and power of style, sets an example, showing the path for future Italian prose. It was a great literary and cultural event, the most important, perhaps, after Machiavelli and before Manzoni".

But why propose Galileo again today? The answer is that his lesson, even if we leave aside its scientific content, has lost none of its validity over the centuries (though it will be surprising to discover just how many pre-Galilean misconceptions survive at the beginning of the twenty-first century). The reading of these passages by Galileo cannot fail to stimulate a taste for observation, a desire to understand, pleasure at giving answers on the basis of experiment and logical deduction. Of this we are convinced, believing that as the third millennium gets under way amidst the decline of reason and unseemly revivals of irrationalism, there can be no more important lesson.

As for those matters about which Galileo was mistaken, they can prove as precious as the most ingenious of his achievements. On the one hand, they help us discern the paths his reasoning took, on the other hand, they teach us how complex and difficult are the mental processes that lead to knowledge.

To spend time living with Galileo has been a great and fortifying privilege for us—an extended journey through the highlands of the mind. It is our wish that the reader will share this experience.

We are grateful to many friends who have read parts of the manuscript and have made valuable comments, in particular to Gabriella Palli Baroni, Anna Maria Pica, Giorgio Salvini, Carlo Bernardini, Giuliano Toraldo di Francia, and especially to Egidio Longo. Thanks go to Anna Degrossi for her help in the electronic handling of the Galilean texts.

Andrea Frova and Mariapiera Marenzana

Rome, September 1, 1998

Notes

- *Edizione Nazionale* refers to the twenty-volume National Edition of Galileo's works by Antonio Favaro, published by Barbèra (Florence, 1890–1907), reprinted in the years 1964–66.

- The ordering of the chapters follows, in the main, a scientific progression—from the fall of bodies to the microscopic structure of matter—rather than the chronological order of Galileo's writings.
- The physics contents of the book should be accessible to senior high-school students. The mathematical notes at the end of each chapter are meant for readers who have a special interest in science, and are in no way indispensable for understanding the Galilean texts.

Persisting Misconceptions—A Test

As we will point out on occasion during the course of this book, many pre-Galilean scientific misconceptions—often about events which are commonplace in our everyday lives—have survived up to the present, and might well persist for centuries to come. In fact, with the exception of those who are professionally involved in physics and technology, even people with a good level of education fall into deadly traps when confronted with elementary physical phenomena. Phenomena for which the explanation, here at the beginning of the third millennium, after four centuries of scientific progress, really ought to be obvious to everybody.

It therefore seems worth presenting readers with a series of questions before they get going on the book, concerning problems for which Galileo and other scientists of his time found the correct answer, but which before them had been wrongly explained. For each question we invite readers to note down, after due consideration, their own view on the problem, and then to go back to these notes after they have finished reading the book. In this way they should be able to get a concrete idea of how much the Galilean revolution has (or has not!) influenced our common way of observing and explaining the events that happen around us.

NOTE: The chapter in which the problem raised by each question is dealt with is indicated in brackets. Brief answers to the questions are provided at the end of the book.

1. Is it really true that inside a container in which a high vacuum has been produced, a steel ball and a cotton flock will fall with exactly the same speed? (Chapter 3)

2. Let us throw upwards, at an equal speed, two balls of equal size, one made of cork and the other of lead: if we assume that friction against air can be ignored, which of the two balls will attain the greater height? (Chapter 13)

3. Let us now suspend these two balls on threads of equal length, and with a push cause them to make small-amplitude oscillations: again ignoring friction, which of the two balls will swing back and forth more rapidly? (Chapter 6)

4. What is the direction of the force acting on a body thrown vertically upwards, at the instant when it comes to a halt and is about to begin its descent? (Chapters 4, 5, 13)

5. Let us throw a stone horizontally at a given height and at the highest possible speed; then let us drop the same stone freely from the same height: assuming the effect of air friction to be negligible, in which of the two cases will the stone take the shorter time to hit the ground? (Chapter 5)

6. What physical phenomenon causes a flying arrow gradually to lose speed as it moves away from the bow from which it was shot? (Chapter 5)

7. Why is a force required to extract the piston from a syringe if the hole for the needle is stopped up? (Chapter 17)

8. What, in precise terms, is the physical mechanism that enables a drink to be sucked up through a straw? (Chapter 17)

9. Why does a weighing-scale not register the weight of the column of air in the atmosphere above it? (Chapter 14)

10. If a small inflated spherical balloon is released many metres below the surface of the sea, what shape will it assume as it moves upwards: elongated vertically, flattened, or still spherical? (Chapter 17)

11. As the earth rotates around its axis, why does the atmosphere, together with birds, clouds, and whatever else is suspended within it, not lose ground with respect to objects that are fixed firmly to the solid earth? (Chapter 10)

12. Suppose that an anchor is thrown from a moving boat, once in the direction of motion, and then again in the opposite direction, but in each case using identical force and at an identical angle with respect to the sea's surface. Ignoring air friction, say whether, at the moment it touches the water, the anchor will be equally distant from the boat in the two cases, or not. (Chapter 5)

13. If we play table-tennis inside a railway carriage travelling at a constant speed, why is there no difference between playing on the side of the table where we are facing the direction of movement and on that where we have our backs to it? (Chapter 4)

14. And why instead, if the locomotive pulling the carriage brakes suddenly, are table, players, and ball pushed forwards in the direction of movement? (Chapters 4 and 5)

15. Newton's law states that under the action of a force a body is subjected to an acceleration: why then does a parachute, though pulled downwards by the weight of the parachutist, descend at a constant, and low speed? (Chapter 3)

16. What is the precise reason why a lead sphere, sinking in a liquid, descends at a higher speed than a marble sphere of equal diameter? (Chapters 14 and 15)

17. Suppose that the surface of the moon, instead of being rugged and opaque, were highly polished, so that it reflected like a mirror: would it appear to us to shine more or less brightly than it does at present? (Chapter 8)

PART I

INTRODUCTION

1

Posthumous Self-Portrait of Galileo Galilei, Philosopher

Io credo nell'uomo, e questo vuol dire che credo nella sua ragione!

I believe in man, and this means that I believe in his reason

This self-portrait was never written by Galileo, of course, but it is what the authors of this book believe he would have sketched, had the times allowed him to express himself without formal respects and cautious attitudes. In compiling this autobiography, the authors have drawn widely upon letters written by Galileo and by his correspondents, as well as on biographies and documents of his time.

My name is Galileo Galilei. I was born in Pisa on the fifteenth day of February, in the year 1564, son of Vincenzo, a musician, and of Giulia Ammannati, a woman of sharp mind but, I cannot avoid saying it, rather malevolent and litigious. From her I must have taken my being easily moved to rage (though I calm down just as easily) and a taste for quarreling, or rather, to put it in the Tuscan Ambassador's words, a stubborn relish for driving monks mad, and for fighting with those by whom I cannot but be defeated. From her I certainly inherited my intolerance of stupidity.

Character and features

As to my character, it has been said that I was all afire in my opinions, because extreme passion I had inside, but not enough strength and prudence in controlling it, and that my bold and resolute speaking much too often could not abstain from stinging. My appearance, it has been said, was serious, but more jovial in old age; bright the eyes and white the complexion, while the hair tended to the reddish. I was of rather tall stature, of strong and well-proportioned body, of robust and healthy appearance. Nevertheless, from forty years of age through to my very last day I was afflicted by very severe or stabbing pains that disturbed me cruelly in different parts of my body whenever the weather changed, and several times I was assailed by grave and perilous illnesses. Often I felt so weakened, owing to fatigues and travails of the spirit as much as of the body, that I ended up in a state of great weariness.

Galileo as an old man

Though eager for glory, I was never ambitious for vulgar honors, but only for that fame which could distinguish me from ordinary people. No vice was more despicable to me than lying, perhaps because, thanks to the mathematical sciences, I knew the beauty of truth so well.

I had more hatred for avarice than for prodigality, and always did my best to succor in their needs my early-widowed mother, my sisters (whom I granted dowries so that they could marry), my brother

Michelangelo, and later his widow and their unlucky children, my sister Virginia's son Vincenzo, whose rent I paid and to whom I warranted a monthly sum of money for many years; and also his sisters, whose dowries I provided when they became nuns. I spent liberally in hosting and honoring foreigners and in helping the poor, especially if they excelled in some art or profession, at times offering them hospitality in my home.

Means and family

In Padua, where I spent the best eighteen years of all my life, I had to toil not a little to provide for the ever-pressing demands of my relatives in Tuscany and also to support my own family that, meanwhile, I had begun. To augment my meager salary as a professor I used to lecture a great many students privately, also foreigners, and to some of them I offered board and lodging. In the small workshop in my home I had my technician Marco Antonio Mazzoleni build astronomical and geometrical instruments for me, that I would then sell. Some income I had from the horoscopes I cast and sold to those unable to find comfort and hope elsewhere. This activity led to me being denounced for heresy, abetting which was the fact that I had a concubine, Marina Gamba, and that I was not in the habit of attending Mass.

Of those years, what I remember best are Marina's silences and her smile. She bore me three children, Virginia, Livia, and Vincenzo. She took care of me, of them, of our large house and of our guests, with industry and joy, never asking for anything. No reproach nor protest on her part, not even when I returned to Tuscany taking our still very young children with me. I found her a husband, I sent her money on several occasions. I wish now, when I recall her big dark eyes and her so firmly sealed lips, that she had spoken, that she had wept.

When Virginia and Livia reached the age of sixteen, nought could I do but consign them to the convent. As "Born of fornication" and of an "unknown father" was written on their birth certificates, to marry them off properly I should have dowered them in a manner I could not afford. Else I should have opposed the obtuse cruelty of the times with that freedom and courage which in other circumstances I found myself capable of. Livia manifested her insufferance for her monastical condition by developing a sour character, while Virginia, sister Maria Celeste, lived through it with resigned sweetness, altruism, and a devoted affection towards her father that was always of great comfort to me.

This daughter of mine, so dear to me on account of her nobility of mind, her wisdom and goodness, was snatched from life when she was only thirty-three years of age, leaving me in loath of myself and in extreme affliction. Thus was I deprived, during the dark years of my internment, of the sweet comfort of talking to her. Out of delicate reserve she would only ever speak little of herself, but she smoothed tensions in family relationships with prudence. And she used to read my books and speak of them with me. I was to receive her gentle letters, the little gifts she would send to me, the citron and rosemary flower jams, the rose bloomed in December, no more. Nor could I fix for her a window frame, nor repair a clock—jobs more suited to a carpenter than to a philosopher, as she used to say, but very dear to me because they were testimony to my paternal affection.

Blindness

In my advanced old age I became blind, and it was not easy to accustom myself to so immense a transmutation, which gave rise to an extraordinary metamorphosis of thoughts and conceptions in my mind. The sky, which my wonderful observations had a thousand and more times widened beyond what scholars of all centuries past had seen, became shrunk to a space no larger than that occupied by my body. I had to rely on the eyes and the pen of other people, I could no longer perform experiments, nor tend my garden, prune and bind the vines—operations that were to me at once a pastime and an opportunity for philosophising. The one consolation in my wretchedness was the thought that, amongst all Adam's children, nobody had ever seen more than me.

Even in the darkness I kept speculating on one or another effect of nature. But I never succeeded, as I would have desired, in quelling my restless mind, which never could cease from keeping on grinding away; an agitation that harmed me greatly as in the last years it kept me in almost perpetual wakefulness.

Sources of pleasure and delight

Among my most pleasing entertainments was the practice of music, touching keys and playing the lute, at times competing with the best professors in Florence and Pisa, and restoring my spirits with musical harmonies, by no means less admirable than those hidden in the phenomena of Nature. Great pleasure I had from drawing, and from the friendship of renowned painters, like Bronzino, Cigoli, Empoli, Passignano, and young Artemisia Gentileschi. I loved good literature, the invention of concepts, and their exposition. I knew by heart very many poems, in particular almost all of *Orlando Furioso* by the divine Ariosto, greatest among all Tuscan and Latin poets. Even today I cannot stand him being compared to that pedant Tasso, as I feel the same difference between the two that my palate senses eating cucumbers after having tasted savorous melons. Moreover, Tasso dealt just with words, whilst Ariosto dealt with things; much better Ruzante's wit and humor, with his free inventions and the argument, as dear to him as to myself, about the difference between life in the town and in the country.

Nowhere could I find better relief for the passions of the soul, nor better medicine for the ailments of the body, than in the open and healthy air of the countryside, far from the great clamor of the city, which in a sense is the prison of speculative minds. It was in the quiet of my country house that for thirty years I could devote myself to my studies and observe the phenomena of Nature that, no matter how minute and unimportant they may appear, must never be scorned by the philosopher. Because the operations of Nature are all equally worthy of being marvelled at; and also because, even from common things, sometimes seemingly worthless, it is possible to draw very intriguing and novel notions, and quite often remote from anything that might have been imagined.

Though by nature inclined to solitude, I always drew comfort and joy from the company of my friends, who used to visit me daily and with whom I would often attend banquets. I loved gracious conversation and being praised for my eloquence, both in serious discussions, for my profound sayings and concepts, and in light-hearted talk, where I did not lack wit and spicy interjection. And I also liked to joke about the most serious and important of matters, which sometimes are no less ridiculous than anything else.

In eating I was moderate, and especially in drinking. Yet I loved to taste different exquisite wines, *ciliegiolo*, *chiarello*, the sweetish *razzese*, the piquant and sweet *bruschetto*, which were also delivered to me from the cellar of the Grand Duke; and to crunch *cantucci*[4] with my friends in the tranquility of my house. I valued good cheer as the best means of preserving health and life.

Studies and activities

My earliest education I received at the convent of St. Mary of Vallombrosa near Florence. In 1581, my father decided to enroll me in the faculty of Medicine at the University of Pisa, where, instead, I became passionate about the investigation of natural phenomena. To Ostilio Ricci I am indebted for introducing me to the world of mathematics and to the works of Archimedes. It was for this reason that I never obtained any academic title. Nevertheless, when not yet twenty years of age I discovered the isochronism of the pendulum and shortly afterwards I invented the hydrostatic balance for the measurement of the density of substances.

Returning to Florence, I spent four years with my family, having no precise occupation, but deepening my studies in science and in literature. The theorems I conceived on the center of gravity of bodies earned me the esteem of Guidubaldo Del Monte and of the renowned astronomer Father Clavius. In the year 1589, the University of Pisa conferred upon me the chair of mathematics, not a very important one and poorly paid. In this period I embarked upon my first studies on motion, inspired by the work of Giovanni Battista Benedetti.

In the decrepit world of the university my juvenile intolerance had to clash against so much hypocrisy and conceit! At this time I wrote

[4] *cantucci*: typical Tuscan crunchy cooky enriched with almonds, to be dipped into a glass of sweet wine. Still very popular today.

a poem against the wearing of the robe. In it I defended the position that it would be better to go around in the nude, since it is clothing, among many other inconveniences, that makes subjects different from masters. Nothing worse than the robe, as it gets in your way, encumbers and hinders you, and suits only those who do everything slowly, not liking to toil, that is to say friars or some fat priest; or those who deem a man more talented or more learned according to whether he wears raw wool or velvet clothes. To me men are made like flasks: some of them have little or nothing around them, but then they are filled with excellent wine; others are elegantly dressed, but they contain just wind or sweet-smelling water, and all they are fit for is pissing into.

The sudden death of my beloved father, and the consequent need for me to earn more in order to support my family, in combination with the hostility of the academic world of Pisa, forced me to look for a position elsewhere. Thanks to the generous recommendation of Guidubaldo Del Monte, I obtained an appointment to the Chair of Mathematics at the University of Padua, and there I moved in December of the year 1592.

What liveliness of minds, what impassioned, free discussions there were to be found in the industrious and vital climate of that university! And how much sharp and unprejudiced conversation in the noble palaces of Venice that I had begun visiting! No less pleasure and opportunity to speculate did I get from frequenting the famous Venice arsenal, where I learned a great deal about mechanics by observing and talking to the craftsmen, quite a few of whom were highly skilled and capable of very fine discourses.

I became a friend, in those years, of Paolo Sarpi, who is too illustrious for me to talk about him here; of the Aristotelian philosopher Cesare Cremonini, later to be tried for atheism by the Inquisition (the inquiry was extended to me, too); of Giovanfrancesco Sagredo, a Venetian nobleman whom I later had pleasure in casting as one of the characters in my *Dialogue*; and of my most devoted pupil, Benedetto Castelli.

At the University I taught the Ptolemaic system in Latin and also architecture and military fortifications in private lectures. I deepened my studies on motion, patented a pump for lifting water, improved and built an instrument which I called the "geometric and military compass", built magnets, and invented a device for measuring temperature.

In the meanwhile I studied the Copernican system. These new ideas, being in contradiction with the way things appeared, forced me to put before my senses that which my reasoning judged contrary to them. How would obstinate or slow brains, unimpassioned minds, have sustained the arduous battle against inveterate habits of thought that I, step by step, was beginning to wage inside myself? One day, with clarity comprehensible to all, I would have given proofs incontrovertible even to the most stubbornly persuaded, as sensible experience is what very clearly draws final conclusions, and settles the account to those who do not want to, or cannot, make use of reason.

Copernicus and the Bible

In the cold and clear winter nights of the year 1609, to the sky I turned the glass I had greatly improved and, first among men, observed through it the wrinkled face of the moon, the clusters of stars, the odd appearance of Saturn; I discovered the satellites of Jupiter and also the phases of Venus, similar to those of the moon.

This was when I became persuaded, with not a doubt remaining, of the validity of the theory of Copernicus, and of the fallacy and inconclusive reasoning of many of the doctrines that for centuries had been taught in schools. Then the propositions of the Bible concerning the motion of the sun around an immobile earth appeared to me obsolete and rendered ridiculous by the new scientific knowledge. Eager to unveil to the world this ancient deceit, thoroughly convinced that the good of mankind resides in the progress of knowledge, and that those who love truth—upon discovering that what they had believed in was mistaken—would thank the man who was showing them what was true, I decided to do my utmost to give the widest diffusion to the Copernican theory.

The time had come for me to return to Florence. My fixed intention was to attain leisure enough and peace so that I could bring to an end a great work on the two world systems, Ptolemaic and Copernican.

Public and private lectures and the students I gave lodging to in my home—from whom I had to draw the maintenance of my family—hindered and delayed my studies. I needed to live free from such burdens, and I longed for spare time more than for gold. But it is not customary to obtain a salary from a Republic, no matter how splendid and generous it may be, without serving the public. The Senate of Venice, having appreciated the use of my glass—or *perspicillo*, or binoculars, or telescope, as some wished to name it—for practical purposes increased my salary to 1000 florins a year; but it could not relieve me of the appointments while leaving to me the emoluments. The granting of such advantage was only to be hoped for from an absolute monarch.

So with joy I accepted the invitation from the Grand Duke of Tuscany Cosimo II, who had been a pupil of mine, to become First Mathematician of the University of Pisa and his Philosopher, with no particular obligation. My decision was cause for resentment and surprise to some, for sorrow to others. I was judged imprudent to leave the freedom I enjoyed in Padua, where I had to serve only myself, to face the infinite and incomprehensible fortuities of a world (the Court and the Curia, I mean) governed by the posturings of ambitious and envious men.

Clash with the Church

I ignored the warnings, being convinced that only in that world, rather than in the enclosed space of the Republic of Venice which, though free inside her boundaries, was isolated in Italy by the hostility of Rome, would I have been able to give wide resonance to my discoveries. I intended to write in the Italian language, rather than in Latin, because it was my wish that truth be received by everybody and in particular by the young; in fact many young people are engaged in studying disciplines unsuitable to them, while others, who would have an aptitude for those disciplines, are kept away from them. I wanted everybody to know that Nature, just as she has given us the eyes to see her works, has also given us the brain to comprehend them. My vanity in wishing to be known and praised everywhere was not extraneous to this decision of mine.

It was my intention to acquaint the Church with the admirable new truths. That same Church which, as the depositary of knowledge, was the vigilant and firm judge of everything that was written in Italy at the time: outside of the Church there was but silence, either by choice, or by imposition. The muzzle that locked Giordano Bruno's mouth shut as he was being dragged, naked, to the stake, was a very significant demonstration to me of the intentions of the Curia. No more words issued from Francesco Pucci's lips once his head had dropped into the basket. Only Tommaso Campanella was intent on breaking the silence, as he continued writing in the confined space of his prison cell, where he was to spend twenty-seven years of his life.

Yet—you will agree with me on this—what effective support could have come from the Church! If the earth moves *de facto*, we cannot change Nature and make the earth stand still: he who follows concrete reasoning follows a guide that does not lead astray. I hoped against hope that the strength of well-founded arguments—once initial resistance to what was new had been overcome—would have prevailed over undemonstrated and unnecessary propositions, whose only efficacy consisted in their being inveterate in the minds of men. That was, perhaps, my most serious mistake: a mistake I would certainly repeat should I have to walk the path of my life again, because I deem reason the only appropriate guide to lead man out of obscurity and to quiet his mind.

In the spring of the year 1611 I went to Rome, where I was admitted as a member to the *Accademia dei Lincei* (which afterwards was always to be of great encouragement and help to me) by its founder, the nobleman Federico Cesi, a young man of passionate and bright talent. But there was more: with great courtesy I was received by Pope Paul V, and also by the Jesuit Fathers, who had made repeated observations of the new Medicean planets that agreed exactly with mine. At the time, I ignored the fact that powerful Cardinal Bellarmine was beginning to become suspicions about the consequences of my discoveries for the traditional view of the world, the one held by the Church.

In the years immediately following I studied bodies floating on water, and gave demonstrations on sun spots. These provided far better evidence than the childish explanations of that malicious great ass, Father Scheiner, who ever sought to deprive me of my merits, and was always envious of me. Moreover, I also wrote some letters on science and faith, in order that the Church might accept the new evidence more readily, and thereby not place its authority in jeopardy by firmly holding principles which one day soon, if not by me then by others, would most clearly and conclusively be shown to be false.

But two Dominican friars declared some of my propositions "suspect or rash", and "improper" certain ways of mine of talking about the Holy Scriptures, especially my statement that, in disputes about the effects of Nature, the Scriptures should be put in the last place. These Dominicans were put up to it by malevolent instigators and out of extreme ignorance, which is the mother of malice, envy, rage, and of all other wicked and ugly vices and sins. Full of holy zeal, they submitted one of these letters of mine to the Holy Office, in order that the necessary remedies might be applied.

Though gravely ill, I did not despair that I could overcome even this difficulty, if I were in a place where I could avail myself of my tongue rather than my pen. Therefore, as soon as I recovered I went to Rome, afeared that an adverse decision might be taken upon the urging of certain malicious men who understood nothing of these subjects. All operations in Rome, especially for a foreigner, were lengthy and laborious, but the firm hope I had of succeeding in a great enterprise—that the Church adhere to Copernicus' opinion and that I depart with my reputation increased threefold—induced me to bear any fatigue patiently.

Saint Robert Bellarmine

Yet hope has very lazy wings, while fortune has the swiftest of ones. Rome was not the place to come to and dispute about the moon, nor, in that century, in which to defend or present new doctrines. Many Jesuits whom I knew to be, in secret, of my own opinion, kept silent—foolish sheep wrapped up in equivocations! Others just did not care much about the matter. And in February 1616, the Holy Office condemned the proposition regarding the motion of the earth around the sun and prohibited the books teaching this suspect doctrine. How many calumnies, frauds, stratagems, and deceits were used to dazzle the superiors and blind their eyes!

I was next summoned by Cardinal Bellarmine, who admonished me not to hold and not to defend the suspect opinion. What arrogance, what hypocrisy in his words! I felt sorry for him, obliged by the will of the Pope to offend his own intelligence. So I kept silent. But, as to what mattered, I did not budge an inch. After all, if my enemies had not dragged me into it, I could not have cared less about the entire affair.

When I returned to Florence, since it seemed absolutely clear to me that it was in my interest to obey and trust the decisions of the superiors, as if they possessed higher knowledge to which my low mind could not reach, I made a show that I considered as poetry, or a dream, certain fancies of mine, certain chimeras going around in my imagination, such as the notes on the ebb and flow of the sea I had written in January in Rome. And I stuck to caution for some time. I studied the periods of Jupiter's satellites with the purpose of finding a reliable method for determining longitude, that I hoped would be useful for fixing the course of ships. Regarding this I entered into negotiations—unhappily without success—with the King of Spain and the government of the Low Countries.

The dispute with Father Grassi

Later I proved that the opinion of the Jesuit Father Orazio Grassi, concerning the nature of comets, was altogether vain and false. This I did by means of a book, *Discourse on comets*, which I persuaded a student of mine, Guiducci, to write. I was wrong in maintaining that comets are only imaginary bodies (that is, optical illusions), yet still today do I defend the principles I adhered to in that dispute, which guided me in writing the pages of the *Assayer*. This work was intended to raise doubts and bring some light into the search for truth. However it earned me, in addition to the esteem of men of lively intellect, the hatred of that gross buffoon Father Grassi whom, I declared openly, lacked logical skills, founding his arguments on nothing more than distinctions, distortions, or other circumlocutions.

Many years later Grassi said that I had been the cause of my own downfall, being so enamored of my own intelligence and having no esteem for the intelligence of others. It is most certainly true that our reputation begins with ourselves, and that anyone who wishes to be valued must be the first to value himself. But it is equally true that to draw ungrateful men out of ignorance is a pain indeed! The venom of that scorpion—who insinuated even that my atomistic idea of matter was repugnant to the real presence of Christ in the Eucharist—was without doubt, along with other snares laid for me by envy and diabolical malice, among the causes of my misfortune.

The election, in the year 1623, of Cardinal Maffeo Barberini as the new Pope Urban VIII gladdened me greatly. Previously, he had demonstrated so high an opinion of my person as to dedicate one of his poems to me, and he was a supporter and generous patron of the arts. With him my hope, by then completely buried, that my suspect ideas might be called back from their long exile, became reawakened. To his most benign protection I turned, and in 1624 I went to Rome to pay homage at his feet. However, it immediately became clear that the Pope had no intention of removing the obstacles. In fact, he limited himself to saying that, since the Holy Church had not condemned the Copernican theory as heretical, but only as rash, there would be no need to condemn it in the future. All the more because there was no fear of anybody ever being able to prove the theory unequivocally true.

I cannot deny that the Pope's words were little to my liking, and yet I did not consider them an obstacle too difficult to surmount. I decided, therefore, to proceed with caution, but with renewed efforts on the path that leads to the discovery of truth. In the years that followed I fashioned a small glass for seeing the smallest objects closely and I went deeper into the investigation of magnets. And above all I embarked, with refound energy, upon writings to show heretics who mock us Catholics because we remain in the certainty of the ancient Truth taught us by the sacred authors, that this is not because we are ignorant about natural philosophy, or because we have not understood the rational arguments they have. And also to show them that they can consider us men of steadfast opinions, not as people that are blind or ignorant about human disciplines.

I wanted to solve age-old problems using well-founded arguments, to the advantage of all lovers of truth. I wanted to persuade the Church that the gravest error for science, and, above all, for the Church itself, was to place faith ahead of all the reasoning and sensible experiences that astronomers and philosophers had produced with overwhelming evidence. And this I wished to do using due caution, since, in the view of my friends, one could never be too cautious. If the Church had accepted my suggestions, the Holy Scriptures, and She Herself, would have benefited greatly through the course of history.

Bigger misfortunes

Trusting in the new Pope, I felt I was allowed to expound the ideas in favor of Copernicus, and I developed them in the extended work I had been intending to write since leaving Padua, the *Dialogue Concerning the Two Chief World Systems, Ptolemaic and Copernican*. This work was printed in Florence in February 1632, with a preface agreed upon by the Florentine Inquisitor, and it marked the beginning of my gravest misfortunes.

Soon afterwards there were rumors that irate theologians were seeking to ban the *Dialogue* and news came that the Pope was wrathful and ill-disposed towards me. I and my friends had foreseen that the *Dialogue* would stir up controversy, because this, it appears, is what commonly happens with new doctrines that distance themselves from customary and inveterate opinions. Yet that the hatred against me and my writings should have been so powerful as to impress upon the most holy minds of the superiors the idea that this book of mine was not worthy of being brought to light was truly unexpected, and it still seems impossible to me.

The order arrived that the book was to be suspended until corrected, and in the end I was summoned to Rome by the Congregation of the Holy Office. I tarried, and the fact that I did not obey the order promptly was judged badly. I was told that a Commissioner would be sent to fetch me and take me to jail, dragged in irons if need be, because I had taken advantage of the benignity of the Congregation.

Heeding this strict injunction—imposed on me without regard for the poor state of my health, nor for the cold winter season, my old age and the plague that still had not abated—I made my will and left for Rome, where I arrived on February 13, 1633. Of solace to me was the purity of my conscience, and this persuaded me that it should not be difficult to prove my innocence.

Yet I cannot deny that the injunction that I must present myself before the Tribunal of the Inquisition was of great affliction to me, when I reflected that the fruits of all my study and toil over many years had transmuted into serious injury to my reputation. This tormented me to the point that I came to detest all the time I had consumed upon those studies, aspiring to depart from the trite, conventional path of scholars. I was tempted, even, to suppress my books myself and set them afire. Nobody seemed disposed to believe that I had written these books for the benefit of the Church, too. My intention had been to help those charged with making decisions about disciplines that by their nature are difficult to grasp, to understand, with less effort and waste of time, in which direction truth lies, so that they could bring the sense of the Holy Scriptures to accord with that.

What was it that impelled the Church to persecute me? What was it that made Her judge my hapless *Dialogue* execrable and more pernicious than the writings of Luther and Calvin? Was it fear of being swept away by the collapse of old truths? Or fear of losing control over a flock that was beginning to think? Or obtuseness and envy on the part of the learned Jesuit Fathers? Or was it, perhaps, just that same myopia which, through the centuries, seems always to have clouded the Church's choices?

During the course of the trial snares were laid for me, with the result that instead of defending my opinions, as I would have wanted, I followed the counsel of those who urged me to put an end to the affair as quickly as possible, to submit to anything my judges wanted me to believe regarding the motion of the earth. They could do whatever they wanted, I was utterly in their hands! I was judged vehemently suspected of heresy, and I was condemned to prison for life and to a salutary penance, and my *Dialogue* was proscribed. After listening to the sentence, dressed in the white robe of the penitent, I read my abjuration. Hatred, impiety, fraud, and falsehood had prevailed. I had been obliged to deny what is clearly written in the great Book of Nature, what is evident to anybody who has eyes to see and ears to hear, what the Bible itself would openly affirm if it were rewritten now. The date was June 22, 1633.

I held out no hope whatsoever for any relief because I had committed no crime. If I had been guilty I might have hoped to obtain mercy and pardon, because faults are a matter over which a sovereign can grant grace and leniency. Instead, in the case of a man innocently condemned it is more fitting for the sovereign to maintain rigor, to show that he has acted with justice. The only concession granted to me was to exchange, for the prison I was assigned to in Rome, the narrow confines of my small house in Arcetri, with the strictest of prohibitions against my going into Florence, having conversations or meetings with many friends at a time, or having them as guests.

There was nothing I could do save curse, in my heart, the authors of my disgrace for centuries to come, keep silent, go back to my beloved studies, and say nothing about my many sorrows. Foremost among these, and reason for inconsolable grief, was the loss of Virginia, which I had to suffer in what remained of my tormented life.

In my decaying old age the wrongs and injustices that had been railed against me by diabolical malice and wicked intent did not afflict me. Quite the contrary, the enormity of the affront was a relief to me, it was a kind of revenge, because the shame had rebounded upon my persecutors.

Bitter reflections

The hypocrisy and reticence I had resolved to practice in order to protect my studies when dealing with men of power and leaders of the Church, had proved futile. And the hiding of my free thoughts and any convictions that might be held as heresies, also had proved futile. Futile it had been to behave like a humble sheep of Saint Peter's flock. It had been wasted effort. Perhaps this was because, at times, astonished by what I had discovered and wishing to share it with others, or struck by somebody or other's idiocy, I had cast aside the caution that it behooved me to use in the circumstances.

Yet I must deeply regret the caution that I did show. In fact there were certain gentlemen, and others who are alive today, who have produced my caution as evidence of what I consider to be the greatest absurdity: that I am one of those who need dogmas. *I*, who had eyes in my head and in my mind; *I*, who always reasoned, drew conclusions, and wrote under the sign of sensible experience and grounded demonstrations; *I*, who if I had any respect for revealed religion, it was only for its role in history and its power over people.

I never wanted to inquire into what God *could be* or *could do*, but only what He *had done*—the Book of Nature, written in mathematical characters, open perpetually before our eyes. And I always enjoyed the simplicity and the easiness of Nature, which does not undertake to do what cannot be done, does not perform by many means what it can by few, and accomplishes with ease what for us is difficult to understand. I always deemed looking for the essence to be no less impossible a task, and no less vain a toil, as regards elementary and nearby substances, than it is as regards the most distant and heavenly ones. And that is that.

Pupils

In the bitterness of my persecutions, the knowledge that I had left traces of my passage in this world and, even more, that I had bred disciples of bright talent and clear mind, was of comfort to me. I wish to mention, among the many, Cavalieri, Torricelli, Viviani, Castelli, who carried on my work eminently well, but, above all, were capable of that freedom of thought that is essential for progress in the sciences and the attainment of conclusive demonstrations. Never were they content with an Authority, or assured of it, whether it be Aristotle, Galileo, the Scriptures, or whoever, and they always opposed to it beautiful and strong considerations, even going so far, in some cases, as to produce new evidence and reasoning against my own results. And for a teacher this, among all possible accomplishments, is the noblest and most gratifying.

Detestable people

I felt repugnance for all those people who:

— for their own advantage, or because of their weak brains, fight against those who introduce novelties
— to better enslave those who depend on them, nourish their ignorance and superstition
— out of vested interest or simple-mindedness, foment the sense of mystery in others
— out of contempt for human rationality, uphold dogma
— out of arrogance, want to set rules for minds and for the arguments they engage with
— out of stupidity, if you dispute a nonsense of theirs, come up with a bigger one
— out of moral baseness, despite knowing the truth, call it a lie
— out of opportunism, pander to men of power
— out of vanity or mediocrity, usurp the work of others
— out of obtuseness, call themselves philosophers, while they are no more than chuckers-together of tales
— out of conceit or laziness, speak obscurely
— out of cowardice, call weakness anything that pleases the body
— for various reasons are made like drums, which when viewed from the front, seem round, and when viewed from the side, square.

Self-criticism

May I call myself free from such faults? Certainly not free from having usurped the work of others, though out of vanity rather than mediocrity. I borrowed the best of Benedetti's ideas, like his admirable argument on the equal free fall velocity of bodies of different weight, and his sharp intuitions about the principle of inertia. I appropriated the theses of my dear friend Guidubaldo Del Monte, from whom I derived, without acknowledging it, the reasoning on the strength of ropes. And I also drew inspiration from Francesco Buonamici's *De motu*, and from Giordano Bruno's writings—may his restless soul sleep in peace. I failed to honor his memory, neither did I dare bring succor to Tommaso Campanella, who wrote in my support even from his prison cell. Yet, this I say, and I have a right to say it: could I have so exposed myself, seeing how gravely at risk I was? Was this a fault? Please favor me with an answer.

In defence of my abjuration

I submitted to the powerful, it is true, and did so to the extent of denying the fulcral ideas of my work. But here I would like you to take note: I am certain I acted not so much to my own advantage—even if I always cared for the integrity of my body—as to the advantage of knowledge.

Nevertheless, in the centuries to come there will be those who wish to sentence me again and on other counts. They will say, I am sure, that in abjuring I sacrificed truth to the violence of the powerful. They will say that I was driven by the idea that the sole obligation of science is to nourish itself, and even that I reduced myself to cultivating science as if it were a vice, in silence, stricken with remorse. And they will also infer that, had I immolated myself, professors of the demonstrative sciences would not have been able to evade their solemn duty to use the sciences exclusively to the benefit of the whole of humankind.[5]

Anyone who, inspired by bad feelings towards me, should suggest fancies such as these would be committing an error that offends me greatly. The matter just does not stand in this way, in fact it is quite the opposite: if it is true that I drew from science the greatest pleasures

[5] This is a reference to Bertolt Brecht's point of view, as expressed in his play *Life of Galileo*.

in life, it is also true that I took science out of the dead libraries and of the academies of the savants and, by writing in my Florentine language, I extended it to all. And this I did because, I say it again, I always greatly trusted in science for a better future for humanity. Not only because from science many advantages come to our life, not less than to justice, but also, and principally, because, as our father Dante admonishes, "we were not made to live like brutes, but rather to follow virtue and knowledge".

Trust in science

It is to me deep cause for bitterness that the weakness of human brains has reached so miserable a state (and particularly in my own country) that hunters and cooks—who by inventing new hunts and recipes strive to satisfy the whims and the palates of men—are so highly rewarded, whereas great hindrances are set up against speculative minds. That greedy and wicked minds have contrived means to enslave science to ignoble purposes is for me cause for sadness and immense melancholy. However, as no hope or real consolation can be found outside of science and reason, my by now feeble voice exhorts you to return to them with limpid mind and renewed vigor.

List of works

For those who would like to examine my work, here follows a list of my main writings.

- *The Sensitive Balance* (*La bilancetta*), written in 1586.
- *Operations of the Geometric and Military Compass* (*Le operazioni del compasso geometrico e militare*), published in 1606.
- *The Starry Messenger* (*Sidereus Nuncius*), written in Latin and published in 1610, in which I desribe the astronomical observations made using the telescope.
- *Bodies that Stay atop Water, or Move in It* (*Discorso intorno alle cose che stanno in su l'acqua o che in quella si muovono*), published in 1612.

- *The Assayer* (*Il Saggiatore*), published in 1623 as a reply to an essay by Father Grassi on comets, and dedicated to Pope Urban VIII. In this work I establish norms and rules for the investigation of Nature. It was praised for its pungent satire and for the elegant style of its writing.
- *Dialogue Concerning the Two Chief World Systems: Ptolemaic and Copernican* (*Dialogo sopra i due massimi sistemi del mondo, tolemaico e copernicano*), begun in 1624 and published in 1632. A book in which, by means of discussions between Salviati, Sagredo, and Simplicio, the Ptolemaic doctrine (accepted by Peripatetic philosophy) is compared to the Copernican theory (according to which the earth rotates around the sun).
- *Discourses About Two New Sciences* (*Discorsi e dimostrazioni matematiche intorno a due nuove scienze*), begun in 1633, immediately after the banning of the *Dialogue* and my abjuration, and published in the Netherlands in 1638 by the Elseviers. In this book, I deal with natural motion due to gravity, and with violent motion due to forces. And I also describe the inner structure of materials and their resistance to rupture.
- Letters to friends and adversaries, some of which are long enough to be called tracts. Among these are:
  - — to Marc Welser, about solar spots (three letters), 1612
  - — to Benedetto Castelli, about the interpretation of the Bible, 1613
  - — to Christina of Lorraine, about the interpretation of the Bible, 1615
  - — to Francesco Ingoli, on the principles of inertia and relativity, 1624
  - — to Leopold of Tuscany, about the silvery-whiteness of the moon, 1640
- Various writings on Dante, Ariosto, and Tasso
- *Against Wearing the Robe* (*Contro il portar la toga*), satirical text written in verse form

2

The Origin of Nerves

Or, Not Believing the Evidence*

quando il testo d'Aristotile non fusse in contrario,
che apertamente dice, i nervi nascer dal cuore,
bisognerebbe per forza confessarla per vera

one would be forced to admit it as true,
if Aristotle's texts were not opposed in saying plainly
that the nerves originate in the heart

An account of a demonstration of an anatomical dissection during which a philosopher, in order not to have his faith in Aristotle undermined, refuses to believe what a surgeon is showing him—namely, that nerves originate in the brain and not in the heart. This provides Galileo with the opportunity for an invective against knowledge based on dogma and tradition (*ipse dixit*), in which his fundamental ideas on the role and methodology of true science emerge.

The principle of authority

"He said it, Aristotle said it, and that is enough"—proclaimed the philosopher's followers in Galileo's time, in an attempt to keep anyone from questioning Aristotelian statements—real or presumed. Even worse, the Aristotelian, or Peripatetic, culture—the official culture in Galileo's time—trusted in Aristotle to the extent of

* From *Dialogue*.

believing that he had the answer to any kind of problem whatsoever (not so different, after all, from what is still the case today with many mass beliefs, especially religious ones). When it came to Aristotle, you had to "understand him, and not only understand him, but also know his books so well that you have a complete picture of them", Simplicio states confidently in the dialogue we shall be looking at shortly. To which Sagredo promptly objects that one can find the answer to any kind of problem in 'a booklet much shorter than Aristotle'—the alphabet—provided, of course, one puts together vowels and consonants in the proper way. Moreover, "this way of all knowledge being contained in a book is very similar to that by which a piece of marble contains within itself a very beautiful statue, or a thousand of them for that matter; but the point is to be able to discover them". The essence of information and of art—Sagredo implies—does not reside in the raw material itself, but rather in the way it is molded.

In the excerpts below,[6] Galileo tackles the theme of knowledge based solely on authority (*ipse dixit*). As a starting point he takes a demonstration of an anatomical dissection, during the course of which a philosopher refuses to acknowledge what he sees—that is, that nerves originate in the brain—simply because it conflicts with what Aristotle says. It is easy to see the dire moral and cultural consequences of this typically late-Aristotelian stance. Not only does it reduce knowledge to passive rote-learning, to sterile formal exercises (often merely a pretext for conversation and prestige), it also permits those who are in bad faith to bend knowledge to their own ends, using Aristotle as a shield. Moreover, and this is an even more serious consequence, it freezes development in science, which by its nature is historical and dynamic. It thus ends up betraying precisely what it purports to represent—the original thought and teachings of Aristotle. Galileo is quite justified, therefore, in declaring that he is often more Aristotelian than his opponents are, and in claiming that

[6] *Galileo on the World Systems*, Second Day, translated by M.A. Finocchiaro, University of California Press, Berkeley, CA and Los Angeles, CA, 1997, p. 119 and following. In Italian: *Dialogo sopra i due massimi sistemi del mondo*, Giornata seconda, Edizione Nazionale, Vol. VII, p. 133 and following.

had the great philosopher lived in their time, he would have altered his opinions and corrected his writings, much to the disappointment of many of his weak-minded followers.

In the specific historical context of seventeenth-century Italy, the principle of authority took the form outlined in the previous two paragraphs. However, taking on different forms and adapting to the times, the principle has survived through the centuries to the present day. It seems, in fact, to satisfy two basic and conflicting human urges: the impulse to dominate and the impulse to obey. It plays an explicit role, for instance, in dictatorships, in religious creeds and sects. But it also plays a hidden, though no less incisive role, even in modern democracies, where consensus is obtained through subtle channels of persuasion rather than rational approaches, and where people are induced to behave in particular ways imposed from above, which they accept uncritically. It is very convenient to accept revealed truths, no matter what they are—it spares one the effort of individual responsibility. Galilean teaching, if followed, can help us live in a worthier way.

The text

Apart from the interest of the argument itself, the dialogue also merits attention for its liveliness and wit (reminiscent of certain passages from Boccaccio), as well as for the vividness of its everyday images and language ("*infilzare i suoi silogismi*", "*trarne il sugo*", "*poveretti di cervello*", "*pecore stolide*", "*uscir un di traverso*", etc., expressions that inevitably lose some of their flavour when translated into English, respectively "spin out his syllogisms", "squeeze their juice", "weak-minded", "silly sheep", "[to see] someone slyly appear").

The character of Simplicio takes on particular prominence as, in his stubborn and pathetic defense of Aristotle, he emerges as a symbol for the declining, untroubled world of knowledge "on paper", based on authority and tradition alone. It is worth noting how the gently mocking humor that pervades the whole passage disappears in Salviati's eloquent closing words, where he passionately invites anyone who "has eyes in his head and in his mind" to proceed to explore the world with the help of these alone.

Galileo liked to call himself a "philosopher", a label which for him was synonymous with "man of science". Salviati's words express well what he meant by the term, and what attributes were required to be worthy of it. For those who debate "about a world on paper", usurping "the honorable title of philosopher", the much more appropriate terms "historians or memory experts" would be enough.

Salviati is Galileo's *alter ego* in most of the dialogues; his interlocutors are Simplicio, who represents all Galileo's Aristotelian opponents united into the one character, and Sagredo, a Venetian man of noble extraction and a sharp mind, who is supposed to act as impartial judge, but almost always ends up extolling the virtues of Copernican theses as compared to those of Peripatetic philosophers. It is worth taking a look at what Tommaso Campanella (an admirer and somewhat inconvenient defender of Galileo[7]) says about the three interlocutors of the *Dialogue*:

> everyone plays his role admirably; Simplicio appears as the stooge of this philosophical comedy, exhibiting at the same time the foolishness of his sect, as well as its manner of speech, its instability, its obstinacy, and everything else about it. Without doubt, it is the equal of anything in Plato. Salviati is a great Socrates, who makes others give birth rather than giving birth himself, and Sagredo is a free mind who has not been adulterated by schools and judges everything with much sagacity.[8]

Italo Calvino writes:

> Salviati and Sagredo represent two different facets of Galileo's character: Salviati is the reasoning man, methodologically rigorous, who proceeds slowly and carefully; Sagredo is characterized by his "very fast discourse",

[7] Tommaso Campanella, 1568–1639, philosopher and writer. A Dominican monk, he was suspected of heresy. He spent nearly twenty-seven years in jail for taking part in a political plot against Spain. He wrote *Apologia pro Galileo*, but his most famous work is *La città del sole*, describing a model of an ideal society where individual happiness and collective welfare converge.

[8] Edizione Nazionale, Vol. XIV, p. 366.

by a spirit more inclined to imagination, to drawing unproved conclusions and to pushing any idea to its extreme consequences, such as when he conjectures on what life could be like on the moon, or on what would happen if the earth stopped.[9]

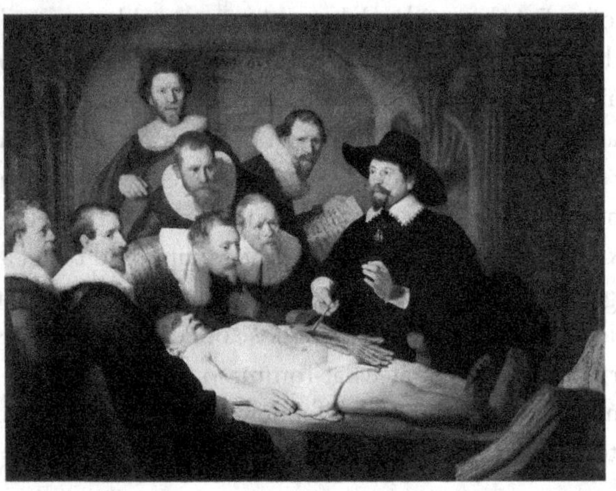

Rembrandt, Professor Tulp's Anatomical Dissection, Mauritshuis, Den Haag.

The anatomical dissection

SIMPL: I confess to you that I thought about yesterday's discussions the whole night, and I really find many beautiful, new, and forceful considerations. Nevertheless, I feel drawn much more by the authority of so many great writers, and in particular . . . You shake your head and sneer, Sagredo, as if I were saying a great absurdity.

SAGR: I merely sneer, but believe me that I am about to explode by trying to contain greater laughter; for you reminded me of a beautiful incident that I witnessed many years ago together with some other worthy friends of mine, whom I could still name.

SALV: It will be good for you to tell us about it, so that perhaps Simplicio does not continue to believe that it was he who moved you to laughter.

[9] I. Calvino, *Lezioni americane*, Garzanti, Milano, 1988, p. 43. English translation by Patrick Creagh, *Six Memos for the Next Millennium/the Charles Eliot Norton Lectures 1985–86* (Vintage International).

SAGR: I am happy to do that. One day I was at the house of a highly respected physician in Venice; here various people met now and then, some to study, others for curiosity, in order to see anatomical dissections performed by an anatomist who was really no less learned than diligent and experienced. It happened that day that they were looking for the origin and source of the nerves, concerning which there is a famous controversy between Galenist[10] *and Peripatetic*[11] *physicians. The anatomist showed how the great trunk of nerves started at the brain, passed through the nape of the neck, extended through the spine, and then branched out through the whole body, and how only a single strand as thin as a thread arrived at the heart. As he was doing this he turned to a gentleman, who he knew was a Peripatetic philosopher and for whose sake he had made the demonstration; the physician asked the philosopher whether he was satisfied and sure that the origin of the nerves is in the brain and not in the heart, and the latter answered after some reflection: "you have made me see this thing so clearly and palpably that one would be forced to admit it as true, if Aristotle's texts were not opposed in saying plainly that the nerves originate in the heart".*
SIMPL: Gentlemen, I want you to know that this dispute about the origin of the nerves is not as settled and decided as some believe.
SAGR: Nor will it ever be decided as long as one has similar opponents. At any rate what you say does not diminish at all the absurdity of the answer of the Peripatetic, who against such a sensible experience did not produce other experiences or reasons of Aristotle, but mere authority and the simple ipse dixit.
SIMPL: Aristotle has acquired such great authority only because of the strength of his arguments and the profundity of his discussions. However, you must understand him, and not only understand him, but also know his books so well that you have a complete picture of them and all his assertions always in mind. For he did not write for the common people, nor did he feel obliged to spin out his syllogisms by the well-known formal method; instead, using an informal procedure, he sometimes placed the proof of a proposition among passages that seem to deal with something else. Thus, you must have that whole picture and be able

[10] *Galenists*—followers of Galenus, the famous physician of Pergamon who lived from 200 to 129 BC.

[11] *Peripatetic philosophers*—followers of Aristotle.

to combine this passage with that one and connect this text with another very far from it. There is no doubt that whoever has this skill will be able to draw from his books the demonstrations of all knowable things, since they contain everything.

The treasures of the alphabet

SAGR: So, my dear Simplicio, you are not bothered by things being scattered here and there, and you think that by collecting and combining various parts you can squeeze their juice. But then, what you and other learned philosophers do with Aristotle's texts, I will do with the verses of Virgil or Ovid, by making patch works of passages and explaining with them all the affairs of men and secrets of nature. But why even go to Virgil or other poets? I have a booklet much shorter than Aristotle or Ovid in which are contained all the sciences, and with very little study one can form a very complete picture of them: this is the alphabet. There is no doubt that whoever knows how to combine and order this and that vowel with this and that consonant will be able to get from them the truest answers to all questions and the teachings of all sciences and of all arts. In the same way a painter, given various simple colors placed separately on his palette, by combining a little of this with a little of that and that other, is able to draw men, plants, buildings, birds, fishes—in short, all visible objects—without having on his palette either eyes, or feathers, or scales, or leaves, or rocks; on the contrary, it is necessary that none of the things to be drawn nor any part of them be actually among the colors, which can serve to represent everything, for if there were, for example, feathers, they would not serve to depict anything but birds and bunches of feathers.
SALV: There are still alive some gentlemen who were present when a professor teaching at a famous university, upon hearing descriptions of the telescope which he had not yet seen, said that the invention was taken from Aristotle. Having asked that a book be brought to him, he found a certain passage where Aristotle explains how it happens that from the bottom of a very deep well one can see the stars in heaven during the day, and he said to the bystanders: "here is the well, which corresponds to the tube; here are the thick vapors, from which is taken the invention of lenses; and lastly here is the strengthening of vision as the rays pass through the denser and darker transparent medium".

SAGR: This way of all knowledge being contained in a book is very similar to that by which a piece of marble contains within itself a very beautiful statue, or a thousand of them for that matter; but the point is to be able to discover them [. . .].

For Simplicio an alternative to Aristotle can only be the written word of some other philosopher, so far removed is he from conceiving of knowledge as an investigation made by individuals who, in the process, put themselves to the test.

Aristotle today

SALV: [. . .] Tell me, if you do not mind, are you so simpleminded that you do not understand that if Aristotle had been present to listen to the doctor who wanted to make him inventor of the telescope, he would have been more angry with him than with those who were laughing at the doctor and his interpretations? Do you have any doubt that if Aristotle were to see the new discoveries in the heavens, he would change his mind, revise his books, accept the more sensible doctrines, and cast away from himself those who are so weak-minded as to be very cowardly induced to want to uphold every one of his sayings? Do they not realize that if Aristotle were as they imagine him, he would be an intractable brain, an obstinate mind, a barbarous soul, a tyrannical will, someone who, regarding everybody else as a silly sheep, would want his decrees to be preferred over the senses, experience, and nature herself? It is his followers who have given authority to Aristotle, and not he who has usurped or taken it. Since it is easier to hide under someone else's shield than to show oneself openly, they are afraid and do not dare to go away by a single step; rather than putting any changes in the heavens of Aristotle, they insolently deny those which they see in the heavens of nature. [. . .]
SALV: I have often wondered how it can be that those who rigidly maintain everything Aristotle said do not notice how much damage they do to his reputation, how much discredit they bring him, and how much they diminish his authority instead of increasing it [. . .].
SIMPL: But, if one abandons Aristotle, who will be the guide in philosophy? Name some author.

A world on paper

SALV: One needs a guide in an unknown and uncivilized country, but in a flat and open region only the blind need a guide; whoever is blind would do well to stay home, whereas anyone who has eyes in his head and in his mind should use them as a guide. Not that I am thereby saying that one should not listen to Aristotle; on the contrary, I applaud his being examined and diligently studied and only blame submitting to him in such a way that one blindly subscribes to all his assertions and accepts them as unquestionable dictates, without searching for other reasons for them. This abuse carries with it another extreme impropriety, namely, that no one makes an effort any longer to try to understand the strength of his demonstrations. Is there anything more shameful in a public discussion dealing with demonstrable conclusions than to see someone slyly appear with a textual passage (often written for some different purpose) and use it to shut the mouth of an opponent? If you want to persist in this manner of studying, lay down the name of philosophers and call yourselves either historians or memory experts, for it is not right that those who never philosophize should usurp the honorable title of philosopher.

However, we should get back to shore in order not to enter an infinite ocean from which we could not get out all day. So, Simplicio, come freely with reasons and demonstrations (yours or Aristotle's) and not with textual passages or mere authorities because our discussions are about the sensible world and not about a world on paper. [. . .]

The three interlocutors

Here are the reasons Galileo himself[12] gave for writing the text in dialogue form and for the choice of the three interlocutors:

Furthermore, I thought it would be very appropriate to explain these ideas in dialogue form; for it is not restricted to the rigorous observation of mathematical laws, and so it also allows digressions which are sometimes no less interesting than the main topic.

[12] *Galileo on the World Systems*, translated by M.A. Finocchiaro, University of California Press, Berkeley, CA and Los Angeles, CA, 1997, p. 81. In Italian: *Dialogo sopra i due massimi sistemi del mondo*, Edizione Nazionale, Vol. VII, p. 29.

Many years ago in the marvelous city of Venice I had several occasions to engage in conversation with Giovanfrancesco Sagredo, a man of most illustrious family and of sharpest mind.[13] *From Florence we were visited by Filippo Salviati,*[14] *whose least glory was purity of blood and magnificence of riches; his sublime intellect fed on no delight more avidly than on refined speculations. I often found myself discussing these subjects with these two men and with the participation of a Peripatetic philosopher, who seemed to have no greater obstacle to the understanding of the truth than the fame he had acquired in Aristotelian interpretation.*

Now, since Venice and Florence have been deprived of those two great lights by their very premature death at the brightest time of their life, I have decided to prolong their existence, as much as my meager abilities allow, by reviving them in these pages of mine and using them as interlocutors in the present controversy. There will also be a place for the good Peripatetic, to whom, because of his excessive fondness of Simplicius's[15] *commentaries, it seemed right to give the name of his revered author, without mentioning his own. Those two great souls will always be revered in my heart; may they receive with favor this public monument of my undying friendship, and may they assist me, through my memory of their eloquence, to explain to posterity the aforementioned speculations.*

[13] Nobleman Giovan Francesco Sagredo (1571–1620) left prestigious appointments with the Republic of Venice to devote himself to scientific studies. He occasionally collaborated with Galileo in carrying out experiments. The dialogues take place in his palace overlooking the Grand Canal.

[14] Florentine nobleman Filippo Salviati (1582–1614), a member of the Lincean Academy, played host to his friend Galileo several times at his Villa delle Selve, where parts of *Letters on Sunspots* (dedicated to him) were written.

[15] Simplicius was a Greek commentator of Aristotle, active in the sixth century AD. The Italian name Simplicio has the connotation "simpleton".

PART II

REVOLUTIONARY MOTIONS

3

Almost Free Fall

Or, Which Reaches the Ground First?*

Simplicio: Io non crederò mai che nell'istesso vacuo . . .
un fiocco di lana si movesse così veloce come un pezzo di piombo

Simplicio: I shall never believe that even in a vacuum . . .
a lock of wool and a bit of lead can fall with the same velocity

Using a piece of *reductio ad absurdum* reasoning which was to become a historical landmark, Galileo proves that, contrary to the Aristotelian conception, all freely falling bodies descend with the same velocity and take equal times to reach the ground. He describes the motion of such bodies as uniformly accelerated. Any divergences in behavior of bodies of different weight and shape falling in media other than a vacuum, he attributes to alterations caused by friction.

How bodies fall

A problem that has long sparked people's interest is that of the real reason why two bodies falling from the same height take different times to reach the ground. And why these times change if the medium through which the fall occurs changes, as can easily be checked by comparing the behavior of a given body falling first through air and then through water (just to take two of the most familiar media).

* From *Discourses*.

The most instinctive answer is the one which Aristotelian philosophers give: the heavier of the two bodies reaches the ground sooner because the fall velocities are proportional to the weights. As for the role of the surrounding medium, according to Aristotle the fall velocity is inversely proportional to its density. This conception, however, does not explain a number of things. For instance, why, when the weight of the two bodies is the same and the density of the medium through which they fall is the same, the times of fall can nevertheless be different if the two bodies have different shapes, that is, if one is more aerodynamic than the other.

Today we know that the net force acting on a falling body is given by the difference between its weight and the friction force due to resistance of the medium. By Newton's law—force is equal to mass times acceleration—the body undergoes acceleration only if the net force acting upon it is not equal to zero, otherwise its velocity remains constant. The reader hardly needs reminding that while the weight of the body does not change during fall, the friction it encounters increases as its velocity increases. Thus if the weight of the body is large, it prevails over the friction force for a large portion of the drop, and so the difference from free fall (i.e. in a vacuum) is small;

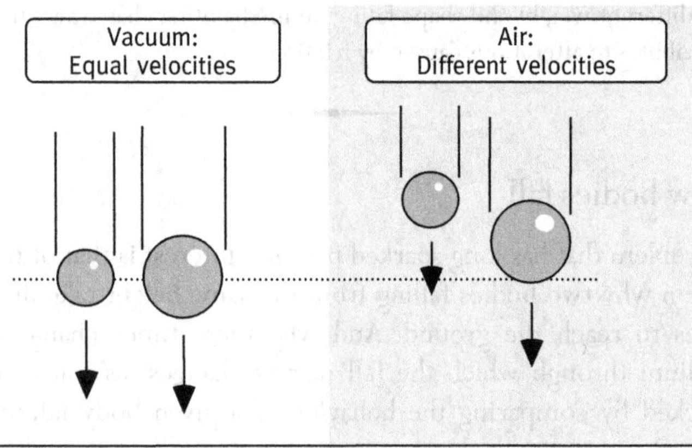

In air: given two spheres made of the same material, the one with the larger radius falls more rapidly because the ratio between its weight and the friction force is greater.

this would be the case, for example, if the body were a sphere made of lead. If it were, instead, a sphere made of cork, the friction force would rapidly equal the weight, thus preventing the body from being further accelerated and permitting it to settle at a limit or terminal velocity (this behavior can clearly be seen in the descent of a parachute). Bodies of intermediate weight would have intermediate behavior.

The weight of the body, therefore, matters only in respect of its magnitude compared to the friction force. Thus, of two spheres made of the same material but having different diameters, the one with the larger diameter falls faster because its weight increases with volume (radius cubed), whereas friction increases with surface area (radius squared), that is to a lesser extent.[16] This implies that bodies whose radius is sufficiently large would, for an appreciable portion of their descent, fall without experiencing the effects of friction.

The Aristotelian mistake

Details of this kind are not taken into account by the Aristotelian approach, which, moreover, also omits the acceleration stage between zero velocity and final velocity, considering it to be a transitory and marginal effect. At the beginning of his studies, in fact, Galileo had not rejected the Aristotelian approach—but his thinking gradually evolved to the point where it completely reversed the entire picture, identifying motion with constant acceleration as the essence

[16] Regarding this, Galileo has Salviati say:

Now you must know, Simplicio, that it is not possible to diminish the surface of a solid body in the same ratio as the weight, and at the same time maintain similarity of figure. For, since it is clear that in the case of a diminishing solid the weight grows less in proportion to the volume, and since the volume always diminishes more rapidly than the surface, when the same shape is maintained, the weight must therefore diminish more rapidly than the surface. But geometry teaches us that, in the case of similar solids, the ratio of two volumes is greater than the ratio of their surfaces [. . .]. Now if the surface varies as the square of its linear dimensions while the volume varies as the cube of these dimensions may we not say that the volume stands in sesquialteral ratio to the surface? (*Two New Sciences*, translated by H. Crew and A. de Salvio, Dover Publications, New York, 1954, p. 89. In Italian: *Discorsi e dimostrazioni matematiche intorno a due nuove scienze*, Edizione Nazionale, Vol. VIII, p. 133). [Note: sesquialteral = 3/2 power].

of the fall mechanism. Indeed, in a vacuum this would be the only mechanism operating. The resisting action by the medium is an "accidental effect", which needs to be examined separately. For Aristotelians, instead, fall in a vacuum—which we now know to be equal for all bodies, regardless of their shape and weight—is inconceivable. They state that where there is a vacuum all motion is impossible. Their argument, in Simplicio's words, is as follows:[17]

SIMPL: So far as I remember, Aristotle inveighs against the ancient view that a vacuum is a necessary prerequisite for motion and that the latter could not occur without the former. In opposition to this view Aristotle shows that it is precisely the phenomenon of motion, as we shall see, which renders untenable the idea of a vacuum [. . .]. [. . .] he assumes that the speeds of one and the same body moving in different media are in inverse ratio to the densities of these media [. . .]. From this second supposition, he shows that since the tenuity of a vacuum differs infinitely from that of any medium filled with matter however rare, any body which moves in a plenum through a certain space in a certain time ought to move through a vacuum instantaneously; but instantaneous motion is an impossibility; it is therefore impossible that a vacuum should be produced by motion.[18]

Fall velocity is indeed proportional to the weight of the body when the terminal velocity is attained (in the Mathematical Note at the end of this chapter it will be shown that the speed of descent of a parachute doubles when it is carrying two persons). With regard to the initial stage of accelerated motion—which is the stage we normally

[17] *Two New Sciences*, translated by H. Crew and A. de Salvio, Dover Publications, New York, 1954, p. 61. In Italian: *Discorsi e dimostrazioni matematiche intorno a due nuove scienze*, Giornata prima, Edizione Nazionale, Vol. VIII, p. 106.

[18] The syllogism is: the time of fall is proportional to the density of the medium; a vacuum has zero density; therefore in a vacuum the time of fall is zero. This implies that a displacement can occur with no time elapsing—an impossible event. Equally unjustified is the assertion that a vacuum is needed in order for motion to occur, as had been claimed by pre-Aristotelian philosophers. The final line in the Crew-de Salvio translation is questionable. A more appropriate translation of the original Italian sentence "adunque darsi il vacuo in grazia del moto è impossibile" might be: "therefore, wherever there is motion there cannot be a vacuum".

observe in everyday life—Aristotle's commentator, Philoponus, had already written: "If you drop two bodies from the same height, one being much heavier than the other, you will find the ratio of the two times of fall does not depend on the ratio of the weights, and that the times differ only slightly".[19] As Stillman Drake points out,[20] this statement had subsequently been ignored for about a millennium.

From the Leaning Tower of Pisa

It is said that in order to clarify his ideas on this question, Galileo dropped objects from the Leaning Tower of Pisa. He is supposed to have used two bodies made of the same material, differing only in size. He had in fact stated that, if the effects of air friction could be ignored, the two bodies would reach the ground at the same instant. We do not know for sure whether he actually performed the experiment,[21] but we do know the argument he proposed in support of his conclusion—a splendid and incontrovertible *gedanken experiment*.[22] In today's language, the argument goes as follows.

Let us assume for a moment that, even in a vacuum, the heavier of the two bodies reaches the ground first, as common sense would suggest. Let us now repeat the experiment with the two bodies joined together, so as to make a single heavier object. This ought to fall

[19] In 1586, Dutch scientist Simon Stevin performed an experiment where two lead spheres, one ten times heavier than the other, were dropped from a height of 30 ft on to a sounding board. He noted: "The two bangs produced result in a single sound perception" (S. Stevin, *The Principal Works*, E.J. Dijksterhuis, Amsterdam, 1955, p. 511).

[20] S. Drake, *Galileo: Pioneer Scientist*, University of Toronto Press, Toronto, 1994, p. 64.

[21] The circumstances are narrated in the imaginative biography of Galileo due to his pupil Vincenzo Viviani (*Racconto istorico della vita di Galileo Galilei*, Edizione Nazionale, Vol. XIX). In a drop as long as that from the top of the tower of Pisa, the different effects of air friction upon two bodies would make the fall time appreciably different even for objects as heavy as lead spheres. Experiments of this kind were actually performed a few years after the publication of the *Discourses*, by Vincenzo Renieri, who let Galileo know the results by letter.

[22] *gedanken experiment*—thought experiment.

to the ground in a shorter time than either of the two bodies separately. However, it is also possible for us to reason differently: the heavier body, having to fall faster, will drag along the lighter body, making it take on a higher velocity. For its part, however, the lighter body will act as a brake on the heavier body, slowing down its fall a little. The fall time will thus end up being intermediate between the times measured for the two bodies separately. Two truths which are in opposition to each other, but equally sound from the point of view of logical reasoning. Therefore, the initial assumption that the heavier body descends faster must be wrong.

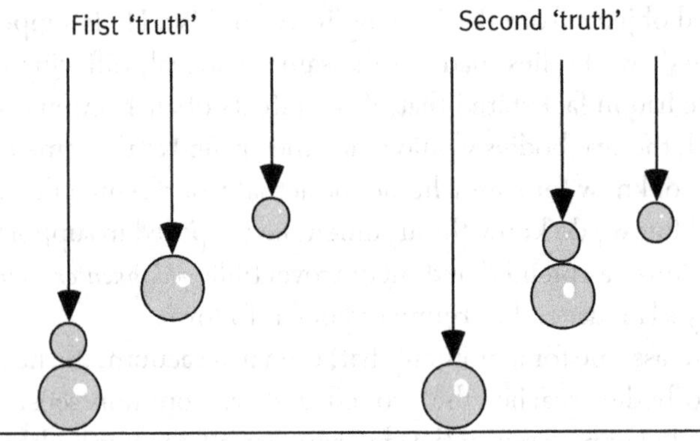

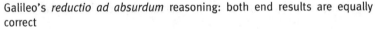

Galileo's *reductio ad absurdum* reasoning: both end results are equally correct

We will now take a look at the argument in the form in which it was originally presented.[23] After Simplicio has expounded (in truth, very competently) the Aristotelian point of view, Salviati replies to him as follows:

SALV: [. . .] I greatly doubt that Aristotle ever tested by experiment whether it be true that two stones, one weighing ten times as much as

[23] *Two New Sciences*, translated by H. Crew and A. de Salvio, Dover Publications, New York, 1954, p. 62 and following. In Italian: *Discorsi e dimostrazioni matematiche intorno a due nuove scienze*, Giornata prima, Edizione Nazionale, Vol. VIII, p. 106 and following.

the other, if allowed to fall, at the same instant, from a height of, say, 100 cubits, would so differ in speed that when the heavier had reached the ground, the other would not have fallen more than 10 cubits.
SIMPL: His language would seem to indicate that he had tried the experiment, because he says: We see the heavier; *now the word* see *shows that he had made the experiment.*
SAGR: But I, Simplicio, who have made the test can assure you that a cannon ball weighing one or two hundred pounds, or even more, will not reach the ground by as much as a span ahead of a musket ball weighing only half a pound, provided both are dropped from a height of 200 cubits.

The paradox of the two falling bodies

SALV: But, even without further experiment, it is possible to prove clearly, by means of a short and conclusive argument, that a heavier body does not move more rapidly than a lighter one provided both bodies are of the same material and in short such as those mentioned by Aristotle. But tell me, Simplicio, whether you admit that each falling body acquires a definite speed fixed by nature, a velocity which cannot be increased or diminished except by the use of force or resistance.
*SIMPL: There can be no doubt but that one and the same body moving in a single medium has a fixed velocity which is determined by nature and which cannot be increased except by the addition of momentum [*impeto[24]*] or diminished except by some resistance which retards it.*
SALV: If then we take two bodies whose natural speeds are different, it is clear that on uniting the two, the more rapid one will be partly retarded by the slower, and the slower will be somewhat hastened by the swifter. Do you not agree with me in this opinion?
SIMPL: You are unquestionably right.
SALV: But if this is true, and if a large stone moves with a speed of, say, eight while a smaller moves with a speed of four, then when they are united, the system will move with a speed less than eight; but the two

[24] Although in the writings of the mature Galileo the term "*impeto*" is mostly used to refer to the speed acquired by a body, in this particular case it is used to mean impulse, that is, stimulus or push.

stones when tied together make a stone larger than that which before moved with a speed of eight. Hence the heavier body moves with less speed than the lighter; an effect which is contrary to your supposition. Thus you see how, from your assumption that the heavier body moves more rapidly than the lighter one, I infer that the heavier body moves more slowly.[25]

SIMPL: I am all at sea because it appears to me that the smaller stone when added to the larger increases its weight and by adding weight I do not see how it can fail to increase its speed or, at least, not to diminish it.

SALV: Here again you are in error, Simplicio, because it is not true that the smaller stone adds weight to the larger.

SIMPL: This is, indeed, quite beyond my comprehension.

SALV: It will not be beyond you when I have once shown you the mistake under which you are laboring. Note that it is necessary to distinguish between heavy bodies in motion and the same bodies at rest. A large stone placed in a balance not only acquires additional weight by having another stone placed upon it, but even by the addition of a handful of hemp its weight is augmented six to ten ounces according to the quantity of hemp. But if you tie the hemp to the stone and allow them to fall freely from some height, do you believe that the hemp will press down upon the stone and thus accelerate its motion or do you think the motion will be retarded by a partial upward pressure? One always feels the pressure upon his shoulders when he prevents the motion of a load resting upon him; but if one descends just as rapidly as the load would fall how can it gravitate or press upon him? Do you not see that this would be the same as trying to strike a man with a lance when he is running away from you with a speed which is equal to, or even greater, than that with which you are following him? You must therefore conclude that, during free and natural fall, the small stone does not press upon the larger and consequently does not increase its weight as it does when at rest.

[25] Galileo had already expounded this *reductio ad absurdum* reasoning in his *De Motu* and in the margin notes to *Esercitazioni filosofiche* by Antonio Rocco (Aristotelian philosopher, opposer of Galileo), published in 1633. Similar arguments, in slightly different but equally elegant terms, were already to be found in Giovanni Battista Benedetti's *Diversarum speculationum mathematicarum et physicarum liber*, published in 1585. Galileo neglected to provide a reference to his source, even though Benedetti's work had played a decisive role in his scientific education.

Salviati's examples seem to anticipate Einstein's famous illustration of a man standing inside a free-falling lift cabin: in the reference system of the lift, the man has no weight, that is, he exerts no force on the floor (a sheet of paper placed under his shoes would offer no resistance to being slipped out). Simplicio seems unpersuaded and comes up with a further objection:

SIMPL: But what if we should place the larger stone upon the smaller?
SALV: Its weight would be increased if the larger stone moved more rapidly; but we have already concluded that when the small stone moves more slowly it retards to some extent the speed of the larger, so that the combination of the two, which is a heavier body than the larger of the two stones, would move less rapidly, a conclusion which is contrary to your hypothesis. We infer therefore that large and small bodies move with the same speed provided they are of the same specific gravity.
SIMPL: Your discussion is really admirable; yet I do not find it easy to believe that a bird-shot falls as swiftly as a cannon ball.
SALV: Why not say a grain of sand as rapidly as a grindstone? But, Simplicio, I trust you will not follow the example of many others who divert the discussion from its main intent and fasten upon some statement of mine which lacks a hair's-breadth of the truth and, under this hair, hide the fault of another which is as big as a ship's cable. Aristotle says that 'an iron ball of one hundred pounds falling from a height of one hundred cubits reaches the ground before a one- pound ball has fallen a single cubit'. I say that they arrive at the same time. You find, on making the experiment, that the larger outstrips the smaller by two finger-breadths, that is, when the larger has reached the ground, the other is short of it by two finger-breadths, now you would not hide behind these two fingers the ninety-nine cubits of Aristotle, nor would you mention my small error and at the same time pass over in silence his very large one. Aristotle declares that bodies of different weights, in the same medium, travel (in so far as their motion depends upon gravity) with speeds which are proportional to their weights; this he illustrates by use of bodies in which it is possible to perceive the pure and unadulterated effect of gravity, eliminating other considerations, for example, figure as being of small importance, influences which are greatly dependent upon the medium which modifies the single effect of gravity alone. Thus we

observe that gold, the densest of all substances, when beaten out into a very thin leaf, goes floating through the air; the same thing happens with stone when ground into a very fine powder. But if you wish to maintain the general proposition you will have to show that the same ratio of speeds is preserved in the case of all heavy bodies, and that a stone of twenty pounds moves ten times as rapidly as one of two; but I claim that this is false and that, if they fall from a eight of fifty or a hundred cubits, they will reach the earth at the same moment.
SIMPL: Perhaps the result would be different if the fall took place not from a few cubits but from some thousands of cubits.
SALV: If this were what Aristotle meant you would burden him with another error which would amount to a falsehood; because, since there is no such sheer height available on earth, it is clear that Aristotle could not have made the experiment; yet he wishes to give us the impression of his having performed it when he speaks of such an effect as one which we see.

Galileo reproaches Aristotle for having made claims without due experimental verification. For Galileo experimentation represents (as, after all, it did for Aristotle, too) the kernel of scientific investigation—and he is keen to point this out again and again: 'It is foolish to look for philosophical arguments that can truly describe an effect better than experiment and our eyes';[26] and then 'among the most reliable means of achieving truth, is to put experience before any kind of discourse . . . as no sensible experience can be contrary to truth'.[27] However, when it came to the free fall of bodies in a vacuum, experience was not feasible—so in this case Galileo, too, was compelled to draw conclusions without being able to verify them. Nevertheless, as we have just seen, he employed cast-iron *reductio ad absurdum* reasoning, the persuasive value of which could be considered as matching that of an actual experiment. This is a classic example of Galileo's working style: preliminary observations, and a theoretical discussion of them, culminate in a main scene where a

[26] Margin note to *Considerazioni* by the Unknown Academic, Edizione Nazionale, Vol. IV, p. 166.

[27] Letter to Fortunio Liceti, September 15, 1640, Edizione Nazionale, Vol. XVIII, p. 247.

striking empirical verification (either actual or *gedanken*) of the basic claim is propounded. After the main scene in the present case, the discussion proceeds to deal with specific details, until it arrives at the question of the differences in fall velocities as the surrounding media differ. Salviati says:[28]

Air friction

SALV: [. . .] We have already seen that the difference of speed between bodies of different specific gravities is most marked in those media which are the most resistant: thus, in a medium of quicksilver, gold not merely sinks to the bottom more rapidly than lead but it is the only substance that will descend at all; all other metals and stones rise to the surface and float. On the other hand the variation of speed in air between balls of gold, lead, copper, porphyry, and other heavy materials is so slight that in a fall of 100 cubits a ball of gold would surely not outstrip one of copper by as much as four fingers. Having observed this I came to the conclusion that in a medium totally devoid of resistance all bodies would fall with the same speed.
SIMPL: This is a remarkable statement, Salviati. But I shall never believe that even in a vacuum, if motion in such a place were possible, a lock of wool and a bit of lead can fall with the same velocity.

Salviati rebuts Simplicio's objection with the following argument: since we cannot have a vacuum, let us experiment in less and less dense media, aiming at describing the behavior in a vacuum by extrapolation. If we find that the fall times of two different bodies get closer and closer the less dense the medium becomes, we will then be able to conclude that in a vacuum they would be the same. Post-Galilean scientific research was subsequently to make widespread use of an extrapolation procedure such as this in determining properties that cannot be observed directly.

[28] *Two New Sciences*, translated by H. Crew and A. de Salvio, Dover Publications, New York, 1954, p. 71 and following. In Italian: *Discorsi e dimostrazioni matematiche intorno a due nuove scienze*, Giornata prima, Edizione Nazionale, Vol. VIII, p. 116 and following.

SALV: A little more slowly, Simplicio. Your difficulty is not so recondite nor am I so imprudent as to warrant you in believing that I have not already considered this matter and found the proper solution. Hence for my justification and for your enlightenment hear what I have to say. Our problem is to find out what happens to bodies of different weight moving in a medium devoid of resistance, so that the only difference in speed is that which arises from inequality of weight. Since no medium except one entirely free from air and other bodies, be it ever so tenuous and yielding, can furnish our senses with the evidence we are looking for, and since such a medium is not available,[29] *we shall observe what happens in the rarest and least resistant media as compared with what happens in denser and more resistant media. Because if we find as a fact that the variation of speed among bodies of different specific gravities is less and less according as the medium becomes more and more yielding, and if finally in a medium of extreme tenuity, though not a perfect vacuum, we find that, in spite of great diversity of specific gravity, the difference in speed is very small and almost inappreciable, then we are justified in believing it highly probable that in a vacuum all bodies would fall with the same speed. Let us, in view of this, consider what takes place in air, where for the sake of a definite figure and light material imagine an inflated bladder. The air in this bladder when surrounded by air will weigh little or nothing, since it can be only slightly compressed;*[30] *its weight then is small being merely that of the skin which does not amount to the thousandth part of a mass of lead having the same size as the inflated bladder. Now, Simplicio, if we allow these two bodies to fall from a height of four or six cubits, by what distance do you imagine the lead will anticipate the bladder? You may be sure that the lead will not travel three times, or even twice, as swiftly as the bladder, although you would have made it move a thousand times as rapidly.*[31]

SIMPL: It may be as you say during the first four or six cubits of the fall; but after the motion has continued a long while, I believe that the

[29] Shortly afterwards, in Magdeburg, Otto von Guericke was to perform his famous experiments with evacuated hemispheres (see Chapter 17).

[30] In other words: the bladder is sufficiently elastic to allow its internal pressure to remain almost equal to the external pressure.

[31] The meaning is: even if the weight of the lead ball were a thousand times larger than that of the inflated bladder, the fall times would differ by at most a factor of two.

lead will have left the bladder behind not only six out of twelve parts of the distance but even eight or ten.[32]

SALV: I quite agree with you and doubt not that, in very long distances, the lead might cover one hundred miles while the bladder was traversing one; but, my dear Simplicio, this phenomenon which you adduce against my proposition is precisely the one which confirms it. Let me once more explain that the variation of speed observed in bodies of different specific gravities is not caused by the difference of specific gravity but depends upon external circumstances and, in particular, upon the resistance of the medium, so that if this is removed all bodies would fall with the same velocity; and this result I deduce mainly from the fact which you have just admitted and which is very true, namely, that, in the case of bodies which differ widely in weight, their velocities differ more and more as the spaces traversed increase, something which would not occur if the effect depended upon differences of specific gravity. For since these specific gravities remain constant, the ratio between the distances traversed ought to remain constant whereas the fact is that this ratio keeps on increasing as the motion continues. Thus a very heavy body in a fall of one cubit will not anticipate a very light one by so much as the tenth part of this space; but in a fall of twelve cubits the heavy body would outstrip the other by one-third, and in a fall of one hundred cubits 90/100, etc.[33]

Salviati's arguments lead to a crystal-clear formulation of the fall mechanisms for a body in a vacuum and in a material medium. In a vacuum, that is, during free fall, the body keeps accelerating indefinitely, its speed increasing in proportion to the time elapsed; in material media, instead, the body accelerates during the initial stage, when its speed is low, but its motion becomes uniform after a certain interval of time, which is the shorter the higher the resistance offered by the medium is. We thus arrive at a definition of terminal velocity as that which the body attains when its gravitational force and the friction force become equaled out (the parachute effect). Galileo's description was to find

[32] When the lead reaches the ground, the bladder will be left behind by half the path, or even by 2/3 or 5/6 of it.

[33] To summarize: the distance gained by the heavy body over the light one is *not* proportional to the difference in their weights.

quantitative expression in Newton's law (see the Mathematical Note at the end of this chapter). In the Aristotelian conception, the transitory stage of acceleration is not taken into consideration.

SALV: [...] I begin by saying that a heavy body has an inherent tendency to move with a constantly and uniformly accelerated motion toward the common center of gravity, that is, toward the center of our earth, so that during equal intervals of time it receives equal increments of momentum and velocity.[34] *This, you must understand, holds whenever all external and accidental hindrances have been removed; but of these there is one which we can never remove, namely, the medium which must be penetrated and thrust aside by the falling body. This quiet, yielding, fluid medium opposes motion through it with a resistance which is proportional to the rapidity with which the medium must give way to the passage of the body; which body, as I have said, is by nature continuously accelerated so that it meets with more and more resistance in the medium and hence a diminution in its rate of gain of speed until finally the speed reaches such a point and the resistance of the medium becomes so great that, balancing each other, they prevent any further acceleration and reduce the motion of the body to one which is uniform and which will thereafter maintain a constant value. There is, therefore, an increase in the resistance of the medium, not on account of any change in its essential properties, but on account of the change in rapidity with which it must yield and give way laterally to the passage of the falling body which is being constantly accelerated. Now seeing how great is the resistance which the air offers to the slight momentum of the bladder and how small that which it offers to the large weight of the*

[34] The fall motion described by Galileo—elsewhere called by him *motus naturaliter acceleratus*, that is, naturally accelerated motion—is a case of uniformly accelerated motion, that is, characterized by a constant value of acceleration. This causes velocity to increase proportionally to the time elapsed from the start (at an earlier stage of his studies Galileo had wrongly proposed—following other scholars—that velocity increases proportionally to the space covered, see Chapter 13). The notion of accelerated motion in the fall of bodies can already be found in Aristotle ("A body heading toward its place of rest always appears to be moving by accelerated motion . . ."), but the various explanations for this given through the centuries had all been far removed from the correct one given by Galileo. Motion with velocity increasing in time had already been described, though in a somewhat unclear and ambiguous fashion, by Leonardo da Vinci, and before him by Jean Buridan and Albert of Saxony.

lead, I am convinced that, if the medium were entirely removed, the advantage received by the bladder would be so great and that coming to the lead so small that their speeds would be equalized. [. . .]

Parachute effect

Much further on in the dialogue[35] Galileo resumes the discussion about the role of the friction force in relation to the weight and the velocity of the falling body, elucidating the meaning of final velocity, beyond which motion becomes uniform:

SALV: [. . .] *For as to velocity, the greater this is, the greater will be the resistance offered by the air; a resistance which will be greater as the moving bodies become less dense. So that although the falling body ought to be displaced in proportion to the square of the duration of its motion, yet no matter how heavy the body, if it falls from a very considerable height, the resistance of the air will be such as to prevent any increase in speed and will render the motion uniform; and in proportion as the moving body is less dense* [men grave] *this uniformity will be so much the more quickly attained and after a shorter fall.*

At the end of the debate, which becomes protracted as a range of further arguments and examples are introduced, Simplicio finally declares himself convinced of Salviati's theses. This causes Sagredo to pronounce a judgment which clearly expresses how justifiable Galileo's resentment of his opponents (or rather, his persecutors) was. The reader should note how, through Sagredo's words, Galileo's painful personal experience acquires wider significance in bitter observations about the nature of human behavior.

SIMPL: The previous experiments, in my opinion, left something to be desired: but now I am fully satisfied.
SALV: The facts set forth by me up to this point and, in particular, the one which shows that difference of weight, even when very great, is

[35] *Two New Sciences*, Fourth Day, translated by H. Crew and A. de Salvio, Dover Publications, New York, 1954, p. 252. In Italian: *Discorsi e dimostrazioni matematiche intorno a due nuove scienze*, Giornata quarta, Edizione Nazionale, Vol. VIII, p. 275.

without effect in changing the speed of falling bodies, so that as far as weight is concerned they all fall with equal speed: this idea is, I say, so new, and at first glance so remote from fact, that if we do not have the means of making it just as clear as sunlight, it had better not be mentioned; but having once allowed it to pass my lips I must neglect no experiment or argument to establish it.

SAGR: Not only this but also many other of your views are so far removed from the commonly accepted opinions and doctrines that if you were to publish them you would stir up a large number of antagonists; for human nature is such that men do not look with favor upon discoveries either of truth or fallacy—in their own field, when made by others than themselves. They call him an innovator of doctrine, an unpleasant title, by which they hope to cut those knots which they cannot untie, and by subterranean mines they seek to destroy structures which patient artisans have built with customary tools. But as for ourselves who have no such thoughts, the experiments and arguments which you have thus far adduced are fully satisfactory; however if you have any experiments which are more direct or any arguments which are more convincing we will hear them with pleasure.

The foregoing dialogue about the fall of bodies illustrates well the dialectical style Galileo typically adopted. This style is wittily described by Canon Querengo,[36] one of Galileo's opponents:

Your Lordship would have great fun at hearing Galileo arguing the way he often does among fifteen or twenty people trying to attack him cruelly . . . But he stays fortified in such a way that he mocks them all; and though the novelty of his opinion does not persuade, he nevertheless demolishes most of the arguments by which his opponents try to knock him down. On Monday . . . before replying to a contrary view, he amplified and strengthened it with new arguments of great appeal, to the purpose of making his adversaries more ridiculous when in the end he made it collapse.

Simplicio and the false science

In spite of his deceptive name, Simplicio is for most of the time a respectable opponent, even if occasionally—more so in the *Dialogue*

[36] Letter by Querengo to Cardinal Alessandro d'Este, January 20, 1616, Edizione Nazionale, Vol. XII, p. 226.

than in the *Discourses*—he comes close to appearing ridiculous. A great expert on Aristotelian thought, he symbolizes, culturally, a long tradition which is no longer capable of renewing itself and, psychologically, the man who dares not think differently. He is therefore the antithesis of Galileo–Salviati. He is intelligent enough to follow a line of reasoning, but at times seems almost pathetic in his stubborn defense of safe traditional ideas ("it seems to me hard to believe", he says when faced with a new concept that has just been demonstrated). If he were to abandon these traditional ideas, however, he would then have to be capable of putting himself up for discussion, of proceeding on his own, and this bewilders him.

Moreover, as Salviati indicates so clearly, Simplicio, in order to defend what is familiar to him and he believes right by principle of authority—*ipse dixit*—picks upon any minor inexactitude in his opponent's arguments, even if this has virtually no influence on the basic mechanism of the phenomenon under observation (in Salviati's words: "a statement of mine that would miss truth by a piece of hair"). In showing that he has been able to catch his opponent at fault, even if on a very minor detail, Simplicio is trying to destroy his credibility and at the same time is also providing cover for the frailty of the entire system upon which his own beliefs are founded. This is a procedure which has nothing of the scientific about it; nevertheless, it is still in widespread practice today and is often employed with stupidity and bad faith far greater than Simplicio's. Think of certain political debates, of poor-quality scientific popularization or, even more conspicuous, of the defense of paranormal phenomena and ufology.

Galileo's great battle, therefore, was fought not only at the level of scientific knowledge, but also at the level of mental habits and ethics. The revival of irrationalism that we are presently witnessing—religious sects, alternative medicines, horoscopes, new age, intolerance and racism, obsessive consumerism, "appearance" rather than "substance", the importance of "image", and so on—unfortunately makes the battle as topical today as it was in Galileo's time.

Salviati's advice to Simplicio illustrates one of the cornerstones of the Galilean scientific revolution: in studying a phenomenon one must first try to single out and investigate the basic mechanism, temporarily ignoring small deviations, the minor effects that occur as a result of marginal factors. For example, given its extremely low density, air, in a first approach can be treated as if it were a vacuum. However, in order for this to be legitimate, the fall path must not be too long, that is, the velocity must remain low and thus the role of friction, small. Subsequently friction can be introduced in a second approximation, as a perturbation over the basic process. Modern physics is based entirely upon this type of procedure.

Free fall above clouds (Skydive WWW, Photo M. Skeffington)

MATHEMATICAL NOTE

The whole problem can be summed up in a single equation. The behavior of acceleration a in time t is described by Newton's law

$$m\, a(t) = P + F_A(t)$$

(mass times acceleration = weight + friction force)

where the weight, that is, the force of gravity, is $P = mg$, with m = mass of the body and g = gravitational acceleration, independent of time.[37] In most cases, provided the velocity is not too high, the friction force F_A is proportional to velocity v, namely $F_A = -bv$, where the constant friction factor b depends on

[37] To be precise, the mass in the first member is the inertial mass, while the mass appearing in the weight is the gravitational mass. As experiments show that the two masses are equal, they have both been indicated by m.

the viscosity of the medium and on the shape of the body (in particular on its cross section), and the minus sign serves to indicate that the friction force opposes the gravitational force. Introducing this expression into the equation of motion, after integration, one obtains an acceleration and a velocity which vary in time according to the equations[38]

$$a(t)=g\left[\exp\left(-\frac{bt}{m}\right)\right] \quad 3.1(a)$$

$$v(t)=\frac{mg}{b}\left[1-\exp\left(-\frac{bt}{m}\right)\right] \quad 3.1(b)$$

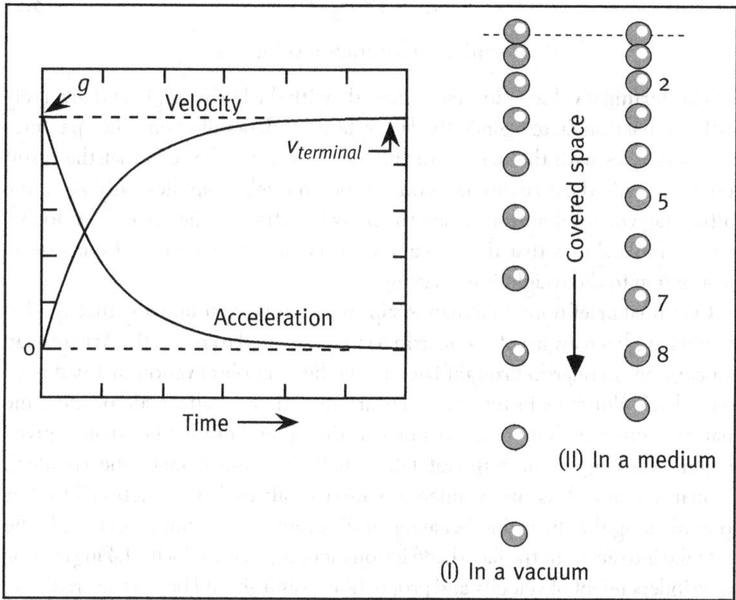

A body falls with constant acceleration—the acceleration of gravity—only in a vacuum. In media that offer resistance, the acceleration decreases gradually during fall and the velocity tends to a saturation value. In the right-hand figure: stroboscopic image of a body falling in a vacuum and in a medium offering resistance.

[38] Instantaneous velocity and acceleration, expressed as derivatives, are briefly discussed in the Mathematical Note of Chapter 12.

The behavior of the two quantities (3.1(a)) and (3.1(b)) is shown in the diagram. For $t = 0$ the acceleration coincides with that of gravity, then it decreases exponentially towards zero, which is reached the more rapidly the smaller the mass of the body is and the larger the friction factor b. After falling for a while, the body virtually stops accelerating and descends with constant speed (*terminal velocity*, parachute effect). If friction is absent—that is, there is free fall—the motion is always uniformly accelerated and the same for all bodies, since Newton's law reduces in all cases to $a(t) = g$. The figure on the right illustrates the first nine intervals of space covered during the fall, as they would appear in a sequence of photographic snapshots equally separated in time: (I) the ideal case in which friction is absent, as hypothesized by Galileo, (II) the real case in which a surrounding medium is present. From equation (3.1(b)), for $t \rightarrow \infty$, it is easily seen that the terminal velocity is given by

$$v_{\text{terminal}} = mg/b \tag{3.2}$$

(terminal or saturation velocity)

This terminal velocity increases directly with the body weight and inversely with the friction factor. Since the latter, besides depending on other parameters, increases with the density of the medium, it can be seen that the result tends towards the Aristotelian conception, though it applies only after the initial stage of acceleration, when the body has attained the terminal velocity. The formula shows that the descent velocity of a parachute will increase in proportion to the weight it is carrying.

One final brief note. In a manuscript which it is most unlikely that Galileo would have been aware of, Leonardo da Vinci seemed to reject the Aristotelian conception. Leonardo brought back to the light an observation that had been made by Philoponus in the sixth century AD: "If two balls made of the same material, one weighing twice as much as the other, fall together from a given height, the bigger one will not fall in half the time it takes the smaller". Leonardo realized (as later Galileo was also to realize) that air friction is the key to explaining the diverging behavior of different bodies, but he evidently did not take into account the fact that friction increases with velocity. Taking friction to be independent of velocity and proportional (as it should be) to the cross section of the body (whereas mass and weight vary according to volume), he arrived at a correction to the Aristotelian law—fall velocity proportional to weight—the sense of which is as follows: "If two spherical bodies unequal in size but made of the same material fall from a given height, the fall times will be in the same ratio to each other as the two diameters are". A statement which is valid only if it is referring to the terminal velocities, given by equation (3.2). In fact, as b is proportional to the cross section of the body (square of the radius) and m to the volume (cube of the radius), v_{terminal} turns out to be proportional to the body radius, in accordance with Leonardo's claim. Likewise, in the more general case of motion due to forces of any kind whatsoever, Leonardo still believes, like Aristotle, that the velocity of a body increases in proportion to the force which is being applied. Aristotle says: "If a given force or power pushes a body at a given speed, to double the speed of the same body a force or power twice as large will be needed". And in Leonardo's words: "If a given force moves a given body over a given distance in a given time, the same force will move another body half the size of the first body over the same distance in half the time".

4

Life Aboard Ship

Or, the Principles of Inertia and Relativity*

pigliatevi anco un gran vaso con acqua,
e dentrovi de' pescetti

bring also a large tank of water,
with some small fish in it

From the way observers see an event taking place, it is not possible for them to tell if they themselves are at rest, or are moving with uniform rectilinear motion. This is confirmed by life aboard ship, and the famous illustration of the motion of various small living creatures and the fall of drops of water inside a cabin below deck, where everything occurs just as it would on dry land. The main arguments of Aristotelian philosophers in support of the thesis that the earth is immobile, are shown to be ineffectual. In this part of the dialogue the revolutionary idea (later to be completed by Newton) of linking the force acting on a body to its acceleration rather than to its velocity, makes its first appearance.

And yet it moves

In this impeccably argued dissertation sent to Francesco Ingoli in reply to one of his essays, Galileo implicitly states the two fundamental principles which, in a sense, were to be the focal point of the Galilean revolution—inertia and relativity. The matter in dispute is whether

* From the *Letter in Reply to Ingoli.*

the earth is immobile—the Aristotelian standpoint—or is in orbit around the sun, while at the same time spinning on its own axis. Here we will just be looking at the latter of these two motions, which Galileo called "diurnal motion". There are basically three arguments which Ingoli puts forward in favor of the immobility of the earth and which are addressed by Galileo: first, the descent of a stone dropped from a tower is perpendicular and is parallel to the tower for the whole of its fall (true); second, the same also holds true for a stone dropped from the mast of a ship which is at rest, but is not the case if the ship is moving (false); third, the range of a cannon is the same whether it shoots to the east or to the west (true). Galileo shows that regardless of whether they are true or false, each of the three arguments is inconclusive.

Galileo pointed out flaws in the observations of phenomena made by certain Aristotelians—observations that were often accepted as valid without direct experimentation. He also pointed out flaws in their reasoning, which could often be used to prove the opposite of what they intended. The argument that the rock hits the ground right at the base of the tower and not some distance to the west of it, proves absolutely nothing. Galileo demonstrated that in either case—earth motionless, or spinning with constant velocity—the fall must take place in exactly the same way. This conclusion amounts to a statement of the two principles, inertia and relativity, at the same time.

The principle of inertia

According to the principle of inertia, a body, once set in motion, maintains its velocity if the resultant of the forces applied to it—both acting and resisting (typically friction)—is zero. The body remains motionless, therefore, if it was so at the initial instant. It is thus no longer necessary to invoke—as Aristotle did—an impressed force (Latin *vis impressa*), or some other *deus ex machina*, to ensure that motion is conserved. If a force intervenes, its effect is either to slow down the body, if it is counter to its motion, or to accelerate the body, if it is in the direction of its motion. The effect of the force is never to maintain the initial velocity of the body. This is a revolutionary concept, since before Galileo it was commonly held that the velocity of a body had to be somehow related to the presence of an applied force. Scholars were still stuck with Aristotle's words: "If a given force or power moves a given body at a given speed, twice as great a force or power will be needed to move the same body at twice as great a speed" (or, to put it another way, two horses pull a coach more rapidly than a single one).[39] Galileo's concept was so revolutionary that even today it is not easily accepted by the lay person. This is shown by surveys conducted among physics freshmen, who are revealed to be holding, in two cases out of three, pre-Galilean conceptions.[40]

Alexandre Koyré, a Frenchman who was a great expert on Galileo and who was also, however, a zealous supporter of Descartes, had the following to say about the respective merits of the two scientists regarding the principle of inertia:[41]

> The greatest glory of Descartes as a physicist is, without doubt, to have provided for the principle of inertia a clearcut formulation and to have inserted it at the right place. One could object, though, that when he formulated it, at the time of the *Principia*[42]—twelve years after Galileo's *Dialogue* and six years after his *Discourses*—there was no particular merit in doing it, nor difficulty. In fact, by 1644 the law of inertia was no longer a novel, unheard-of idea. Quite the contrary, thanks to the works and writings of Gassendi,[43] Torricelli and Cavalieri, it was becoming a universally

[39] In truth, Leonardo da Vinci had already expressed, even if in somewhat ambiguous form, ideas that can be seen as implying the principle of inertia. In his *Codice Atlantico* he writes: "Every motion will keep its course rectilinearly as long as the nature of the violence impressed by its motor will last in it". He does not say, unfortunately, whether such "impulse [*violenzia*]" dies out by itself, or by the action of external resisting forces. Giovanni Battista Benedetti, instead, was more explicit on this issue, and here, as on other matters, was a great source of inspiration for Galileo. It was Benedetti who first talked of the persistence of the speed of a body, even when the causes that have produced that speed come to an end—and to assign to the force the role of giving rise to acceleration (G.B. Benedetti, *Diversarum speculationum mathematicarum et physicarum liber*, Torino, 1585).

[40] A. Frova, *Perché accade ciò che accade* (Supersaggi BUR Rizzoli, Milano, 1995).

[41] A. Koyré, *Études galiléennes*, Hermann, Paris 1966. English translation *Galileo Studies* by J. Mepham, Humanities Press, Atlantic Highlands, NJ, 1978.

[42] R. Descartes, *Principia Philosophiae*, Amsterdam, 1644.

[43] Pierre Gassendi was a French scientist, a contemporary of Torricelli and Cavalieri who were both disciples of Galileo.

acknowledged truth. One could also add that, even if Galileo himself had not declared it *expressis verbis*, or at least had not stated it as the fundamental law of motion, his physics was so imbued with it that Baliani, a scientist whose mind in no way bore comparison to the minds of the aforementioned scholars, was able to derive it quite easily. One could cite Newton's judgment, which gave the entire credit for the discovery to Galileo, completely ignoring the name of Descartes . . . .

Indeed it was Newton who perceived in Galileo's work all the fundamental premises of a revolutionary conception of motion. A conception that had been centuries in preparation, but that only in Galileo became defined in its essential and systematic aspects.

Circular motion

If one really wanted to express a reservation on Galileo's merits, it may be said that he did not put all aspects of inertia in perfect focus. For instance, nowhere did he state explicitly that uniform circular motion is not inertial, that is, a type of motion that occurs without the action of forces. Circular motion, in fact, is an accelerated motion, because it implies a continuous change in direction (and for this to happen a force directed towards the center of the trajectory is needed, for example, the gravitational force in the case of an orbiting planet). Instead, in the *Dialogue* Galileo has Salviati say:[44] "and since in circular motion the moving body is continually going away from and approaching its natural terminus, the repulsion and the inclination are always of equal strengths in it. This equality gives rise to a speed which is neither retarded nor accelerated; that is, a uniformity of motion". On the other hand, no statement made by Galileo would permit it to be argued that he held the contradictory idea of circular inertia. From the excerpts which follow, both later in this chapter and in the next chapter, readers will be able to judge for themselves to what extent Newton was right in his judgment.

[44] *Dialogue Concerning the Two Chief World Systems*, First Day, translated by S. Drake, The Modern Library, New York, 2001, p. 35. In Italian: *Dialogo sopra i due massimi sistemi del mondo*, Giornata prima, Edizione Nazionale, Vol. VII, p. 56.

The principle of relativity

The principle of relativity states that a given physical phenomenon is described by one and the same law, whether observed by someone at rest or by someone who is in motion, provided this motion is rectilinear and uniform, that is, it is at a constant velocity (in scientific terms one says that the observer is in an *inertial reference system*). More generally: physical laws are the same for two observers who are moving with rectilinear uniform motion with respect to each other. Or, in other words, from the way observers see an event taking place, it is not possible for them to tell if they themselves are at rest, or are moving with uniform rectilinear motion. Only a non-inertial observer—that is, one who is in a state of acceleration—would give a different account of the event. Strictly speaking, observers located on the earth's surface are non-inertial, because the rotation of the earth carries them along an arc (or, we may say, bestows them with radial acceleration). However, given the size of the earth's radius, their motion deviates very little from uniform rectilinear motion and the error incurred by treating them as inertial observers is almost negligible (an estimate is given in the Mathematical Note at the end of this chapter). This is why Galileo does not reject, indeed he welcomes, the similarity Ingoli establishes between the fall of a body from a tower on dry land and from the mast of a moving ship.

The horse and coach

Before coming to Galileo's words themselves, it may be helpful to explain why the fact that two horses can pull a coach faster than one horse on its own, does not prove that it is necessary to maintain an applied force in order to produce motion. The exact contrary is the case—in order for a body (the coach) to cruise with constant velocity, it is essential that the overall force acting upon it be null: otherwise, by Newton's law—force equals mass times acceleration—the body would keep accelerating. How, then, does the power of the horses enter into the game? Why, if the force produces acceleration and not velocity, is a motor or some other device always required to keep a body moving?

The answer is: because in events in the real world the resistance of friction has to be overcome. The absence of acceleration simply indicates that the pulling force of the horses is such as to balance the resisting force due to friction (of the air, of the wheels on the road surface, etc.). It is well known that friction tends to increase as the speed of a body increases,[45] and so the power of two horses rather than one enables greater friction to be overcome and a higher speed attained. It can, instead, be stated without hesitation that two horses will be able to accelerate a coach more quickly than a single one (in a motor car: more power, more acceleration). This novel vision of the relationships among force, velocity, and acceleration constitutes the fulcrum of Galileo's new way of thinking. Several decades later Newton was to express these concepts in quantitative form, stating the fundamental law that bears his name.

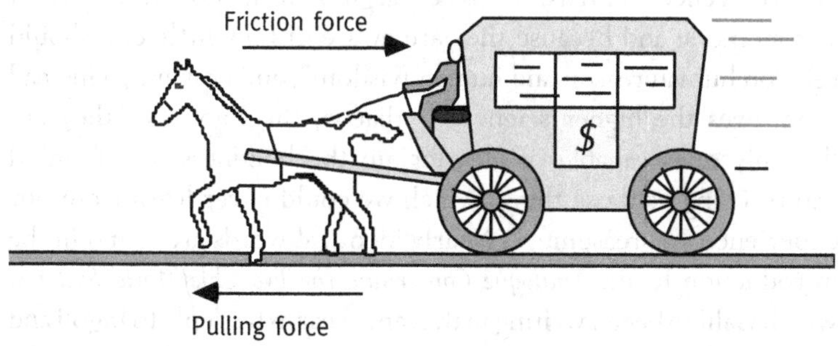

Cruising speed is attained when the horse develops a pulling force equal to the resisting force due to friction

Caution and irony

The opening paragraph of the letter is as ceremonious, baroque, and exaggeratedly refined in style, as it is pungent in content. Galileo says that he has not confuted Ingoli's theses earlier out of respect for his reputation. But, since his silence had led to many believing he had

[45] See Mathematical Note at end of Chapter 3.

been persuaded by arguments that were totally untenable, he intended to take steps to remedy the situation. In this first paragraph, moreover, Galileo briefly mentions his recent visit to Rome to pay homage to the newly elected Pope Urban VIII, Maffeo Barberini, who, it seems likely, was the real addressee of the *Reply*. The following observations would support this hypothesis: the text is crystal-clear, detailed perhaps even to excess, as when one is addressing a person who, though competent, is not an expert in the field, but whom one intends absolutely to convince; the style is particularly over-polished, especially where the argument is not scientific; Galileo is careful to stress that he does not wish to rebut any theological question, but stay "within the boundaries of human and natural discourses". Above all, there is a brief reference—in the second paragraph—to the so-called "Urban VIII argument" (see Chapter 9), where Galileo claims that if Catholics, as opposed to heretics, do not accept Copernican theses, it is not because they are ignoring them, but because they put "reverence and trust" ahead of "arguments and observations" of astronomers; and because they are aware of how little one should rely "on human reason and human wisdom", and how much instead "one owes the higher sciences"—that is, theology—as "they are the only ones capable of clearing up the blindness of our mind and teaching us those things which we could never learn from our experience and reasoning". Nearly identical words are found in the introduction to the *Dialogue Concerning the Two Chief World Systems*, which Galileo began writing in the same year as his reply to Ingoli and in which he returns, in greater scientific depth, to many of the themes dealt with here (see Chapter 5).

Now, anyone who knows even a little of Galileo cannot fail to notice, in this open declaration of submission, all the bitterness of a man who is being forced to write the exact opposite of what he thinks, and forms the basis of his life's work. Plus the profound disappointment of a scientist at finding he belongs to so backward a social and cultural milieu, and also the irony, bordering on the rash, that in places leaps out of the text. Equally imprudent appear: the relentless force of the argumentation he uses in demolishing each of the anti-Copernican theses (justified though this may have been by the troubles that had preceded the *Reply*—see the Historical Note towards the end of this chapter); the accusation of "lie(s)" and "deception" he makes against his opponents; and finally, his stubbornness in pointing out his opponents' two main methodological mistakes ("one is always to commit equivocations, assuming as known what is in question; the other is that when you think of experiments which could be made to discover the truth, without having made them you take them as made and report them as resulting in favour of your conclusion").

All told, then, in its substance the *Reply* is a splendid example of trust in rationality and of loving attention to the world of experience, and, as such, stands in contradiction to all the declarations made in its introduction. Anyhow, the Pope apparently read it and was satisfied with it. Let us read it too.[46]

LETTER TO FRANCESCO INGOLI IN REPLY TO THE *DISPUTATIO DE SITU ET QUIETE TERRAE*[47]

Eight years have already passed, Mr. Ingoli, since while in Rome I received from you an essay written almost in the form of a letter addressed to me. In it you tried to demonstrate the falsity of the Copernican hypothesis, concerning which there was much turmoil at that time. In particular, you dealt with the location and motion of the sun and the earth, maintaining that the latter is at the center of the universe and completely motionless and the former in motion and as far from the said center as from the earth itself. To confirm this you advanced three types of arguments: the first astronomical, the second philosophical, the third theological. Then, very courteously, you urged me to reply if I had noticed any fallacy or incorrect reasoning in them. Moved by your sincerity and other courteous feelings I had detected in the past, and completely sure that you had offered your thoughts to me with a candid heart and no envy, I examined them more than once and was desirous of reciprocating as best I could the sincerity of your soul. I concluded inwardly there was no more appropriate means of putting my desire into practice than to remain silent. For in this way I would not be spoiling the satisfaction which

[46] *Galileo's reply to Ingoli*, translated by M.A. Finocchiaro, in *The Galileo Affair: A Documentary History*, University of California Press, 1989, p. 154. In Italian: Edizione Nazionale, Vol. VI, p. 509.

[47] *DISPUTATIO... TERRAE*: Discussion on the location and immobility of the Earth.

I imagine you felt in coming to think that you had convinced a man like Copernicus; at the same time I would be leaving intact your reputation with those who had read your essay, insofar as it was within my power. I will not say that the regard for your reputation made me disregard my own, for I never believed mine is that tenuous; indeed I felt no one who has carefully examined your objections to that opinion which I then considered true could infer from my silence that I have less intelligence than is sufficient to refute them all. I say all, except the theological ones, concerning which it seems to me one must proceed differently from the others; in fact, they are not subject to refutations but are only liable to interpretations. However, I have now discovered very tangibly that I was completely wrong in this belief of mine: having recently gone to Rome to pay my respects to His Holiness Pope Urban VIII, to whom I am tied by an old acquaintance and by many favors I have received, I found it is firmly and generally believed that I have been silent because I was convinced by your demonstrations; some even regard them as necessary and unanswerable. Although their being so regarded is of some relief for my reputation, nevertheless experts as well as non experts have formed a very low opinion of my competence: the former because they understand the ineffectiveness of the objections and yet they see me being silent, the latter because they are capable of judging only by the results and so from my silence have inferred the same conclusion. Thus I have found myself being forced to answer your essay, though, as you see, very late and against my will.

Note, Mr. Ingoli, that I do not undertake this task with the thought or aim of supporting as true a proposition which has already been declared suspect and repugnant to a doctrine higher than physical and astronomical disciplines in dignity and authority.[. . .] Moreover, I am thinking of treating this topic very extensively, in opposition to heretics, the most influential of whom I hear accept Copernicus's opinion; I would want to show them that we Catholics continue to be certain of the old truth taught us by the sacred authors, not for lack of scientific understanding, or for not having studied as many arguments, experiments, observations, and demonstrations as they have, but rather because of the reverence we have toward the writings of our Fathers and because of our zeal in religion and faith. Thus, when they see that we understand very well all their astronomical and physical reasons, and indeed also others much more powerful than those advanced till now, at most they will blame us as men who are steadfast in our beliefs, but not as blind to and ignorant of the human disciplines; and this is something which in the final analysis should not concern a true Catholic Christian—I mean that a heretic laughs at him because he gives priority to the reverence and trust which is due to the sacred authors over all the arguments and observations of astronomers and philosophers put together. To this we should, finally, add another benefit, that is, to understand how little one should rely on human reason and human wisdom, and therefore how much one owes the higher sciences; they are the only ones capable of clearing up the blindness of our mind and teaching us those things which we could never learn from our experience and reasoning.

Diurnal motion

After this obsequious, but at the same time ironical, preface, Galileo gets to the contents. The extract which follows deals with the rotation of the earth about its axis.

[...] As regards the diurnal motion, namely the motion of rotation every twenty-four hours from west to east, there are many reasons and experiments produced by Aristotle, Ptolemy, Tycho,[48] *and others. However, you get by very lightly by mentioning only two, namely the much used one about heavy bodies falling vertically on the surface of the earth and the other one about projectiles, which without any difference move through equal distances toward the east as well as toward the west and toward the south as well as toward the north. Maybe you get by so quickly, I believe, because of the great obviousness and necessity with which they seem to convince you. However, these and others were well known to Copernicus, and they were examined by him and much more attentively by myself; and I know they are all such that either nothing can be proved on the affirmative or negative side, or else if any consequence can be deduced it favours the Copernican opinion. Moreover, I say I have other evidences not previously observed by anyone, which are necessarily convincing about the certainty of the Copernican system (as long as we remain within the limits of human and scientific inquiry). As all these things need more extended analyses to explain them, I reserve them to*

[48] Tycho Brahe—the great sixteenth century Danish astronomer, Kepler's teacher.

some other time.[49] *In the meantime, to answer what is sufficient for the things you touched upon, I repeat that you and all the others begin by having the earth's stability firmly impressed upon your minds, and because of this you fall into two errors: one is always to commit equivocations, assuming as known what is in question; the other is that when you think of experiments which could be made to discover the truth, without having made them you take them as made and report them as resulting in favor of your conclusion. I shall try as concisely as possible to make you grasp these two errors; some other time you will be able to see this point treated at great length, with the answers to all the objections which seem prima facie to have some probability, but have none at all.*

The earth and the ship

With Aristotle and others, you say the following. If the earth were turning on itself every twenty-four hours, then rocks and other heavy bodies falling from on high downward, for example, from the top of a tower, would not hit the earth at the foot of the tower; for, during the time the rock spends in the air going down toward the center of the earth, the earth would proceed at great speed toward the east, carrying along with it the foot of the tower, and so the rock would necessarily be left behind by a distance equal to how much the earth's turning would have advanced forward in the same time, which would be many hundreds of feet. They confirm this reasoning with an example taken from another experiment, where they say the same thing can be clearly seen: if one lets a rock fall freely from the top of the mast of a ship standing still in harbor, it falls vertically and hits the foot of the mast precisely at that point which is perpendicularly below the place from which the rock was dropped; this effect does not happen (they say) when the ship moves at a fast speed; for in the time the rock takes to go from top to bottom, and during which it falls freely and perpendicularly, the ship moves forward and leaves the rock behind toward the stern many feet from the foot of the mast; the corresponding thing should happen to the rock falling from the top of

[49] Galileo was to return to these arguments eight years later, with the publication of the *Dialogue Concerning the Two Chief World Systems*.

a tower, if the earth were turning with such a great speed. This is the argument, in which I see very clearly the mistakes I am talking about.

My reasons are as follows. When you and Aristotle infer that the rock falling from the top of the tower moves in a straight line perpendicular to the earth's surface, you do this, and can do this, only from seeing how during its fall it goes on licking (as it were) the surface of the tower, erected perpendicularly on the earth's surface; this makes one perceive the line traced by the rock as being also straight and perpendicular. However, here I say that from this appearance one cannot infer that conclusion unless one assumes the earth is motionless while the rock falls, which is the question at issue. For, if with Copernicus I say the earth turns and consequently carries with it the tower and also us who are observing the rock, then we can say the rock moves with a motion composed of the general circular motion toward the east and the accidental straight motion toward the whole to which it belongs, and the result of these is a motion inclined toward the east;[50] *in this case, the motion which is common to me, the rock, and the tower is for me imperceptible, as if it did not exist, and only the other remains observable, which the tower and I do not share, namely the motion of getting nearer the earth. Here, then, is the misunderstanding exposed, if I have been able to explain myself sufficiently. Let me add another point. Arguing from parts to whole, you and Aristotle elsewhere say that, since the parts of the earth are seen to move naturally straight downward, one can infer that such is the natural inclination of the whole earth, namely to desire the center and, having already reached it, to stop therein. Therefore, arguing much more correctly from whole to parts, I shall say that, since the natural inclination of the terrestrial globe is a circular motion around its center every twenty-four hours, this is also the inclination of the parts, and so by their nature they circle the earth's center every twenty-four hours, and this is their congenital, proper, and completely natural behaviour; if by some violence they are*

[50] A vectorial sum readily illustrates the statement: the motion of the rock is in an oblique direction, with an eastward component identical to that of the earth's surface. Terrestrial observers do not perceive this component as it is common to them, so they describe the fall as being vertical.

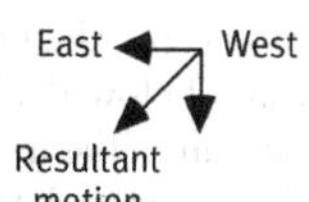

removed from their whole, then as an accident the other motion of fall is added to that. [...]

The other error involves your putting forth experiments as having been made and corresponding to your needs, without ever having carried them out or made the observations. To begin with, if you and Tycho wanted sincerely to confess the truth, you should say that you have never tested whether or not any difference is observable in shooting artillery toward the east and toward the west, or toward the north and toward the south (especially in regions near the pole, where according to you the effect should be more noticeable); I am moved to believe this, indeed to be sure of it, by the fact that you put forth as certain and unequivocal other experimental observations which are much easier to make and which I am so sure you did not make as to be able to say that whoever makes them will find the effect to be contrary to what you claimed with excessive confidence. One of these experiments is precisely that of the rock falling from the top of the mast on a ship; the rock always ends up hitting the same spot, whether the ship is standing still or moving forward fast; it does not strike away from the foot toward the stern,[51] *as you believed (on account of the ship moving forward while the rock comes down through the air). Here I have been a better philosopher than you in two ways: for, besides asserting something which is the opposite of what actually happens, you have also added a lie by saying that it was an experimental observation; whereas I have made the experiment, and even before that, natural reason had firmly persuaded me that the effect had to happen the way it indeed does. Nor was it difficult for me to discover your deception. For you first imagine someone on a motionless ship at the top of the mast who drops a rock while all is still, and then you fail to notice that when the ship is in motion the rock no longer starts from rest, since the mast, the man at the top, his hand, and the rock are also moving with the same speed as the whole boat; I still frequently meet*

[51] Strictly speaking, Galileo's statement is valid when there is no movement of air in relation to the ship, such as below deck, as in the situation described later in this chapter beginning with these words: "Shut yourself with a friend in the largest cabin below deck". When air movement is present, including that occurring as a result of the motion of the ship, some displacement of the rock occurs, however this is minimal if the rock is very heavy.

people with such a thick skull that I cannot put it into their head that, because the man on the mast keeps his arm still, the rock does not start from rest. Thus, I say to you, Mr. Ingoli, that when the ship is moving forward, the rock also moves with the same impetus; this is not lost just because the person who held it opens his hand and drops it, but is permanently conserved in the rock, and so is sufficient to allow it to follow the ship; no longer held by the person, the rock falls down on account of gravity, thus combining the two motions into a transversal and inclined (and perhaps also circular) motion toward where the ship is going; and so it strikes the same spot on the ship where it fell when all was still. From this you can understand how the same experiments advanced against Copernicus by his opponents speak much more in his favor than in theirs; for the motion transmitted by the ship is certainly accidental to the rock, and so, if it is conserved to such an extent that the effect is observed to be exactly the same whether the ship is still or moving, then what doubt can remain that the rock brought to the top of the tower and sharing the same speed as the terrestrial globe would conserve it while falling? Note, in fact, that such a speed would be a natural, primary, and eternal inclination, unlike that of the ship, which is accidental.

As Drake pointed out,[52] the anti-Copernican argument of the rock dropped from a tower was not Aristotle's, although his followers liked to attribute it to him. Galileo uses it extensively because it allows him to introduce simultaneously the principles of relativity, of conservation of motion and, as will be shown in the next chapter, of independence of motions. In earlier writings Galileo had dismissed the argument of the tower—addressed here in such rich detail—in a much simpler manner. A body always falls along a line that connects it to the center of the earth; since the side of the tower is parallel to this line, the body must fall alongside the tower, both if the earth is spinning and if it is standing still. This idea was, indeed, partly derived from Aristotle, except that the Greek philosopher had put it forward with reference to the center of the universe (which in

[52] S. Drake, *Galileo: Pioneer Scientist*, University of Toronto Press, Toronto, 1990, p. 82.

his view coincided with the center of the earth—a detail which, however, makes no difference to the argument).

Let us now return to Galileo's words, where he is considering the second argument, the one regarding the range of artillery projectiles.

Artillery shots

As regards the projectile motions of artillery, I have no doubt that things happen precisely as Tycho and, following him, you say, although I have not made the experiments; that is, no difference is seen, and the shots range the same in any direction you wish. However, I add (a point Tycho did not understand) that this happens because it must happen this way whether the earth moves or stands still, and that no difference can be conceived, as you will understand with clear reasons at the proper time. In the meantime, there is another way to remove these and all other difficulties of this kind, such as the flight of birds and how they could keep up with so much motion, and also the suspension of clouds in the air and the fact that they are not always running toward the west, as you think it should happen if the earth were in motion. To solve all these apparent difficulties I say that, as long as water, land, and air (which are their environment) do the same thing together, namely either jointly move or jointly stand still, then necessarily all phenomena appear to us exactly the same in the one as well as in the other state. I say all, meaning those involving the mentioned motions of falling bodies, projectiles upward or sideways in this or that direction, birds flying toward the east or the west, clouds, etc.

The cabin below deck

Now comes the famous and evocative passage regarding the motion of small creatures and the fall of water drops inside a ship's cabin—motion which occurs in exactly the same way whether the ship is moving, or is standing still. In these pages, Galileo introduces the principle of relativity and implicitly suggests its link with the principle of inertia (actually, the former implies the latter). The enchanting realism of the passage is a sign of how Galileo's inquiring eye fixes upon the most humble of everyday objects and phenomena with the same curiosity and attention with which it explores the heavens. The laws which govern the very small and the immense are the same: the flight of an insect is no less worthy of observation than the movement of a star in the sky. And in fact, with strict logic, Galileo extends the behavior of what he has seen aboard ship to the earth itself, which rolls through space, dragging along its envelope of air, all things and ourselves too, without us being able to perceive it.[53] Galileo's attitude is worth highlighting to the many young people who go into science today because they are attracted by the remote charm of galaxies or black holes, while they remain blind and indifferent to the myriad questions that spring out of everyday experience.

Shut yourself with a friend in the largest cabin under the deck of a large ship, and make sure you have with you some flies, butterflies, and similar small flying animals; bring also a large tank of water, with some small fish in it; arrange also some vase up high dripping into a lower one with a narrow neck. When the ship is standing still observe carefully how those small flying animals go with equal speed toward all parts of the room; you will see the fish swim indifferently toward any side of the tank whatever; the falling drops will all go into the vase below; if you throw any object to your friend you need not exert greater effort in one direction than in another, as long as the distances are equal; and if you jump with your feet tied together, you will reach equal distances in all directions. After carefully observing all these things, let the ship move with as much speed as you wish; as long as the motion is uniform and does not fluctuate here

[53] An analogy had been drawn between the earth and a ship prior to Galileo, by Giordano Bruno (*La cena de le ceneri*). The bodies that are on the earth participate in its movement because they are part of it, they *are in it*—in precisely the same way as the bodies on a ship participate in the ship's motion: "if this were not true it would follow that, when a ship was sailing on the sea, nobody could ever throw something straight from one part of it to another, and it would not be possible to jump vertically and fall back with one's feet to the same spot they had left from". Bruno had been ahead of Galileo also as regards the fall mechanism of a rock from the mast of a moving ship: "If from the foot of the mast somebody on the ship throws a rock straight up to the top of the mast, or to the crow's nest, the rock will fall back along the same line no matter how fast the ship is traveling, as long as it does not heel over".

and there, you will not notice the least change in all the things mentioned, nor from any of them or anything within yourself will you be able to ascertain whether the ship is advancing or standing still.[54] *If you jump on the floor, you will reach the same distances as before, nor, because the ship is moving very fast, will you jump farther toward the stern than toward the bow, even though while you are in the air the floor advances in a direction opposite to your jump; if you throw a fruit to your friend, you will not need to throw it harder to reach him when he is toward the bow and you astern than when your positions are reversed; the drops will fall into the lower vase without any being left behind astern, even though while the drop is in the air the ship advances many inches; the fish in their water will not have to work harder to swim toward the front than toward the rear side of the tank, but with equal ease they will go to catch the food you give them from any part of the rim of the tank; the butterflies and flies will be able to fly indifferently in all directions, nor will they ever be stuck to the wall toward the stern, as if they were tired of keeping up with the fast course of the ship, from which they are separated when they remain suspended in air; and if you produce some smoke by burning a little incense, you will see it rise on high and linger there, and like a small cloud move indifferently no more in one direction than in another. If you ask me for the cause of all these effects, at the moment I answer: "Because the general motion of the ship is indelibly conserved in all those things contained in it, being transmitted to the air and to them, and not being contrary to their natural inclination".*[55] *Some other time you will hear detailed answers explained at length. Now, once you have made all these observations, and seen how these motions (though accidental and temporary) appear exactly the same when the ship moves as well as when it stands still, will you not abandon all doubt that the same must happen in regard to the terrestrial globe, as long as the air moves together with it? This is especially true since that general motion, which for the case of the ship is accidental, is taken by us as natural and proper for the case of the earth and terrestrial things. Add to this that for the case of the ship we cannot determine from the things within it what it does, even though with ease we can make it move or make it stand still; how can it be possible to determine this for the earth, which in our experience has always been in the same state?*

[54] Statement of the principle of relativity.

[55] Statement of the principle of inertia.

HISTORICAL NOTE

"Eight years have already passed, Mr. Ingoli, since while in Rome I received from you an essay written almost in the form of a letter addressed to me. In it you tried to demonstrate the falsity of the Copernican hypothesis, concerning which there was much turmoil at that time . . .".

The letter addressed in 1616 to Galileo by jurist Francesco Ingoli, secretary to the Congregation of *Propaganda Fide*, followed up a series of conversations which had taken place between him and Galileo in Rome in December 1615. In the letter Ingoli tried to confute Copernicus' theories using arguments taken from Aristotle, Ptolemy, and Tycho Brahe, the famous Danish astronomer whom Galileo particularly disliked because of his conciliatory theses (all planets go around the sun, but the earth stands still and the sun, together with the planets, goes around it).

Early in 1616 events took a dramatic turn which forced Galileo to postpone his reply to more favorable times. In February that year, in fact, the consultors of the Holy Office declared the two propositions "The Sun is the center of the world and hence immovable of local motion" and "The Earth is not the center of the world, nor immovable, but moves according to the whole of himself, also with a diurnal motion", philosophically erroneous and heretical, because they were in contradiction with the Holy Scripture. Following this, in March the same year, the Decree of the Congregation of the Index suspended the Copernican writings until they had been corrected. There was no explicit condemnation of Galileo, though a few days earlier he had been summoned by Cardinal Robert Bellarmine—the most authoritative theologian in the Jesuit order, who personified the spirit of the Counter-Reformation; he was made a saint in 1930 and named Doctor of the Church—and warned that he should relinquish altogether the condemned opinion, and not hold, teach, or defend it in any way whatsoever, verbally or in writing.

Galileo had gone to Rome with an optimistic attitude, sure that the strength of his arguments would enable him to counter the anti-Copernican and anti-Galilean campaign which had been started by two Dominican friars, Niccolò Lorini and Tommaso Caccini, and had reached the attention of the Holy Office. At the Vatican he had achieved considerable personal success, thanks to his brilliant talents in logic and speech and to the breadth of his culture, which ranged across music and the humanities. Nevertheless he left defeated, and defeated along with him was that internal faction of the Church that would have liked to open up to scientific culture. "this is not a place to come to and dispute about the moon, nor to try, in the current century, to defend or introduce new doctrines": thus, with prophetic scepticism, had written Francesco Guicciardini, the Florentine ambassador to Rome, in December 1615, upon learning of Galileo's forthcoming visit to town.[56]

It was already quite something that Galileo had been able to leave Rome, as Cardinal del Monte writes, "with his reputation intact and praised by everybody who dealt with him", and with a certificate, given to him by Cardinal Bellarmine, acknowledging that he had not abjured his opinions, "nor has he received any penances salutary or otherwise".

Eight years later times seemed to have changed and become more favorable to a resumption of the discussions concerning Copernican issues. In 1623, in fact, Cardinal Maffeo Barberini—who on several occasions had shown his liking for Galileo and a strong interest in his ideas—had been elected Pope, under the name Urban VIII. The new Pope was well-versed in the arts and sciences and seemed the ideal man to lead the Church towards a position of tolerance and cultural opening (so much so that the Lincean Academicians decided to dedicate the *Assayer*, which Galileo had finished writing in 1622, to him).

Galileo did not waste any time and in April 1624 he journeyed to Rome. He hoped to persuade the Pope to relax the "salutary edict" which banished Copernicanism, and which, according to what the Pope himself had apparently confessed to Campanella,[57] "had it depended on him, would not have been issued". However, Barberini, probably rendered particularly cautious by his new position, made it clear during no less than six audiences granted to Galileo that he did not intend to go beyond a few friendly gestures, even less to authorize him to ignore the 1616 decree. Moreover, the Pope still seemed convinced of a thesis he had repeatedly expressed in the past,

[56] Edizione Nazionale, Vol. XII, p. 207.

[57] Tommaso Campanella was a Dominican monk and a philosopher. See more details in Chapter 2.

which meant he had no fear that Copernicanism could ever be proven true. This thesis, which bears his name, is surprising to say the least: no matter how many proofs can be arrayed in favour of Copernicus' theses, God, in his infinite might, can achieve the same effects by making the sun rotate about the earth, as claimed by the Holy Scripture.

Even if the primary objective of his journey had failed, Galileo somehow felt entitled to resume the discussion, owing to the Pope's benevolent attitude towards him. He finally composed his reply to Ingoli and began to write a book on the world systems, the *Dialogue*, which he had been thinking about for years.

MATHEMATICAL NOTE

As an example of the principle of relativity, let us confirm that the fundamental law of dynamics—Newton's law—as written by a given observer S

$$\boldsymbol{F} = m\boldsymbol{a} \tag{4.1}$$

is the same for another observer S′ who is moving with uniform rectilinear motion with respect to S (i.e. *the fundamental laws of physics are invariant in two reference systems that can be related by a Galilean transformation*). $\boldsymbol{F}$ and $\boldsymbol{a}$ are vectors that give—in value, direction, and orientation—the force applied to a body and its acceleration, respectively, m is the inertial mass, a quantity characterizing the body, which establishes the proportionality between force and acceleration. We need to show that Newton's law for observer S′:

$$\boldsymbol{F}' = m\boldsymbol{a}' \tag{4.1′}$$

can be brought back to (4.1).

Let $\boldsymbol{V}$ be the relative velocity of the two observers. If $\boldsymbol{v}$ is the velocity of the body as measured by the first of the two, the transformation that enables the velocity $\boldsymbol{v}'$ of the body as simultaneously seen by the other observer to be obtained—the Galilean transformation—is

$$\boldsymbol{v}' = \boldsymbol{v} + \boldsymbol{V}$$

and by taking the time derivative,[58] for the accelerations one has:

$$\boldsymbol{a}' = \boldsymbol{a} + \boldsymbol{A}$$

where $\boldsymbol{a} = d\boldsymbol{v}/dt$, $\boldsymbol{a}' = d\boldsymbol{v}'/dt$, and $\boldsymbol{A} = d\boldsymbol{V}/dt$ is the acceleration of the second observer with respect to the first. Since $\boldsymbol{V}$ is constant, one has $\boldsymbol{A} = 0$ and $\boldsymbol{a}' = \boldsymbol{a}$. On the other hand, the force turns out to be the same for both

[58] A short presentation of instantaneous velocity and acceleration, expressed in terms of derivatives, is given in the Mathematical Note of Chapter 12.

observers, because it can be determined statically by means of a dynamometer. Therefore $\boldsymbol{F} = \boldsymbol{F'}$ and (4.1′) is brought back to (4.1).

This is not so, instead, if one of the two observers is in a state of acceleration, because $\boldsymbol{V}$ does not remain constant in time. Newton's law according to S′ is therefore different from 4.1′:

$$\boldsymbol{F} = m(\boldsymbol{a} + \boldsymbol{A})$$

Galilean transformation proves to be valid as long as the relative velocity $\boldsymbol{V}$ between the two observation systems remains low with respect to the velocity of light, that is, within the limit of non-relativistic velocities. If this is not the case, the description of the motion of the body must be done using Einstein's relativity laws, and the Galilean transformations must be replaced by the less obvious Lorentz transformations.

A comment for the more expert reader. How far is Galileo's reasoning (a rock in free-fall from a tower hits the ground at exactly the same point that it would if the earth were immobile) affected by the assumption that the earth is an inertial system? In a rotating system a body in motion is subjected to the Coriolis force, the very same force that induces a rotation in the oscillation plane of the Foucault pendulum. Leaving aside the calculations, this force would make the path of the rock deviate slightly from a perfectly vertical line, giving rise to an eastward displacement of the impact point equal to (approximate result, valid at the equator):[59]

$$\Delta x \cong \frac{1}{3}\omega g t^2 \approx 2 \text{ cm}$$

where the angular rotation velocity of the earth is $\omega = 7.27 \cdot 10^{-5}$ rad/s, the acceleration of gravity is $g = 9.81$ m/s^2, and the free fall time for a

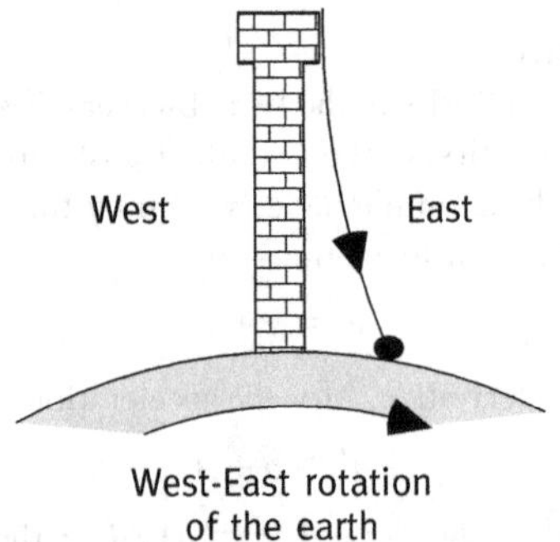

How a rock would fall by effect of the Coriolis force.

[59] See most general physics or mechanics textbooks.

100 m-high tower is $t \approx 4.5$ s. In terms of simple physical intuition, this is due to the fact that at the top of the tower the rock is traveling (horizontally) faster than it is at the foot of the tower, owing to the fact that it is at a slightly greater distance from the earth's rotation axis. Therefore, as a consequence of the principle of inertia, as it descends the rock always has a slightly greater horizontal velocity than each successive point on the tower that it passes, and thus it ends up gaining space towards the east, the direction in which all objects are moving. The effect, as the calculation shows, is so small as to be hardly measurable.

As to the fact that observers on the earth's surface are "nearly" inertial, but not quite, it is worth estimating, by way of example, the error they would make in measuring their own weight, compared to the value they would obtain if the earth were still. The law of gravitation states that, for inertial observers, their weight p would be given by their mass m times the acceleration of gravity g

$$p = mg$$

In reality, terrestrial observers—who for the sake of simplicity we shall place at the equator—feel the action of two forces upon their bodies: the gravitational force and the centrifugal force, associated with the rotational motion of the earth and directed upwards. The value of the latter is

$$F_c = m\omega^2 R$$

where ω is again the angular rotation velocity of the earth and R is its radius. Observers would therefore measure a smaller (apparent) weight, given by $p' = p - F_c$.

Let us make an estimate of the relative error, knowing that $R = 6.360$ km:

$$\frac{p - p'}{p} = \frac{m\omega^2 R}{mg} = \frac{(7.27 \cdot 10^{-5})^2 \cdot (6.36 \cdot 10^6)}{9.81} = 3.4 \cdot 10^{-3}$$

which amounts to a few tenths of a per cent, but is definitely measurable. The weight measured gradually gets larger as the poles are approached, both because the centrifugal force diminishes and because the flattening of the terrestrial globe causes an increase in g.

5

Towers Arrows Cannons and Birds

More on Inertia and Relativity*

Quando gli uccelli avessero
a tener dietro al corso de gli alberi
con l'aiuto delle loro ali, starebbero freschi

If birds had to keep up with the course
of the trees by means of their wings,
they would soon fall behind

Using a wealth of arguments Galileo reaffirms his assertion that the phenomena invoked by Aristotelians to demonstrate that the earth is immobile do not prove their claim. He rejects the need for an "impressed force" and again states that it is inertia which ensures that motion continues—and explains the observed behavior—when a number of events are examined: the fall of a rock from a tower or from the mast of a moving ship; trajectories of artillery shots; the motion of projectiles shot from a moving carriage; the flight of birds—and others. Among the pearls we have the discussion about the independence of mutually perpendicular motions and that most famous of examples—the time it takes for cannon balls to fall to earth.

Inertia inertia inertia

In Chapter 4 we saw how Galileo, in his reply to Ingoli, said repeatedly that he intended to return to the arguments on inertia and relativity

* From *Dialogue*.

and to deal with them more thoroughly. In fact, in his *Dialogue Concerning the Two Chief World Systems*, published in 1632, he returned to the comparison between a rock falling from a tower situated on dry land and a rock falling from the mast of a moving ship. It is upon the presumed difference between the behavior of the stone in the two cases that the Aristotelians based their claim that the earth does not rotate. Point by point Galileo proves that this difference does not exist. Although his discussion is repetitive in places, it contains a wealth of new examples, each one of which surpasses its predecessor in strengthening his arguments.

The subject, therefore, more than merits being revisited at some length. It may be useful to mention in advance some of the themes which we think are likely to capture the attention of the reader. First—the delightful way in which Salviati traps Simplicio, forcing him to state, albeit reluctantly, that in the absence of "external impediments" a moving body will retain its state of motion indefinitely. From here it is but a small step to the conclusion that the rock must fall at the foot of the mast, both when the ship is moving and when it is stationary. Thus, whether the Aristotelians like it or not, their argument that the vertical fall of a rock from a tower is proof that the earth is immobile, just fades away.

No less delightful is the theme which follows this: the confutation of the Aristotelian belief that a body thrown into the air continues in its forward motion because it is carried by the air itself—which, it was claimed, received an impulse from the arm of the thrower—rather than as a consequence of inertia. Once again Simplicio finds himself in trouble when Sagredo shows him that, if such were the case, an arrow "shot sideways" instead of "point foremost" would travel further.

Motion of projectiles

Another passage of particular importance is the one in which Galileo asserts that two mutually perpendicular motions have no influence on one another—each motion occurs precisely as it would if the other were absent. Here we have the famous example, once again entrusted to Sagredo, which opens with the words "Similarly, if a perfectly level cannon on a tower . . .". Sagredo proves that a cannon ball shot horizontally at very high speed takes the same time to fall to the ground as another cannon ball which is merely dropped from the mouth of a cannon, having received no forward propulsion (ignoring, of course, any slight differences due to air friction). The reason for this is that the vertical motion of descent of the cannon ball is not influenced by its horizontal motion. Even today many people are likely to be surprised by this result, so far removed from common sense is the idea of the independence of the two motions (a mathematical treatment of the problem is given in the Mathematical Note at the end of this chapter).

In fact, the Aristotelian prejudice that two motions experienced by one and the same body are not independent of each other was firmly rooted even among some of Galileo's pupils and followers. Antonio Nardi and Giovan Battista Baliani, to mention but two, held that horizontal velocity gradually decreased as a result of simultaneous vertical fall, and that the latter occurred more slowly because of the former. Descartes declared himself utterly opposed to the description put forward by Galileo. It is disconcerting, however, that today—at the beginning of the third millennium—the typical response to this problem remains (just as it does to other problems which are even more common to everyday experience) substantially pre-Galilean.

In 1658, the members of the Accademia del Cimento decided to put Galileo's idea to the test in an experiment conducted from the tower of the old Fortress in Livorno. Although the results were close to Galileo's hypothesis, they were not entirely unequivocal—probably owing to air friction and to inaccuracies in the experimental operating conditions. As a consequence the experimenters wrote in the Academy diaries that the matter could not be settled in a decisive manner. Of interest is the method they devised to ensure the two cannon balls started at the same time: the ball to be merely dropped was suspended just below the mouth of the cannon on a cord; the fired ball broke this cord as it exited from the cannon, thereby setting the suspended ball free.

Dispute with a pupil

As regards the trajectory described by a projectile—that is any object which is thrown—a rather curious event is worth recounting. Galileo's description contains all the elements needed to deduce the exact shape of this trajectory, that is, a parabola (prior to Galileo it had been thought that the shape of a trajectory was the arc of a circle). Even though he was well aware of this parabolic behavior, Galileo had never stated it explicitly. However, he reports that he had spoken of it in private with his friend and pupil Benedetto Castelli. It is reasonable to assume that he intended to deal with the matter more thoroughly in a further paper. In fact, he was eventually to do so in the fourth day of the *Discourses*, started in 1633 and published in 1638. Whatever, the fact is that his pupil Bonaventura Cavalieri[60] took the liberty of publishing this news of the parabola—which by then had evidently become common knowledge within the circle of close friends—in his book *Specchio Ustorio* of 1632. Cavalieri informed Galileo of this by letter only afterwards, when it was already a *fait accompli*. This made Galileo—whose morale had already been considerably shaken by the disturbing news he was getting from Rome[61]—furious. In a letter to Marsili[62] he wrote:

I cannot hide from Your Lordship that this news has much disappointed me, seeing as I am now deprived not only of the priority of a 40 year long study of mine (the larger part of which I have described in confidence to the aforementioned Father[63]), but also of that glory which I longed for so keenly and might have expected from such hard work. Because, in truth, my primary intention when I started speculating about motion was to find just such a trajectory. Once found, this trajectory is not too difficult to

[60] Holder of the Chair of Mathematics at the University of Bologna from 1629.

[61] Just a few days earlier the Grand Duke of Tuscany's Ambassador to the Vatican had reported of the Pope's anger at the contents of the *Dialogue* and of his suspicion that it was he who was being portrayed in the figure of Simplicio. Shortly afterwards Galileo was to be summoned to Rome by the Inquisition and brought to the trial culminating in his abjuration.

[62] Letter to Cesare Marsili of September 11, 1632, Edizione Nazionale, Vol. XIV, p. 368.

[63] Benedetto Castelli.

demonstrate—but it requires a lot of effort to first devise it, as I well know having gone through it. If Father Bonaventura had made his intention known to me before going ahead with publishing (as, perhaps, good manners required) I would have immediately asked him to let me first publish my book, after which he could have then added as many further findings as he wished.

Thus admonished, poor Cavalieri, who clearly held his teacher in the highest regard and had not the slightest wish to defraud him, tried to justify himself as follows:[64]

What I said about motion I said as a disciple of yours and of Father Benedetto's.[65] You are quite right in saying that I should have been a little clearer that the idea of the parabolic curve was Your Excellency's. But you must know that my suspicion that I was not, perhaps, in full agreement with your conclusion, kept me from explicitly ascribing to you something that you might later have to reject as not belonging to you.

He added that, having seen Castelli "perform experiments with other students and heard the same conclusion reported by these students, too" it had seemed to him

that both the conclusion and you being its author were such common knowledge that there could be no doubt that I could arrogate it to myself . . . and this also has contributed to the delay in writing to you about it, as I deemed that you actually would not care, on the contrary that you would be pleased that a disciple of yours, through such an appropriate action, would show himself a follower of your doctrine, admitting moreover to have learned it from you.

For the record, it has to be said that, shortly afterwards, Galileo was to acknowledge that his pupil had acted in good faith.

Shots towards the east and shots towards the west

From the motion of projectiles the discussion next turns to the problem of the range of artillery shots, which is the same regardless of whether the shots are made towards the east or towards the west.

64 Edizione Nazionale, Vol. XIV, p. 394.
65 Castelli.

This, however, is still not a proof that the earth is immobile. Galileo argues that exactly the same would be the case if the earth were rotating and for this purpose he draws an analogy between the earth and a "running carriage"—similar to the earlier analogy he drew between the earth and a ship. A bolt shot from the carriage would fall to the ground at the same distance from it, whether it was shot in the direction of motion of the carriage, or against it.

Many other issues are dealt with in the discussion—the nature of wind, the fall of objects from moving vehicles, the flight of birds, the art of taking aim. The examination of each of the phenomena in this section of dialogue illustrates clearly how Galileo's new scientific thought represents a total overturning of Aristotelian ideas ("this whole thing takes place just exactly opposite to what Aristotle says"), in terms not only of the principles that are being expounded, but also of the continuous reference to tangible experience as the starting point for all logical thinking.

The text

The technique Salviati uses to lead Simplicio, little by little, to the discovery of some fundamental concepts, is worth noting. The discussion never starts out with these concepts, partly because this is not in the style of the new science, but also because Salviati is well aware that if he were to start with them he would never succeed in dislodging Simplicio from his creed. Instead he prefers it to be Simplicio himself—appropriately prodded and spurred—who, almost inadvertently, hits upon some incontrovertible principles, drawing upon experiences that either he has already had, or he could have, but upon which he has never reflected. In other words, he wants to show him that "he has the solutions at his fingertips though he does not notice them". Simplicio traces out the line of reasoning all by himself, or at least this is what he is led to believe. It is reminiscent of the Socratic maieutic art, but here applied to scientific investigation—in the place of a young man who does not know that he knows and discovers his knowledge gradually with joy, we have here an adult who presumes that he knows. He, too, can attain knowledge, using instruments he has within himself, provided he starts by dismantling bad mental habits and freeing himself from the dead weight of his old beliefs.

As for the psychology of the interlocutors, note how masterfully Salviati lays siege to Simplicio ("You say . . . now tell me . . . very good . . . I do not want you to declare or reply anything that you do not know for certain . . . now tell me . . . pay careful attention to what you are saying . . .") and how he makes sure, before moving on, that there is no risk of Simplicio slipping back ("And you take this for granted not because I have taught it to you . . . but all by yourself, by means of your own common sense"). Possessing a much more refined dialectical ability than his opponent, Salviati does not hesitate at times to resort to some "trick", such as presenting as a truth something he knows to be false, in order to put Simplicio's new-born deductive abilities to the test, or to speed things up.

In this dialogue Simplicio is more boastful and know-all than elsewhere. Most of the time he is sure he has the situation under control ("I completely understood you") and even believes initially that he can catch his opponent at fault ("Well, Salviati, so long as you make use of assumptions of this sort I shall cease to be surprised that you deduce such false conclusions"). He drops sentences in Latin and almost has the air, when answering precise and demanding questions, that it is he who is teaching Salviati. As soon as he suspects he has been sent in a direction other than the one he would have wished, he plays for time ("Here I must think a moment about my reply"), he hesitates ("It seems so to me . . . It seems that it ought to be"), he reveals himself as suspicious of what he thinks he is perceiving ("All this is true, but what next?"), to the point where he makes Salviati lose patience and exclaim "Go on and draw the final consequence by yourself, if by yourself you have known all the premises". And indeed it is Simplicio by himself who in the end goes on to express clearly, for instance, the mechanism of the fall of a body simultaneously undergoing a horizontal motion. Immediately afterwards, however, he is quick to raise further objections, as if he has reached the conclusion not with joy, but with reluctance. Simplicio is a character that most people will have happened upon at some time in their lives. He is the prototype of the Aristotelian thinker, and he is also almost certainly a portrait of a particular opponent of Galileo, so realistically is he drawn.

Beneath the liveliness of the dialogue and the fascinating contrast of the characters, Galileo's lucid thinking unfolds with marvelous clarity for the reader. When confronting a problem he aims directly at solving the central question and only once this has been resolved does he tackle secondary aspects—in the end building up a coherent whole, whose parts mutually support and strengthen one another.

A final remark on style.[66] In narrative sections, Salviati's and Sagredo's speech is wide-ranging and digressive, rich in subordinate sentences. However the more scientific the discussion becomes, the quicker and sharper becomes their speech, resolving matters 'on the surface', that is, the main sentences are joined tightly together by linkers (then, therefore, nevertheless, etc.) and each sentence seems to stem out of necessity from the one that precedes it. Vocabulary is never dryly technical, but is drawn from the world of everyday experience, attributing human qualities and feelings to the inanimate world, almost as if to point out that the two worlds have a common nature (literal translation from Italian: "the moving body has neither a *repugnance* nor a *propensity*", "the force that causes motion . . . begins to *languish*").

Simplicio's speech, instead, is tortuous and self-satisfied—his choice of words abstract and technical. Except, that is, when he is cornered and is forced to utter a scientific statement. Here, too, Galileo is warning us: too many words, when not actually intended to deceive, are often a cover for paucity of ideas. Simplicity and clarity of expression are a primary goal and a moral obligation for anyone who genuinely wants to share their knowledge with others, rather than just trying to make it appear that they know. The language used today by many politicians and experts should cause us to reflect upon this.

[66] The comments which follow, on the style of the language Galileo is using, refer in particular to the Italian language of his original text. Such stylistic aspects may not have been preserved in the English translation.

It is now time to take a look at the original text.[67] Salviati is addressing Simplicio:

A fall from the top of the mast

SALV: [. . .] You say, then, that since when the ship stands still the rock falls to the foot of the mast, and when the ship is in motion it falls apart from there, then conversely, from the falling of the rock at the foot it is inferred that the ship stands still, and from its falling away it may be deduced that the ship is moving. And since what happens on the ship must likewise happen on the land, from the falling of the rock at the foot of the tower one necessarily infers the immobility of the terrestrial globe. Is that your argument?

SIMPL: That is exactly it, briefly stated, which makes it easy to understand.

SALV: Now tell me: If the stone dropped from the top of the mast when the ship was sailing rapidly fell in exactly the same place on the ship to which it fell when the ship was standing still, what use could you make of this falling with regard to determining whether the vessel stood still or moved?

SIMPL: Absolutely none; just as by the beating of the pulse, for instance, you cannot know whether a person is asleep or awake, since the pulse beats in the same manner in sleeping as in waking.

SALV: Very good. Now, have you ever made this experiment of the ship?

SIMPL: I have never made it, but I certainly believe that the authorities who adduced it had carefully observed it. Besides, the cause of the difference is so exactly known that there is no room for doubt.

SALV: You yourself are sufficient evidence that those authorities may have offered it without having performed it, for you take it as certain without having done it, and commit yourself to the good faith of their dictum. Similarly it not only may be, but must be that they did the same thing too—I mean, put faith in their predecessors, right on back without ever arriving at anyone who had performed it.[68] *For anyone*

[67] *Dialogue Concerning the Two Chief World Systems*, translated by S. Drake, The Modern Library, New York, 2001, p. 167 and following. In Italian: *Dialogo sopra i due massimi sistemi del mondo*, Edizione Nazionale, Vol. VII, p. 169 and following.

[68] An example of how what starts out as an unsubstantiated claim can become accepted history, or an object of faith.

who does will find that the experiment shows exactly the opposite of what is written; that is, it will show that the stone always falls in the same place on the ship, whether the ship is standing still or moving with any speed you please.[69] *Therefore, the same cause holding good on the earth as on the ship, nothing can be inferred about the earth's motion or rest from the stone falling always perpendicularly to the foot of the tower.*

At this point Simplicio declares himself doubtful that Salviati, either, has ever performed the experiment of the falling rock—and he is not mistaken in this. So, asks Simplicio, how can he be so sure of the outcome? While admitting he has never performed the experiment, Salviati states that when the principles governing a phenomenon are completely clear, the conclusion can be reached through "argument", that is logical reasoning, which in such cases has the same value as experimental evidence (take note: on the basis of an "argument", not of a received truth, as Simplicio had claimed a little earlier). It is now that Salviati's subtle trap is sprung, inducing Simplicio to arrive, on the basis of reasoning, at conclusions which are counter to his most rooted convictions.

A perfectly smooth plane

SALV: Without experiment, I am sure that the effect will happen as I tell you, because it must happen that way; and I might add that you yourself also know that it cannot happen otherwise, no matter how you may pretend not to know it—or give that impression. But I am so handy at picking people's brains that I shall make you confess this in spite of yourself.
[. . .]

I do not want you to declare or reply anything that you do not know for certain. Now tell me: Suppose you have a plane surface as smooth as a mirror and made of some hard material like steel. This is not parallel to the horizon, but somewhat inclined, and upon it you have placed a ball which is perfectly spherical and of some hard and heavy material like

[69] Provided, of course, that effects due to air movement can be ignored.

bronze. What do you believe this will do when released? Do you not think, as I do, that it will remain still?
SIMPL: If that surface is tilted?
SALV: Yes, that is what was assumed.
SIMPL: I do not believe that it would stay still at all; rather, I am sure that it would spontaneously roll down.
SALV: Pay careful attention to what you are saying, Simplicio, for I am certain that it would stay wherever you placed it.
SIMPL: Well, Salviati, so long as you make use of assumptions of this sort I shall cease to be surprised that you deduce such false conclusions.
SALV: Then you are quite sure that it would spontaneously move downward?
SIMPL: What doubt is there about this?
SALV: And you take this for granted not because I have taught it to you—indeed, I have tried to persuade you to the contrary—but all by yourself, by means of your own common sense.
SIMPL: Oh, now I see your trick; you spoke as you did in order to get me out on a limb, as the common people say, and not because you really believed what you said.
SALV: That was it. Now how long would the ball continue to roll, and how fast? Remember that I said a perfectly round ball and a highly polished surface, in order to remove all external and accidental impediments. Similarly I want you to take away any impediment of the air caused by its resistance to separation, and all other accidental obstacles, if there are any.

Salviati says "a perfectly round ball and a highly polished surface . . .", he is proposing, that is, a world composed of abstractions, where anything that is marginal is set aside in order to make a mathematical formulation of a phenomenon in its essence possible. This is the typical Galilean approach, the fundamental feature of the new science: it is no longer a matter of confuting the Aristotelian viewpoint, but instead of affirming the idea that reality can be described mathematically. Simplicio, still unaware of the way the dialogue is developing, readily accepts.

SIMPL: I completely understood you, and to your question I reply that the ball would continue to move indefinitely, as far as the slope of the surface extended, and with a continually accelerated motion. For such is

the nature of heavy bodies, which vires acquirant eundo;[70] *and the greater the slope, the greater would be the velocity.*
SALV: But if one wanted the ball to move upward on this same surface, do you think it would go?
SIMPL: Not spontaneously, no; but drawn or thrown forcibly, it would.
SALV: And if it were thrust along with some impetus impressed forcibly upon it,[71] *what would its motion be, and how great?*
SIMPL: The motion would constantly slow down and be retarded being contrary to nature, and would be of longer or shorter duration according to the greater or lesser impulse and the lesser or greater slope upward.
SALV: [. . .] Now tell me what would happen to the same movable body placed upon a surface with no slope upward or downward.
SIMPL: Here I must think a moment about my reply. There being no downward slope, there can be no natural tendency toward motion; and there being no upward slope, there can be no resistance to being moved, so there would be an indifference between the propensity and the resistance to motion. Therefore it seems to me that it ought naturally to remain stable. [. . .]
SALV: I believe it would do so if one set the ball down firmly. But what would happen if it were given an impetus in any direction?
SIMPL: It must follow that it would move in that direction.
SALV: But with what sort of movement? One continually accelerated, as on the downward plane, or increasingly retarded as on the upward one?
SIMPL: I cannot see any cause for acceleration or deceleration, there being no slope upward or downward.
SALV: Exactly so. But if there is no cause for the ball's retardation, there ought to be still less for its coming to rest;[72] *so how far would you have the ball continue to move?*
SIMPL: As far as the extension of the surface continued without rising or falling.

[70] *vires acquirant eundo*—Latin for "they gain energy as they go".

[71] Drake uses the word "impetus" to translate *impeto*, which here Galileo is using to indicate impulse or momentum. This meaning of the word is somewhat different from the one Galileo gives it elsewhere in the *Dialogue* and in later writings, where he uses *impeto* generically to indicate the velocity acquired by a body, either spontaneously—for example, due to gravity—or as a result of an applied force.

[72] Rest is being equated to extreme retardation (motion of "infinite slowness").

SALV: Then if such a space were unbounded, the motion on it would likewise be boundless? That is, perpetual?
SIMPL: It seems so to me, if the movable body were of durable material.
SALV: That is of course assumed, since we said that all external and accidental impediments were to be removed, and any fragility on the part of the moving body would in this case be one of the accidental impediments.
Now tell me, what do you consider to be the cause of the ball moving spontaneously on the downward inclined plane, but only by force on the one tilted upward?
SIMPL: That the tendency of heavy bodies is to move toward the center of the earth, and to move upward from its circumference only with force; now the downward surface is that which gets closer to the center, while the upward one gets farther away.
SALV: Then in order for a surface to be neither downward nor upward, all its parts must be equally distant from the center. Are there any such surfaces in the world?
SIMPL: Plenty of them; such would be the surface of our terrestrial globe if it were smooth, and not rough and mountainous as it is. But there is that of the water, when it is placid and tranquil.

Returning to the falling rock

SALV: Then a ship, when it moves over a calm sea, is one of these movables which courses over a surface that is tilted neither up nor down, and if all external and accidental obstacles were removed, it would thus be disposed to move incessantly and uniformly from an impulse once received?
SIMPL: It seems that it ought to be.
SALV: Now as to that stone which is on top of the mast, does it not move, carried by the ship, both of them going along the circumference of a circle about its center? And consequently is there not in it an ineradicable motion, all external impediments being removed? And is not this motion as fast as that of the ship?
SIMPL: All this is true, but what next?
SALV: Go on and draw the final consequence by yourself, if by yourself you have known all the premises.

SIMPL: By the final conclusion you mean that the stone, moving with an indelibly impressed motion, is not going to leave the ship, but will follow it, and finally will fall at the same place where it fell when the ship remained motionless. And I, too, say that this would follow if there were no external impediments to disturb the motion of the stone after it was set free. But there are two such impediments; one is the inability of the movable body to split the air with its own impetus alone, once it has lost the force from the oars which it shared as part of the ship while it was on the mast; the other is the new motion of falling downward, which must impede its other, forward, motion.

We finally get to the point Salviati had been waiting for, that is, the exposition of the erroneous Aristotelian idea that two motions—one along the horizontal and the other along the vertical—must influence each other. He is quick to tear it to shreds.

SALV: As for the impediment of the air, I do not deny that to you, and if the falling body were of very light material, like a feather or a tuft of wool, the retardation would be quite considerable. But in a heavy stone it is insignificant [. . .]. All the same, as I said, I concede to you the small effect which may depend upon such an impediment, just as I know you will concede to me that if the air were moving at the same speed as the ship and the rock, this impediment would be absolutely nil.

As for the other, the supervening motion downward, in the first place it is obvious that these two motions (I mean the circular around the center and the straight motion toward the center) are not contraries, nor are they destructive of one another, nor incompatible. As to the moving body, it has no resistance whatever to such a motion, for you yourself have already granted the resistance to be against motion which increases the distance from the center, and the tendency to be toward motion which approaches the center. From this it follows necessarily that the moving body has neither a resistance nor a propensity to motion which does not approach toward or depart from the center, and in consequence no cause for diminution in the property[73] *impressed upon it. Hence the*

73 *Property*—that is, velocity.

cause of motion is not a single one which must be weakened by the new action, but there exist two distinct causes. Of these, heaviness attends only to the drawing of the movable body toward the center, and impressed force only to its being led around the center, so no occasion remains for any impediment.

Forcibly impressed motion

The scenario Salviati is describing has now touched upon a critical point—the mechanism whereby a body, once thrown, retains its motion after separation from the thrower ("forcibly impressed motion"[74]). Simplicio immediately reacts:

SIMPL: This argument is really very plausible in appearance, but actually it is offset by a difficulty which is hard to overcome. You have made an assumption throughout which will not lightly be granted by the Peripatetic school, being directly contrary to Aristotle. You take it as well known and evident that the projectile when separated from its origin retains the motion which was forcibly impressed upon it there. Now this impressed force is as detestable to the Peripatetic philosophy as is any transfer of an accidental property from one subject to another. In their philosophy it is held, as I believe you know, that the projectile is carried by the medium, which in the present instance is the air. Therefore if that rock which was dropped from the top of the mast were to follow the motion of the ship, this effect would have to be attributed to the air, and not to the impressed force; but you assume that the air does not follow the motion of the ship, and is quiet. Furthermore, the person letting the stone fall does not need to fling it or give it any impetus with his arm, but has only to open his hand and let it go. So the rock cannot follow the motion of the boat either through any force impressed upon it by its thrower or by means of any assistance from the air, and therefore it will remain behind.
[. . .]

[74] The term "forcibly impressed motion" (*moto violento* in Galileo's words) refers to the outcome of forces applied by man or his instruments and is to be contrasted to "natural motion" (*moto naturale*) which takes place spontaneously owing to gravity.

The mistaken Aristotelian view

Before moving on to Salviati's reply, it seems appropriate to define a little more clearly the concept Simplicio is in the process of expressing. Let us anticipate what he will be going on to say. Aristotle, who had no inkling of inertia, held that motion is associated with an "impressed force". He assigned to the medium the role of ensuring that, once an object has been thrown, its motion continues. Air is pushed forward at the moment of the throw and would have a tendency to refill the vacuum created behind the projectile, thus providing it with an additional push (it is worth noting that, in reality, the depression which is created behind a moving body has the opposite effect—it slows it down[75]). According to the Aristotelian view, therefore, motion of a projectile would not be possible in a vacuum.

SALV: [. . .] Tell me: Seeing that your objection is based entirely upon the nonexistence of impressed force, then if I were to show you that the medium plays no part in the continuation of motion in projectiles after they are separated from their throwers, would you allow impressed force to exist? Or would you merely move on to some other attack directed toward its destruction?
SIMPL: If the action of the medium were removed, I do not see how recourse could be had to anything else than the property impressed by the motive force.
SALV: It will be best, so as to get as far away as possible from any reason for arguing about it forever, to have you explain as clearly as you can just what the action of the medium is in maintaining the motion of the projectile.

In the reply that Simplicio is about to give it is plain that, while he has made the language Salviati uses his own, he has assimilated just the formal appearance of the scientific method based on observation.

[75] As had also been the case for other problems, this idea had already been put forward by Giovanni Battista Benedetti, cited in the previous two chapters: "the more condensed is the air in the front of a body, the more rarefied it becomes behind it . . . For such reason, when the air is dragged by the body, the body itself is kept back by the air". (G.B. Benedetti, *Diversarum speculationum mathematicarum et physicarum liber*, Turin, 1585).

His style is punctilious and dry—like a pupil who has learned a lesson by rote. It is a pity that immediately afterwards, in drawing his conclusions, he ignores experience and relies entirely upon ancient 'truths' that he has absorbed acritically. This passive attitude to learning is summed up well by Salviati when he points out to "credulous" Simplicio that he has not persuaded himself, but has allowed himself to "be persuaded"—even though he possesses his "own senses", that is, experimental evidence that can be used for rejecting errors and reaching valid conclusions.

SIMPL: Whoever throws the stone has it in his hand; he moves his arm with speed and force; by its motion not only the rock but the surrounding air is moved; the rock, upon being deserted by the hand, finds itself in air which is already moving with impetus, and by that it is carried. For if the air did not act, the stone would fall from the thrower's hand to his feet.
SALV: And you are so credulous as to let yourself be persuaded of this nonsense, when you have your own senses to refute it and to learn the truth? Look here: A big stone or a cannon ball would remain motionless on a table in the strongest wind, according to what you affirmed a little while ago. Now do you believe that if instead this had been a ball of cork or cotton, the wind would have moved it?
SIMPL: I am quite sure the wind would have carried it away, and would have done this the faster, the lighter the material was. For we see this in clouds being borne with a speed equal to that of the wind which drives them.
SALV: And what sort of thing is the wind?
SIMPL: The wind is defined as merely air in motion.
SALV: So then the air in motion carries light materials much faster and farther than it does heavy ones?
SIMPL: Certainly.
SALV: But if with your arm you had to throw first a stone and then a wisp of cotton, which would move the faster and the farther?
SIMPL: The stone, by a good deal; the cotton would merely fall at my feet.
SALV: Well, if that which moves the thrown thing after it leaves your hand is only the air moved by your arm, and if moving air pushes light

material more easily than heavy, why doesn't the cotton projectile go farther and faster than the stone one? There must be something conserved in the stone, in addition to any motion of the air. [. . .]

Arrows shot sideways

Sagredo now intervenes, bringing into play the amusing story of arrows "shot sideways" to back up the principle of inertia and demolish the Aristotelian idea of an impressed force. The tone he uses in addressing Simplicio is ceremonious and falsely humble—it will help in making him lower his guard. Everything happens very quickly. Two verbal exchanges and Simplicio is induced to state that "the arrow shot point foremost has to penetrate only a small quantity of air, and the other has to cleave as much as its whole length". Enough for Sagredo to observe, rather brusquely, that in reality matters are the exact opposite of what Aristotle says. And here Salviati rejoins the discussion to deliver the *coup de grace* ("How many propositions I have noted in Aristotle . . . that are not only wrong, but wrong in such a way that their diametrical opposites are true").

SAGR: What an incredible stroke of luck it is that when an arrow is shot against the wind, the slender thread of air driven by the bowstring goes along with the arrow! But there is another point of Aristotle's which I should like to understand, and I beg Simplicio to oblige me with an answer.

If two arrows were shot with the same bow, one in the usual way and one sideways—that is, putting the arrow lengthwise along the cord and shooting it that way—I should like to know which one would go the farther? Please reply, even though the question may seem to you more ridiculous than otherwise, forgive me for being, as you see, something of a blockhead, so that my speculations do not soar very high.

SIMPL: I have never seen an arrow shot sideways, but I think it would not go even one-twentieth the distance of one shot point first.

SAGR: Since that is just what I thought, it gives me occasion to raise a question between Aristotle's dictum and experience. For as to experience, if I were to place two arrows upon that table when a strong wind was blowing, one in the direction of the wind and the other across it, the wind would quickly carry away the latter and leave the former. Now apparently the same ought to happen with two shots from a bow, if Aristotle's doctrine were true, because the one going sideways would be spurred on by a great quantity of air moved by the bowstring—as much as the whole length of the arrow—whereas the other arrow would receive the impulse from only as much air as there is in the tiny circle of its thickness. I cannot imagine the cause of such a disparity, and should like very much to know it.

SIMPL: The cause is obvious to me; it is because the arrow shot point foremost has to penetrate only a small quantity of air, and the other has to cleave as much as its whole length.

SAGR: Oh, so when arrows are shot they have to penetrate the air? If the air goes with them, or rather if it is the very thing which conducts them, what penetration can there be? Do you not see that in such a manner the arrow would be moving faster than the air? Now what conferred this greater velocity upon the arrow? Do you mean to say that the air gives it a greater speed than its own?

You know perfectly well, Simplicio, that this whole thing takes place just exactly opposite to what Aristotle says, and that it is as false that the medium confers motion upon the projectile as it is true that it is this alone which impedes it. Once you understand this, you will recognize without any difficulty that when the air really does move, it carries the arrow along with it much better sideways than point first, because there is lots of air driving it in the former case and little in the latter. But when shot from the bow, since the air stands still, the sidewise arrow strikes against much air and is much impeded, while the other easily overcomes the obstacles of the tiny amount of air that opposes it.

SALV: How many propositions I have noted in Aristotle (meaning always in his science) that are not only wrong, but wrong in such a way that their diametrical opposites are true, as happens in this instance! But keeping to our purpose, I believe that Simplicio is convinced that from seeing the rock always fall in the same place, nothing can be guessed about the motion or stability of the ship. [. . .] Now if in this example no difference whatever appears, what is it that you claim to see in the stone falling from the top of the tower, where the rotational movement is not adventitious and accidental to the stone, but natural and eternal, and where the air as punctiliously follows the motion of the

earth as the tower does that of the terrestrial globe? Do you have anything else to say, Simplicio, on this particular?
SIMPL: No more, except that so far I do not see the mobility of the earth to be proved.
SALV: I have not claimed to prove it yet, but only to show that nothing can be deduced from the experiments offered by its adversaries as one argument for its motionlessness, as I believe I shall show of the others.

The astounding cannon

Sagredo now intervenes once more, with the observation that the independence of the two motions—horizontal and vertical—of a rock dropped from the mast of a ship, as demonstrated by Salviati, has an astounding implication—that projectiles shot horizontally by a cannon will always take the same time to fall to the ground, regardless of the length of the shot. The description of the phenomenon offers one of the most evocative images to be found in Galileo's writings.

The passage again shows the complicity that develops between Salviati and Sagredo when the joke is at Simplicio's expense. From its opening lines—where Sagredo is affectionately making fun of his friend using delightfully graphic language—it seems we already glimpse a smile of understanding pass between the two. And this explains the piqued tone with which Simplicio turns first to the one ("I do not feel that all my doubts are removed") and then to the other ("but perhaps the fault is mine for not being as alert and quick-witted as Sagredo"), as well as his efforts to come up with a brilliant objection.

SAGR: Excuse me, Salviati, but before going on to the others let me bring up a certain difficulty that has been going round in my head while you were so patiently going into such detail with Simplicio on this ship experiment.
SALV: What we are here for is to discuss things, and it is good for everyone to raise his objections as they occur to him, for that is the road to knowledge. So speak up.
SAGR: If it is true that the impetus of the ship's motion remains indelibly impressed on the stone after it has separated from the mast, and that

furthermore this motion occasions no hindrance or slowing in the straight-downward motion which is natural to the stone, then an effect of a remarkable nature must take place.

Let the ship be motionless and the fall of the stone from the mast take two pulse beats. Then cause the ship to move, and drop the same stone from the same place; from what has been said, it will still take two pulse beats to arrive at the deck. In these two pulse beats the ship will have gone, say, twenty yards, so-that the actual motion of the stone will have been a diagonal line much longer than the first straight and perpendicular one, which was merely the length of the mast; nevertheless, it will have traversed this distance in the same time. Now, assuming the ship to be speeded up still more, so that the stone in falling must follow a diagonal line very much longer still than the other,[76] *eventually the velocity of the ship may be increased by any amount, while the falling rock will describe always longer and longer diagonals, and still pass over them in the same two pulse beats.*

Similarly, if a perfectly level cannon on a tower were fired parallel to the horizon, it would not matter whether a small charge or a great one was put in, so that the ball would fall a thousand yards away, or four thousand, or six thousand, or ten thousand, or more; all these shots would require equal times, and each time would be equal to that which the ball would have taken in going from the mouth of the cannon to the ground if it were allowed to fall straight down without any other impulse.

[76] A composition of the two motions of the rock—fall and advance along with the ship—helps clarify why the path of the rock, though covered in the same interval of time, gets longer and longer as the velocity of the ship increases (note that, since the vertical motion is accelerated, the trajectory is parabolic).

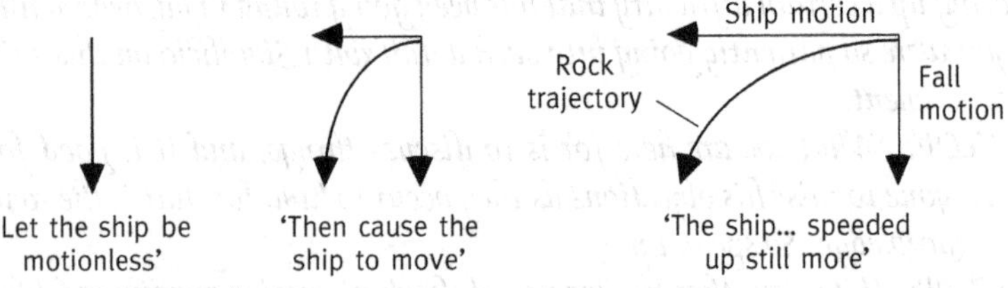

The trajectories illustrated are those that would be seen by an observer standing still on dry land.

Now it seems a marvelous thing that in the same short time of a straight fall from a height of, say, a hundred yards to the ground, the same ball driven by powder could go now four hundred, now a thousand, again four thousand, or even ten thousand yards, so that all shots fired point-blank would stay in the air for an equal time.
SALV: [...] Now if you are satisfied with this, let us get to the solutions of the other arguments, since, so far as I know, Simplicio is persuaded of the uselessness of this first one taken from bodies falling from heights.

Objects thrown from horseback

SIMPL: I do not feel that all my doubts are removed, but perhaps the fault is mine for not being as alert and quick-witted as Sagredo. It seems to me that if this motion which the stone shares while on top of the ship's mast were, as you said, conserved in it also after it is separated from the ship, then it would likewise be necessary for a ball dropped to earth by the rider of a galloping horse to continue to follow the horse's path without lagging behind. I do not believe that this effect is seen except when the rider throws the ball forcibly in the direction in which he is riding. Outside of that, I believe that it will remain where it strikes the ground.

The parallel drawn by Simplicio reveals him to be, in this instance, a good reasoner—but in equal measure a poor observer, since the hypothesis that he proposes as an absurdity is, in fact, precisely what one finds occurs. His lack of attention derives from his lack of courage in the face of the new, from which he tries to avert his eyes. Note, in Salviati's reply, the identity that becomes established between the case where the "impressed force" on the projectile is due to a throw, and the case where it is due to the fact that the whole system from which the projectile departs (i.e. the horse and rider) is not at rest.[77]

SALV: I think you are much deceived, and I am sure that experience will show you on the contrary that the ball, having hit the ground, does run

[77] Just think of a javelin throw, where the run and the action of the arm combine to give the overall effect.

along with the horse and does not drop behind, except as the roughness and unevenness of the path impedes it. And the reason seems clear to me, too. For if you, standing still, were to throw the same ball along the ground, would it not continue the motion also after it was out of your hand? And the distance would be the longer according as the surface was the more even; on ice, for example, it would go a long way.

SIMPL: No doubt it would, if I gave it an impetus with my arm; but in the other example it was assumed that the horseman merely let it fall.

SALV: That is what I want to have happen. When you throw it with your arm, what is it that stays with the ball when it has left your hand, except the motion received from your arm which is conserved in it and continues to urge it on? And what difference is there whether that impetus is conferred upon the ball by your hand or by the horse? While you are on horseback, doesn't your hand, and consequently the ball which is in it, move as fast as the horse itself? Of course it does. Hence upon the mere opening of your hand, the ball leaves it with just that much motion already received; not from your own motion of your arm, but from motion dependent upon the horse, communicated first to you, then to your arm, thence to your hand, and finally to the ball.

I should add that if the rider threw the ball in the direction opposite to his course, when it struck it would sometimes still follow the horse's route, and sometimes it would lie still on the ground; it would move away from him only if the motion received from the arm exceeded that of the rider in velocity. [. . .]

Shooting towards the east and towards the west

Salviati's various arguments about the fall of a rock seem at last to have made a good impression on an eager Simplicio, and he honestly admits it. Nonetheless, he tries to defend his position by shifting the discussion to the other presumed fundamental proof of the earth's immobility—artillery shots—which he considers incontrovertible. The discussion starts with shots along lines of latitude.

When Sagredo interjects saying that he finds it difficult to understand how flying birds manage to keep up with the earth's movement, it should not go unnoticed that, with polite complicity, he is hereby offering Salviati yet another argument to deal with—to the further detriment of Simplicio's theses. In fact, Simplicio's outburst ("But, good heavens . . .") in defense of the "manifest sense" (i.e. of the experimental evidence) is decidedly comic: once again he has learned the form well, but the substance has escaped him.

SIMPL: As to this first argument, I really must admit I have been listening to various subtleties that I have not thought about, and since they are new to me I cannot answer them right now. But I have never taken this argument based upon vertically falling bodies to be one of the strongest arguments in favor of the immobility of the earth. I am wondering what is going to happen to the argument from[78] *cannon shots, especially those opposite to the diurnal motion.*[79]
SAGR: If only the flying of the birds didn't give me as much trouble as the difficulties raised by cannons and all the other experiments mentioned put together! These birds, which fly back and forth at will, turn about every which way, and (what is more important) remain suspended in the air for hours at a time—these, I say, stagger my imagination. Nor can I understand why with all their turning they do not lose their way on account of the motion of the earth, or how they can keep up with so great a velocity, which after all much exceeds that of their flight.[80]
SALV: As a matter of fact, your point is well taken. Perhaps Copernicus himself was unable to find a solution which entirely satisfied him, and for that reason he remained silent on it. Though indeed he was very brief in his examination of the other adverse arguments; by reason of the profundity of his mind, I suppose and his preoccupation with the most abstruse reflections, just as a lion is but little impressed by the insistent baying of small dogs. Therefore let us save the objection of the birds for the last and meanwhile try to satisfy Simplicio as to the others by showing him that, as usual, he has the solutions at his fingertips though he does not notice them.

First, let us take the flight of shots made with the same cannon powder, and ball, now toward the east and now to the west. Tell me

[78] *From*—that is, based upon.
[79] *The diurnal motion*—the rotation of the earth.
[80] The velocity of bodies due to the earth's rotation is, at the equator, 1665 km/h.

what it is that moves you to believe that, if the diurnal revolution were the earth's, the westward shot would have to carry much farther than the eastward one?

SIMPL: I am inclined to believe this because on the eastward shot the ball is followed by the cannon while it is outside the cannon. The latter, carried by the earth, travels rapidly in the same direction; hence the fall of the ball to earth takes place but a short way from the cannon. In the westward shot, on the other hand before the ball hits the earth the gun is removed far to the east wherefore the space between the ball and the cannon—that is the length of this shot—will appear greater than the other by the length of the cannon's path (that is, the earth's) during the time the two balls are in the air.

From a moving carriage

SALV: I should like to find some way of setting up an experiment which corresponds to the motion of these projectiles as that of the ship corresponded to the motion of falling bodies. I am trying to think how to do so.

SAGR: I believe it would turn out very satisfactorily to take a little open carriage, place a crossbow in it with the bolt at half-elevation[81] *(since in that way the shot goes farthest of all), and then, while the horses are running, to shoot once in the direction of their motion and again the opposite way. Taking careful note where the carriage is at the moment the arrow strikes the ground in each case, it could be seen exactly how much farther the one carried than the other.*

SIMPL: It seems to me that such an experiment would be very suitable, and I have no doubt that the shot (that is, the space between the arrow and the place where the carriage was when the arrow struck the ground) would be much less when it went in the direction of the carriage than when it went the other way. Let the shot in itself be 300 yards, for example, and the travel of the carriage while the arrow is in the air, 100 yards. Then, when the shooting is with its course, the carriage will pass 100 of the 300 yards of the shot, so that at the time the

[81] *Half-elevation*—with a tilt of 45°.

arrow strikes the ground the space between it and the carriage will be only 200 yards. But on the other hand in the shot with the carriage running opposite to the arrow, when the arrow shall have passed over its 300 yards and the carriage its 100 additional the other way, the distance between them will be found to be 400 yards.
SALV: Would there be any way to make these two shots travel equally?
SIMPL: I don't know of any other way than to make the carriage stand still.
SALV: That, of course; but I mean with the carriage going full speed.
SIMPL: Only by bending the bow harder with the course and more weakly against the course.
SALV: Then there is another way, and this is it. But how much would you need to strengthen your bow, and later to weaken it?
SIMPL: In our example, in which we have assumed that the bow would shoot 300 yards, it would be required for the shot along the course to strengthen the bow so as to shoot 400 yards, and the other way to weaken it so as to shoot no more than 200. Thus each shot would go out 300 yards with respect to the carriage, which, with its travel of 100 yards which is to be subtracted from the shot of 400 and added to that of 200, would reduce both to 300. [. . .]

Simplicio's reasoning is very precise and is perfectly consistent with the Aristotelian view that the forward movement of the projectile is maintained by the reaction of the medium, and is completely uninfluenced by any motion of the shooter. However, his reasoning is also conspicuously erroneous.

After Simplicio has specified the speeds at which the bolt must start in order for the ranges of the shots in the two cases to be equal (i.e. speeds of 4 in the direction of motion and of 2 against it, when the speed of the carriage is 1), we pass the word back to Salviati for him to draw the conclusions on Galileo's behalf.

SALV: [. . .] But, tell me, when the carriage is running, don't all the things in the carriage move with that same speed?
SIMPL: No doubt about it.
SALV: Also the bolt, and the bow, and the string with which this is strung?

SIMPL: That is right.

SALV: Then when the bolt is discharged in the direction of the carriage, the bow impresses its three degrees of speed upon a bolt which already possesses one degree, thanks to the carriage which carries it at that speed in that direction. Thus when the nock leaves the string it does so with four degrees of speed. And on the other hand, shooting the other way, the same bow confers its three degrees upon a bolt moving with one degree in the opposite direction, so that at its separation from the string only two degrees of speed remain with it. But you yourself have already declared that in order to make the shots equal it is required that the bolt leave with four degrees in one case, and with two in the other. Hence, without changing the bow, the course of the carriage itself regulates the flights, and this experiment clinches the matter for those who would not or could not be convinced of it by reason.

Now apply this argument to the cannon, and you will find that whether the earth moves or whether it stands still, shots made with the same force must always carry equally no matter in what direction they are sent. Aristotle's error, and Ptolemy's, and Tycho's, and yours, and that of all the rest, is rooted in a fixed and inveterate impression that the earth stands still; this you cannot or do not know how to cast off, even when you wish to philosophize about what would follow from assuming that the earth moved. Thus in the other argument, without reflecting that when the stone is on the tower it does whatever the terrestrial globe does about moving or not moving, and having it fixed in your mind that the earth stands still, you always argue about the fall of the rock as if it were leaving a state of rest, whereas you ought to say:

If the earth is fixed, the rock leaves from rest and descends vertically; but if the earth moves, the stone, being likewise moved with equal velocity, leaves not from rest but from a state of motion equal to that of the earth. With this it mixes its supervening downward motion, and compounds out of them a slanting movement.

SIMPL: But, good heavens, if it moves slantingly, why do I see it move straight and perpendicular? This is a bald denial of manifest sense; and

if the senses ought not to be believed, by what other portal shall we enter into philosophizing?
SALV: With respect to the earth, the tower, and ourselves, all of which all keep moving with the diurnal motion along with the stone, the diurnal movement is as if it did not exist; it remains insensible, imperceptible, and without any effect whatever. All that remains observable is the motion which we lack, and that is the grazing drop to the base of the tower. You are not the first to feel a great repugnance toward recognizing this nonoperative quality[82] *of motion among the things which share it in common. [. . .]*

Shooting at flying birds

And now, in the midst of this discussion full of "pearls", we come to a curious mistake which Galileo made concerning the mechanism hunters use to take aim when shooting at flying birds. It is rather surprising that Galileo made a slip like this in applying a concept—the principle of inertia—that he was doing everything he could to establish. He had evidently been wrongly informed about hunting. If one were to shoot in the way he described, the prey would escape unharmed, as the projectile would pass behind it.

On the other hand the explanation which follows this in the dialogue, dealing with artillery shots, is correct—at least in its first approximation. This time the shots are made along the lines of latitude and it is argued that they are not, as Aristotelians would have it, altered by the earth's rotation (however, contrary to Galileo's claim, a fundamental difference does, in fact, exist between the physics of artillery shots and that of hunters' shots). Some quantitative aspects of the mistaken aim-taking mechanism described by Galileo, addressed to readers interested in physical problems, will be examined in the Mathematical Note at the end of this chapter. The precise behavior of south–north artillery shots in the rotating

[82] *Non-operative quality*—not producing any perceptible effect (think of two trains running parallel at the same speed—either train will appear at rest to an observer traveling on the other).

earth system, described with the benefit of present-day knowledge, will also be discussed briefly.

SALV: Let us give in, Simplicio, for the matter stands just as he says. And now from this argument I begin to understand the hunter's problem—that of those marksmen who kill birds in the air with their guns. I once thought that because of the birds' flight, aim must be taken some distance from the bird, anticipating it by a certain interval, more or less according to the speed of flight and the distance of the bird, to the end that the ball when fired would go along the direct line of sight and arrive at the same time and the same point as the bird would in its flight, and they would meet. Therefore I asked one of these men whether that was their practice, and he told me no, that the device used was much easier and surer. They work in exactly the same way as if shooting at a stationary bird; that is, they fix their sights on a flying bird and follow it by moving the fowling piece, keeping the sights always on it until firing; and thus they hit it just as they would a motionless one. So the turning motion made by the fowling piece in following the flight of the bird with the sights, though slow, must be communicated to the ball also; and this is combined with the other motion, from the firing. Thus the ball would have from the firing a motion straight upward, and from the barrel a slant according to the motion of the bird [. . .]. Therefore, to hold the sights continually directed at the mark makes the shot carry properly. In order to hold the sights on the target if the mark is standing still, the barrel must be held still; and if the target is moving, the barrel will be held on the mark with that motion.

Upon this depends the proper answer to that other argument, about shooting with the cannon at a southerly or northerly mark.[. . .]. I reply, then, by asking whether it is not true that once the cannon was aimed at a mark and left so, it would continue to point at that same mark whether the earth moved or stood still. It must be answered that the sighting changes in no way; for if the mark is fixed, the cannon is likewise fixed; and if it moves, being carried by the earth, the cannon also moves in the same way. And if the sights are so maintained, the shot always travels true, as is obvious from what has been said previously. [. . .]

Birds in pursuit

And so, after the questions of the hunter's aim and north–south cannon shots, we finally come to the discussion about birds' flight, which was mentioned earlier. This is dealt with by Salviati using scientific rigour, but at the same time that humorous tone which has already come to the surface here and there in the dialogue. Now it stems from his ability to "see", that is to translate into an image, a paradoxical situation ("If birds had to keep up with the course of the trees by means of their wings, they would soon fall behind"). His typically Tuscan taste for wit shows up again in the closing words of the passage ("So the birds do not have to worry about following the earth, and so far as that is concerned they could remain forever asleep"). This gentle humor softens the seriousness of the demanding problems being faced—but at the same time is perfectly in harmony with them. It is as if Galileo is suggesting we should not trust anyone who mistakes solemnity for seriousness and prohibits light-heartedness in order to cloak themselves with importance, thus depriving themselves of an essential tool for reasoning. To be able to laugh—at oneself, too—means to be capable of putting what one knows and what one is, up for discussion.

SAGR: For my part I am fully satisfied, and I understand perfectly that anyone who will impress upon his mind this general communication to all terrestrial things of the diurnal motion (which suits them all naturally, just as in the ancient idea it was considered that rest with respect to the center suited them) will discern without any trouble the fallacy and the equivocation that make the arguments appear conclusive.

There remains for me only that doubt which I hinted at before, about the flight of birds. Since these have the lively faculty of moving at will in a great many ways, and of keeping themselves for a long time in the air, separated from the earth and wandering about with the most irregular turnings, I am not entirely able to see how among such a great mixture of movements they can avoid becoming confused and losing the original common motion. Once having been deprived of it, how could they make up for this or compensate for it by flying, and keep up with all

the towers and trees which run with such a precipitous course toward the east? I say 'precipitous', because for the great circle of the globe it is little less than a thousand miles an hour, while I believe that the swallow in flight makes no more than fifty.
SALV: If birds had to keep up with the course of the trees by means of their wings, they would soon fall behind; and if they were deprived of the universal rotation, they would remain so much behind and their westward course would be so furious that, to anyone who could see it, it would surpass that of an arrow by a great deal. But I think we should not be able to perceive it, just as cannon balls are not seen when they race through the air, driven by the energy of the charge. Now the fact is that the birds' own motion—I mean that of flight—has nothing to do with the universal motion, from which it receives neither aid nor hindrance. What keeps that motion unaltered in the birds is the air itself through which they wander. This, following naturally the whirling of the earth, takes along the birds and everything else that is suspended in it,[83] *just as it carries the clouds. So the birds do not have to worry about following the earth, and so far as that is concerned they could remain forever asleep.*
[. . .]

MATHEMATICAL NOTE

Independence of motions

We will now demonstrate that the fall time of a ball fired horizontally and that of a ball merely dropped from the mouth of a cannon is always the same, no matter how high the horizontal speed of the fired ball may be. Ignoring air friction, the motion of the projectile along the horizontal axis x occurs because of inertia at a constant velocity v_x—that which it had

[83] Elsewhere, the same concept is expressed in a cruder way: "it is a wonder we can urinate, while running so quickly behind our urine; I would expect at least we should urinate down our knees". (Note Galileo made in the margins of his copy of *Contro il moto della Terra* by Lodovico delle Colombe, Edizione Nazionale, Vol. III, p. 255). Lodovico delle Colombe was a mediocre Aristotelian scholar and a fierce opponent of Galileo.

when it exited from the mouth of the cannon. Along the vertical axis, instead, the ball falls with uniformly accelerated motion as an effect of

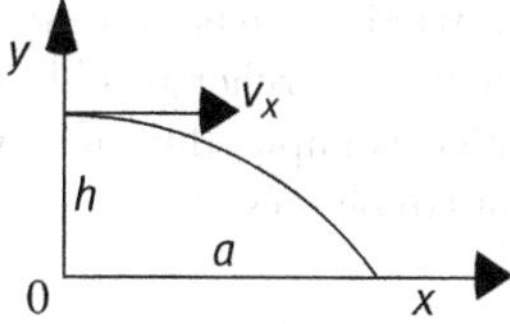

gravity, the initial height being h. Let g be the gravitational acceleration. If we take the origin of the axes at ground level on the vertical from the cannon, we can write for the co-ordinates of the projectile as a function of time:

$$x = v_x t \tag{5.1a}$$

$$y = h - \frac{1}{2} g t^2 \tag{5.1b}$$

By eliminating time, one gets the equation for the parabolic trajectory:

$$y = h - \frac{1}{2} g \left(\frac{x}{v_x} \right)^2$$

The range a, which is found from the value of x when $y = 0$, turns out to be proportional to the initial velocity v_x:

$$a = v_x \sqrt{\frac{2h}{g}}$$

The time T required to fall to the ground is obtained from the value of t when $x = a$. From equation (5.1a) one has

$$T = \sqrt{\frac{2h}{g}}$$

which is actually independent of the initial velocity v_x and therefore equal to that of a projectile simply dropped vertically.

A simple problem (almost a game) which is closely related to the problem of the fall of the projectile, is worth recalling. Suppose an archer wants to shoot an arrow to hit a coconut hanging from a palm-tree at a certain distance from him. He aims directly at the nut. At the very moment that he releases the bow-string the nut breaks off the branch and starts to fall. Alas, the archer thinks, a wasted shot! But he is wrong: much to his surprise the arrow hits its target, breaking the nut in flight before it reaches the ground. The explanation is obvious: the vertical displacements of the nut and the arrow are identical, since both drop by effect of gravity (provided, as usual, that air friction can be considered negligible). In aiming straight at the nut, therefore, the archer would have missed it, had it not broken off the branch! A nice example, this, of how useful it can be to have some knowledge of the laws of physics.

Taking aim

When we turn to the mechanism Salviati describes for taking aim at a bird in flight, however, we find that things do not correspond to the facts. Galileo has evidently been wrongly informed about aim-taking techniques and, perhaps spellbound by the prospect of providing a revolutionary explanation, runs into a conceptual error, allowing Salviati to apply incorrectly the very principle of inertia he is bent on demonstrating. If the fowling-piece is always kept aimed at the target, the bullet, at the moment of exiting from the barrel of the gun, has a transverse speed which is markedly lower than that of the bird. This is because it is much closer to the hunter's shoulder—the rotation fulcrum of the gun–bird system as a whole. Let us consider the simplest case, that where the bird is flying in a transverse direction with respect to the rifle. The transverse velocity

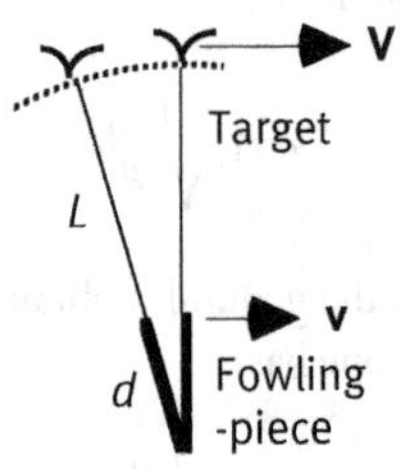

of the bullet is lower in comparison to the bird's speed by a factor d/L, where d is the length of the gun and L is the distance hunter–target. Once the bullet is free its transverse velocity can no longer change, precisely because of the principle of inertia. Thus, the path covered transversally by the bullet is shorter than that covered by the bird and so the target is missed. Here is a numerical example: length of gun $d = 50$ cm; distance to bird $L = 100$ m; speed of bird $V = 30$ km/h; longitudinal speed of bullet $u = 1000$ km/h. The transverse velocity of the mouth of the gun (and of the bullet) turns out to be $v = Vd/L = 0.15$ km/h, and the time the bullet takes to go from the gun to the target is $t = L/u = 0.36$ s. During this interval of time the bird advances by 3 m, while the bullet is displaced by only 1.5 cm and it thus passes way behind the target. In practice, this problem is resolved by means of a number of tricks, which require a certain expertise with guns to appreciate. In any event, to improve the odds, hunters shoot at birds with a burst pattern rather than with single bullets.

Artillery shots

Finally, on the same theme, a note for the more meticulous reader who may wish—with the benefit of present-day knowledge—to object to Galileo's claim that cannon shots from south towards north, with sights set on a target, are not influenced by terrestrial rotation. In fact, in the Northern Hemisphere a target situated further to the north than a cannon will travel from west to east more slowly than that cannon. By the principle of inertia a ball would miss the target, passing ahead of it. This effect is small at the latitudes of Europe, but can become quite marked as the extreme north is approached. To consider the limit case, let us place the target at the North Pole and the cannon at a certain distance from it. The earth's rotation causes the cannon to travel in a circle around the target, which for its part remains fixed. On exiting from the mouth of the cannon the motion of the ball has a component tangent to the circle: this is conserved and prevents the ball from hitting the target. Here is a numerical example: let the cannon be at $r = 10$ km from the target, which means that it describes around this target a circular orbit at a speed $v = \boldsymbol{\omega} r = 0.727$ m/s ($\boldsymbol{\omega} = 7.27 \cdot 10^{-5}$ rad/s is the angular velocity around the terrestrial rotation axis). If the longitudinal velocity of the ball is 1000 km/h, the time required for the ball to cover the distance to the target is 36 s and in this interval it deviates towards the right by $0.727 \times 36 = 26$ m.

In modern mechanics this kind of effect is more conveniently described in terms of the Coriolis force, which causes a bending of the trajectory of any body moving in a non-inertial system such as the rotating earth. The bending occurs to the right in the Northern Hemisphere, to the left in the Southern Hemisphere. This same effect is responsible for the rotation of the oscillation plane of the Foucault pendulum about the vertical, the formation of spiral vortices in water flowing towards the plug-hole in a sink, or the deviation of a falling body from the perfect vertical line, as we discussed in the Mathematical Note of Chapter 3. The mathematical expression for the Coriolis force is the following

$$\boldsymbol{F}_{\text{Coriolis}} = -2m(\boldsymbol{\omega} \times \boldsymbol{v})$$

where $\times$ is the sign of vector product, m is the mass of the body, $\boldsymbol{v}$ is its velocity, and $\boldsymbol{\omega}$ is the vector which expresses the angular rotation velocity of the rotating system (with the same orientation as the rotation axis). It can be seen that the force is transverse to both the direction of motion of the body and to the rotation axis of the system in which the body is to be found. Moreover, since the vector product of $\boldsymbol{v}$ and $\boldsymbol{\omega}$ is zero if the two vectors are parallel, and at a maximum if the two vectors are perpendicular, one realizes that in the above example of the cannon the Coriolis force is zero at the equator and at a maximum at the poles.

The Foucault pendulum, hanging from the dome of the Panthéon in Paris, rotates by effect of the Coriolis force and is the primary proof of the rotation of the earth

PART III

THE PENDULUM AND MUSIC

6

The Divine Harmonies

From Pendulums to Musical Consonance*

Consonanti, e con diletto ricevute,
saranno quelle coppie di suoni che verranno
a percuotere con qualche ordine sopra 'l timpano

Agreeable consonances are pairs of tones
which strike the ear with a certain regularity

As a music-lover Galileo asked himself why certain chords sound consonant, while others sound dissonant. His explanation was rooted in the physical nature of sound (an oscillation of mechanical parts of a source system that becomes transferred to the air and eventually reaches our eardrums). Galileo perceived the similarity that exists between sound waves, which are oscillatory in nature, and the motion of pendulums—and, indeed, it was from the latter that his discussion started. His hypothesis that consonance arises from more synchronous and regular stimulations of the eardrum was a forerunner of modern psychoacoustics, which attributes consonance to a simpler configuration of nerve impulses reaching the brain from the ear.

Consonance and dissonance

Galileo was an accomplished lute player and, like his father, Vincenzo, greatly appreciated the art of music.[84] It was perfectly natural, therefore,

* From *Discourses*.

[84] Galileo's pupil Viviani wrote in his biographical work *Racconto istorico della vita di Galileo Galilei* (Edizione Nazionale, Vol. XIX, p. 602):

Meanwhile, among the entertainments he enjoyed most were practicing music and playing the keyboard and the lute, upon which, thanks to the example and to the that he should seek an answer to the ancient question: what is it that makes a particular ensemble of notes consonant, rather than dissonant? His answer, as will be shown in a while, was firmly rooted in physics, instead of mathematics. Pythagoras had already observed that two notes played together are the more pleasing to the ear, the closer their fundamental frequencies are to forming a ratio of small integers (e.g. in order of decreasing consonance, 2/1 for the octave interval, 3/2 for the perfect fifth *doh-soh* (C-G[85]), 4/3 for the perfect fourth *doh-fah* (C-F), 5/4 for the major third *doh-me* (C-E), and so on). On the basis of this observation, which he had made with the help of a monochord,[86] he arrived at his famous conclusion: "The secret of harmony lies in the magical power of numbers".

Galileo did not believe that there could be magical or abstract effects in Nature and focused, instead, upon the concrete physical aspects of consonance. He established a parallel between consonance and the periodic synchrony that exists in the movements of pendulums whose oscillation periods are in ratios of small whole numbers to each other. As far as he was concerned, there had to be a basic similarity between pendulums and vibrating musical strings, since in both cases oscillating mechanical systems are involved. The only difference between them is the acting force—gravity in the case of pendulums, elastic force in the case of vibrating strings. From here, Galileo, with remarkable insight, derived an explanation of consonance in terms of the eardrum being stimulated in a coordinated way, whereas dissonance corresponds to a random, uncoordinated stimulation. Two pendulums, said Galileo, whose oscillation periods are respectively 3 and 2 s, will be found to be in synchrony with each other every second

instructions of his father, he attained such a level of excellency that he could often compete with the most famous music professors of his time in Florence and Pisa. With the lute, he was very rich of invention and had such a gentle and graceful touch as to be superior to his own father. And this suavity of style he kept to the last days of his life.

[85] In the C-major key: we refer to this key throughout the chapter.

[86] An instrument which has a single string whose length can be varied at will by means of a sliding *ponticello*. Both ends of the vibrating string are fixed, but one end can be moved either closer to, or farther from the other. The wavelength of the fundamental tone is equal to twice the length L of the string; it follows that the fundamental frequency f is inversely proportional to the string length through the relation $f = v/2L$, where v is the sound velocity in the string.

oscillation of the slower pendulum, that is, every 6 s—indicating that the two motions have an element of periodicity in common. Likewise, two musical strings whose lengths are in a ratio of 3 to 2 will produce waves which act synchronously on the eardrum every second oscillation of the longer string. In contrast, two strings whose lengths are not in a simple ratio will never produce waves which act synchronously on the eardrum—and this results in a disturbing rather than pleasing effect. Correspondingly, two pendulums whose oscillation periods are not in a simple ratio will never swing in tandem and will present confusing, uncorrelated motions to the eye.[87]

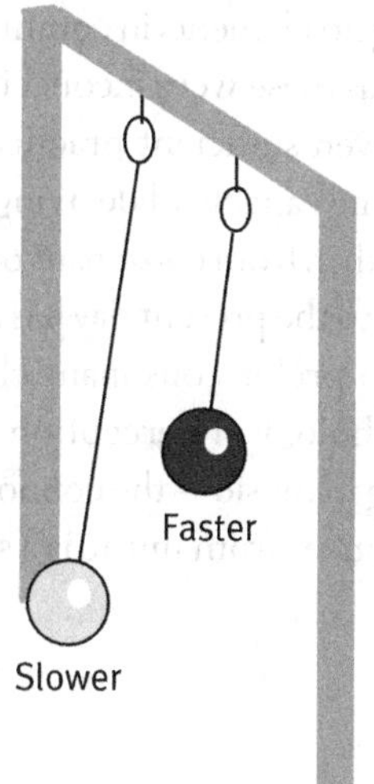
For an instant–every second oscillation of the longer one–the two pendulums appear to swing in tandem
Faster
Slower

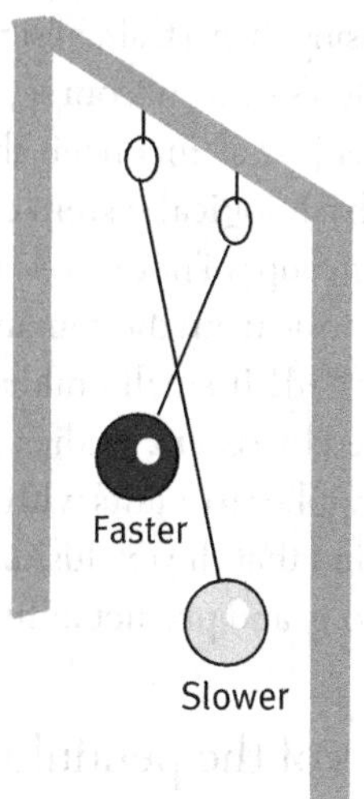
At all other instants the swing of the two pendulums is totally uncoordinated
Faster
Slower

If the oscillation frequencies of the pendulums are in ratios of small integers, for instance 3 to 2, they periodically pass through a phase of perfect synchronization

[87] In this "physical" approach to the problem of consonance, Galileo was inspired by his father Vincenzo, who had been a pupil of the famous music theorist Gioseffo Zarlino. Zarlino had introduced the *zarlinian comma* and was a believer, after Pythagoras, in the importance of numbers in music. Vincenzo, however, was later to fight systematically against his teacher's conceptions.

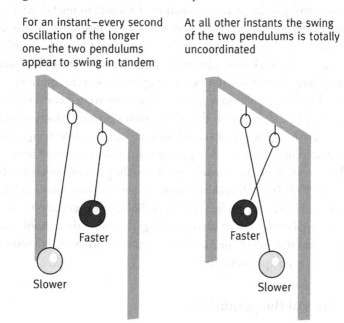

At this point it should be underlined that the criteria used by modern psychoacoustics to define consonance are, in some respects, traceable back to Galileo's hypothesis. His discussion of the fundamental tone was extended to include the higher harmonics, some of which, it turns out, are the same for groups of notes that are consonant. For instance, in the perfect fifth chord C–G, the third harmonic of C coincides in frequency with the second harmonic of G, and the same happens with the sixth harmonic of C and the fourth of G.[88] There is thus an excitation of the eardrum which, for the two notes, contains a sequence of synchronized stimuli, a unifying track for the two sounds. Psychoacoustics, for its part, does not attribute the pleasant sensation of a sound to the behavior of the eardrum, but rather to the nervous system; in the conditions of stimulation described above, the train of signals sent from the cochlea to the neural network is simpler, more easily decoded, and is therefore more pleasing than a train of signals having no elements in common.

In Galileo's time, of course, ideas such as these were inconceivable. And it was just as inconceivable that, given sufficient practice, our psycho-physiological system could become capable of deriving from dissonant groups of notes no less pleasure than from consonant ones—as all the music from the romantic period to the present day has amply demonstrated! If small children seem to prefer consonant chords, as indicated by recent studies in the psychology of perception—and the same applies to adults with no training in music—the connoisseur tends to find that the exclusion of dissonances from music is a source of monotony and predictability.

The laws of the pendulum

Galileo passed from pendulums to musical consonance as a natural development of his thought processes, in much the same way as earlier he had arrived at pendulums starting from falling bodies. He had

[88] If 1 is the frequency of the fundamental tone of C, the frequencies of G and of the various harmonics are:

	fundamental	II	III	IV	V	VI	...
C	1	2	3	4	5	6	...
G	3/2	3	9/2	6	15/2	9	...

been studying these with a completely different aim in mind—namely, to complement his reasoning against Aristotelian beliefs with experimental evidence. Galileo set out with the notion that, in order to analyze fall experimentally, it was opportune to have the body—a well polished sphere—run down an inclined plane, since in this way the times of fall became longer. The duration of the event was measured by counting heart beats, or by weighing the amount of water dropped from a leaking container (the so-called "water chronometer", see Chapter 13)—both of which are methods subject to large errors if used to time phenomena that are short-lived. This was why everything had to be slowed down as much as possible. Moreover, since the weight of a pendulum oscillates as a consequence of the action of gravity, this mechanical system appeared to him even more suitable for the study of gravitational phenomena. In fact, a pendulum's motion can be slowed down at will—the longer the string which its weight is hanging from, the slower its swing will be.

The gradual transition of reasoning from the fall of bodies to musical consonance offers a wonderful example of logical concatenation in advancing a scientific argument. Each individual phenomenon is tightly linked to the others, so as to form a homogeneous and consistent whole. This way of penetrating deeply into the hidden recesses of matter—which is one of the essential features of modern physics—permeates the entire *corpus* of Galileo's works.

Isochronism of oscillations

It is perhaps worth adding a few brief observations regarding Galileo's discovery of the isochronism of the pendulum, that is, equal duration of oscillations for equal length of string, irrespective of their amplitude and of the magnitude of the oscillating weight. Today we know that this holds only for small oscillations (i.e. vertex angles of a few degrees) and then only approximately. In Galileo's example, two balls of equal diameter, one made of lead and one of cork, oscillate with almost identical behavior. If there is any difference, it is in the damping time of the oscillations, which is much longer for the lead ball.

Viviani tells us that his teacher had deduced the isochronism of pendulums when still a young man, by observing the motion of chandeliers in Pisa cathedral. Although there are scholars who hold that isochronism was already known to the Arab astronomer, Ibn Yunus several centuries before Galileo was born, there is no doubt that Galileo was the first to study the phenomenon in a systematic way and to tackle the subject with scientific precision. In Galileo's reasoning, the notion of isochronism leads to the notion that the oscillation speeds of two pendulum weights—one of lead and one of cork—are equal if the lengths of the threads and the amplitudes of the oscillations are the same.[89] This is a property which enters within the general mechanism of falling bodies, since the descent of a pendulum weight is nothing other than a fall along a circular arc. In Galileo's view this mechanism is identical for all bodies, provided incidental effects, such as the resistance offered by the medium, can be ignored.[90] Nevertheless, in the case of the pendulum a new fact emerged, one that is conceptually unpredictable (and was so even for Galileo) and is detectable only through experimentation: the fall time is independent of the height from which the pendulum starts—isochronic oscillation, as defined above—a feature which does not hold true for a vertical fall, or for falls along inclined planes.[91]

[89] Actually, today we know that the two speeds would differ slightly, owing to the different influence of air friction in the two cases: a lead ball would gain over a cork ball of equal diameter something like one oscillation in every hundred.

[90] In Chapter 13, a more detailed comparison between "pendulum-like" fall motion and fall along an inclined plane will be made. This will lead to the definition of a "Galileo constant", establishing a relationship between the two types of motion.

[91] The apparent contradiction between the two behaviors is best illustrated graphically:

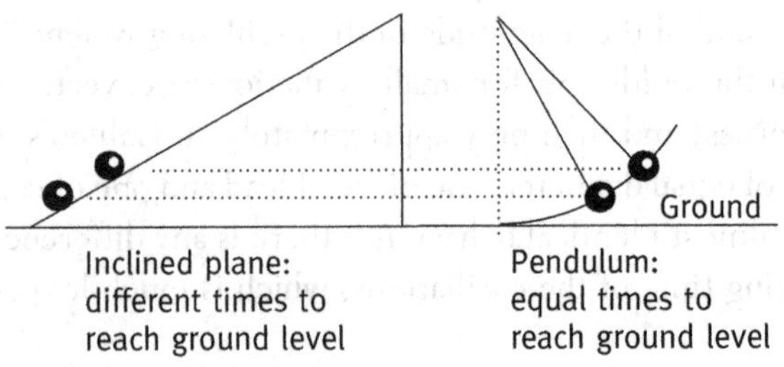

Galileo never claimed that isochronism was exact. We know from Viviani that he had observed very small differences between the periods of small and large oscillations. However, since he was well aware of the effects of air resistance (which are the more significant, the higher the velocity of the body is, that is, the higher the oscillation amplitude) he was led to attribute these differences to this 'impediment', rather than to the fact that the physical law of isochronism ceases to be valid when the oscillations become wide.[92] On this issue, however, Galileo was very cautious. He did not attempt to justify isochronism using logical arguments, nor by mathematics, and was content to accept it simply as an empirical fact.

With regard to this last point, we quote a remark made by Toraldo di Francia:[93]

> The definition of small oscillations obviously depends on convention. What matters is to define *the level of precision* one wants and can measure. Galileo, luckily enough, was not able to obtain exceedingly high *levels of precision*, such as are attainable today, or will be attainable in the distant future, otherwise it is unlikely that he would have been able to discover anything at all about the fall of bodies (gravitational acceleration depends on the location on the earth's surface, on the distance from the earth's center, on the position of the moon and of the planets; the earth attracts the body and the body attracts the earth; the fall time is different if measured while standing on the ground or falling along with the body; the light reflected by the body is bent by the earth's attraction, etc. . . .).

Furthermore, the still rudimentary means of measuring time available to Galileo—which allowed a considerable margin to precision—reinforced him in his conviction (shared by all scientists

[92] This is the reason why he never made a thorough study of the argument—unlike Mersenne, who succeeded in detecting the tiny differences in the oscillation times for different amplitudes using a simple technique. He launched two pendulums simultaneously along different arcs (one small and one large) and let them hit a wooden block placed at the common vertical. He then just noted the time delay between the sounds produced by the collisions.

[93] Giuliano Toraldo di Francia is an Italian physicist and author, formerly President of the Italian Physical Society.

not only of his time, but also of the following two centuries) that physics was a science of certainties, governed at all levels by determinism, that is, by unique and certain links between causes and effects. It was not until the nineteenth century that the concept of probability in the realization of a specific physical macroscopic system became established, thanks to Clerk Maxwell and Ludwig Boltzmann. And it was only in the twentieth century that quantum mechanics was to remove causality effects in physical events at a microscopic level. When will the next step be taken, and in what direction?[94]

But let us return to the isochronism of Galileo's pendulum. It is interesting to note that Galileo's mistake in treating this isochronism as perfect led to some important scientific and technological advances. Without this error, for example, perhaps no one would ever have thought of using the pendulum as a device for measuring time. For a detailed description of this topic, and in particular of Galileo's attempts to build a pendulum-clock with the help of his son Vincenzo, see the book by Ferdinando Flora.[95]

Foucault's pendulum

One last remark concerning the oscillation of the chandeliers in Pisa cathedral. We have to ask ourselves, if Galileo really did devote long periods of time to observing their motion, how is it that he never

[94] We report a comment by Evandro Agazzi:

the great revolutions in contemporary physics have not consisted in recognizing that it was no longer possible 'to do physics' the way shown by Galileo, but only in the fact that we have faced phenomena of completely new orders of magnitude . . . characterized by properties and laws quite different from those previously known. However, these phenomena, too, have been approached and studied according to the major methodological lines and the conceptual system that were developed by the great man from Pisa. (E. Agazzi, 'Fisica galileiana e fisica contemporanea', in *Nel quarto centenario della nascita di Galileo Galilei*, Società Editrice Vita e Pensiero, Milan, 1966, p. 51).

[95] F. Flora, *Galileo e gli scienziati del Seicento*, Tomo I, Ricciardi Editore, Milan—Naples, 1953, p. 699 and following.

happened to notice the rotation of their oscillation plane due to the earth's rotation, as occurs with the Foucault pendulum? Of course, the chandeliers would not have been ideal Foucault pendulums.[96] It is likely that, if they rotated, they would have done so in a somewhat haphazard way. So Galileo would probably have attributed the phenomenon to some "accidental impediment", such as different starting conditions, air streams, or friction of various origins—all effects of marginal scientific interest. Too bad! because this would have been the direct and incontrovertible proof of the earth's rotation, that he was to spend his whole life looking for, in order to silence his Aristotelian opponents. The proof which, instead, he believed he had found in the phenomenon of tides—for which his interpretation was as ingenious as it was mistaken (see Chapter 9).

Chandeliers, however, were not the only source of information on pendulums available to Galileo. Outside the window of his study at the University of Padua he had set up a long pendulum and it is possible that this would have been capable of revealing the Foucault effect to him.

A short digression: in 1995, Matteo Bissiri, a Physics freshman at the University of Rome, "La Sapienza", demonstrated experimentally that it is possible to confirm the earth's rotation (and to measure the velocity of this rotation) by observing the displacement of the shadow of a pendulum over 20 min—the time available before the amplitude of the oscillation becomes too small.[97] The thread of the pendulum was 22 m long and it carried a 1 kg weight.

[96] In order to operate properly, a Foucault pendulum must be built using a few special features: a very long thread to ensure low speed and a heavy weight to minimize the effect of air friction—otherwise, the pendulum's oscillation does not last long enough to allow the rotation of the plane of oscillation to be observed, and the trajectory, which is initially linear, can evolve into an elliptical curve.

[97] M. Bissiri, "L'attimo fuggente: ovvero come verificare la rotazione della Terra in meno di venti minuti", *La Fisica nella Scuola*, Vol. 3, p. 144, 1997. With the help of modern electronic devices, Bissiri has also built a miniature Foucault pendulum, which oscillates without interruption and does not require periodic restarting or maintenance. This pendulum has been in continuous operation since December 23, 1996 in the Physics Department of the University of Rome "La Sapienza" and is described in: M. Bissiri and A. Frova, "Pendolo di Foucault in miniatura da esposizione", *La Fisica nella scuola*, XXXI, 1, 1998, p. 47.

Returning to Galileo, we know that his pendulum was not less than 10 m long[98] and it is likely that its weight would have been rather heavier than 1 kg, as he was perfectly aware that it is best to use a heavy weight in order to minimize the effects of air friction. Therefore, it seems reasonable to suppose (considering, also, his great abilities as an experimentalist) that the observations Bissiri made would have been possible for Galileo, too. With hindsight, of course.

Detail of the miniature Foucault pendulum built by Matteo Bissiri at the Physics Department of the University of Rome 'LA SAPIENZA'

The text

Coming now to the text itself, the reader should note how "grounded" Galileo's reasoning is, how reference is constantly made to phenomena that everybody can observe. And also how natural the style of argumentation is. It builds towards its objective with clarity and rigor, but there are frequent interruptions as the interlocutors make known their moods, involvement, and expectations—just as happens in ordinary conversation. Galileo's "digressions", as commentators have often called them, when they do not serve the artistic purpose of adding liveliness to the dialogue (a format which was chosen in place of the more formal one of the essay) are, in reality, digressions only at

[98] S. Drake, *Galileo: Pioneer Scientist*, University of Toronto Press, Toronto, 1994, p. 18.

a surface level. As he passes from one phenomenon to the next Galileo picks out what it is the two phenomena have in common and in doing so he isolates, from the seeming heterogeneity of the physical world, the elements which unify. He thereby performs an operation of illuminating simplification—a forerunner of what has become one of the prominent features of modern physics.[99] Finally, it is worth noting how ably Salviati (here as elsewhere in the discussions) uses Sagredo's remarks, integrating them into his own discourse in such a way as to make his interlocutor feel that he has made an effective contribution to the explanation of the phenomenon.

Taking it as given that (as he had established earlier) two bodies having different weights fall in the same manner, Galileo opens the dialogue with a discussion in which it is argued that the oscillation period of a pendulum does not depend on the mass hanging from its thread (or string, or cord—Galileo uses all three terms interchangeably); this is followed by the argument concerning the isochronism of oscillations.[100]

The mass of the pendulum

SALV: The experiment made to ascertain whether two bodies, differing greatly in weight will fall from a given height with the same speed offers some difficulty; because, if the height is considerable, the retarding effect

[99] Some people did not appreciate Galileo's digressive style. These are Descartes' words concerning the *Discourses*: "it seems to me that a serious fault is that of continuously making digressions, never dwelling enough on a given argument to explain it in depth. This is a sign that he had not examined it with the required care and that, without considering the primary causes of Nature, he has only looked for the reasons of some particular effects, and placed his foundations on these" (*Œuvres de Descartes*, Vol. I, Paris, 1897, p. 380). Antonio Banfi's opinion, instead, was altogether different: "digressions which are sometimes no less intriguing than the main subject, through which emerged not only the methodological non-systematic unity of the new science, but also its process of self-development . . ." (A. Banfi, *Galileo Galilei*, Milan, 1949, p. 185).

[100] *Two New Sciences*, First day, translated by H. Crew and A. de Salvio, Dover Publication, New York, 1954, p. 84 and following. In Italian: *Discorsi e dimostrazioni matematiche intorno a due nuove scienze*, Giornata prima, Edizione Nazionale, Vol. VIII, p. 128 and following.

of the medium, which must be penetrated and thrust aside by the falling body, will be greater in the case of the small momentum of the very light body than in the case of the great force of the heavy body; so that, in a long distance, the light body will be left behind; if the height be small, one may well doubt whether there is any difference; and if there be a difference it will be inappreciable.

It occurred to me therefore to repeat many times the fall through a small height in such a way that I might accumulate all those small intervals of time that elapse between the arrival of the heavy and light bodies respectively at their common terminus, so that this sum makes an interval of time which is not only observable, but easily observable. In order to employ the slowest speeds possible and thus reduce the change which the resisting medium produces upon the simple effect of gravity it occurred to me to allow the bodies to fall along a plane slightly inclined to the horizontal. For in such a plane, just as well as in a vertical plane, one may discover how bodies of different weight behave: and besides this, I also wished to rid myself of the resistance which might arise from contact of the moving body with the aforesaid inclined plane. Accordingly I took two balls, one of lead and one of cork, the former more than a hundred times heavier than the latter, and suspended them by means of two equal fine threads, each four or five cubits long. Pulling each ball aside from the perpendicular, I let them go at the same instant, and they, falling along the circumferences of circles having these equal strings for semi-diameters, passed beyond the perpendicular and returned along the same path. This free vibration repeated a hundred times showed clearly that the heavy body maintains so nearly the period of the light body that neither in a hundred swings nor even in a thousand will the former anticipate the latter by as much as a single moment, so perfectly do they keep step. We can also observe the effect of the medium which, by the resistance which it offers to motion, diminishes the vibration of the cork more than that of the lead, but without altering the frequency of either; even when the arc traversed by the cork did not exceed five or six degrees while that of the lead was fifty or sixty, the swings were performed in equal times. [. . .]

Isochronism

SALV: [. . .] But observe this: having pulled aside the pendulum of lead, say through an arc of fifty degrees, and set it free, it swings beyond the perpendicular almost fifty degrees, thus describing an arc of nearly one hundred degrees; on the return swing it describes a little smaller arc; and after a large number of such vibrations it finally comes to rest. Each vibration, whether of ninety, fifty, twenty, ten, or four degrees occupies the same time: accordingly the speed of the moving body keeps on diminishing since in equal intervals of time, it traverses arcs which grow smaller and smaller.

Precisely the same things happen with the pendulum of cork, suspended by a string of equal length, except that a smaller number of vibrations is required to bring it to rest, since on account of its lightness it is less able to overcome the resistance of the air; nevertheless the vibrations, whether large or small, are all performed in time-intervals which are not only equal among themselves,[101] *but also equal to the period of the lead pendulum. Hence it is true that, if while the lead is raversing an arc of fifty degrees the cork covers one of only ten, the cork moves more slowly than the lead; but on the other hand it is also true that the cork may cover an arc of fifty while the lead passes over one of only ten or six; thus, at different times, we have now the cork, now the lead, moving more rapidly. But if these same bodies traverse equal arcs in equal times we may rest assured that their speeds are equal.*

Despite the touching image conjured up by Viviani's description of his teacher pensively observing the chandeliers in Pisa cathedral, it seems more likely that Galileo would have deduced the isochronism of the pendulum on the basis of the similarity existing between the pendulum and the musical string, as two mechanical oscillating systems. As mentioned earlier, this similarity was quite obvious to him. He was certainly fully aware of the fact that as the sound of a string fades, which means that the amplitude of the oscillation is getting smaller, the pitch (i.e. frequency) of the note remains the same.

The discussion continues with details and examples, until Salviati, warmly encouraged by Sagredo, begins to speak about the acoustic resonance effect and establishes the connection between the phenomenon

[101] It was stated earlier in this chapter that this becomes less the case, the more the amplitude of the oscillations grows. It is surprising that averaging over one hundred vibrations—as Galileo claimed to have done—did not bring the differences out.

of pendulum oscillations and the mechanisms that lead to the definition of consonance and dissonance—the fundamental issue in musical harmony.

SALV: We come now to the other questions, relating to pendulums, [. . .] I may give you some of my ideas concerning certain problems in music, a splendid subject, upon which so many eminent men have written: among these is Aristotle himself who has discussed numerous interesting acoustical questions. Accordingly, if on the basis of some easy and tangible experiments, I shall explain some striking phenomena in the domain of sound, I trust my explanations will meet your approval.
SAGR: I shall receive them not only gratefully but eagerly. For, although I take pleasure in every kind of musical instrument and have paid considerable attention to harmony, I have never been able to fully understand why some combinations of tones are more pleasing than others, or why certain combinations not only fail to please but are even highly offensive. Then there is the old problem of two stretched strings in unison; when one of them is sounded, the other begins to vibrate and to emit its note; nor do I understand the different ratios of harmony and some other details.

And now arrives the moment when Salviati states, albeit implicitly, the law which links the period of oscillation of pendulums to the length of the thread from which the oscillating weight is suspended. Here the concept of resonance emerges clearly: an external perturbation can cause a pendulum to oscillate only if it excites it in a synchronous way, that is, if its frequency coincides with the characteristic frequency of the pendulum—uniquely defined for it by the length of its thread (think of a garden-swing, for example).[102] Sagredo's words ("Thousands of times I have observed vibrations

[102] It has to be acknowledged that Girolamo Fracastoro had already discussed this phenomenon in rather similar terms (as well as other aspects related to the vibration of strings) in his *De sympathia et antipathia rerum* of 1584. Similar concepts had also been expressed in manuscripts entitled *Meditatiunculae*, which Galileo would certainly have known about, as they were written by his friend Guidubaldo Del Monte. However, Galileo's description was, as usual, more lucid and closer to the facts. For more details, see for instance F. Flora, *Galileo e gli scienziati del Seicento*, Tomo I, Ricciardi Editore, Milan—Naples, 1953, p. 710 and following.

especially in churches where lamps, suspended by long cords, had been inadvertently set into motion") echo the young Galileo's personal experience.

The concept of resonance is then extended to musical strings, the fundamental frequency of which is defined by their length (given equal tension, thickness, and material). The oscillation of a string, the transmission of this oscillation to air-particles (or other media) and finally to the eardrum: here in a nutshell is the mechanism of sound—the sensation we have of a mechanical perturbation of periodic oscillatory character. As usual, more quantitative details will be found in the Mathematical Note at the end of this chapter.

The length of the thread

SALV: Let us see whether we cannot derive from the pendulum a satisfactory solution of all these difficulties. [. . .] As to the times of vibration of bodies suspended by threads of different lengths, they bear to each other the same proportion as the square roots of the lengths of the thread; or one might say the lengths are to each other as the squares of the times; so that if one wishes to make the vibration-time of one pendulum twice that of another, he must make its suspension four times as long. In like manner, if one pendulum has a suspension nine times as long, this second pendulum will execute three vibrations during each one of the first; from which it follows that the lengths of the suspending cords bear to each other the [inverse] ratio of the squares of the number of vibrations performed in the same time.[103]

SAGR: Then, if I understand you correctly, I can easily measure the length of a string whose upper end is attached at any height whatever even if this end were invisible and I could see only the lower extremity. For if I attach to the lower end of this string a rather heavy weight and give it a to-and-fro motion, and if I ask a friend to count a number of its vibrations, while I, during the same time-interval, count the number of vibrations of a pendulum which is

[103] Namely the squares of the oscillation frequencies, if the reference time is the second. Here Galileo forgets to specify that the proportionality is inverse.

exactly one cubit in length, then knowing the number of vibrations which each pendulum makes in the given interval of time one can determine the length of the string.[104] *[. . .]*

SALV: Nor will you miss it by as much as a hand's breadth, especially if you observe a large number of vibrations.

SAGR: You give me frequent occasions to admire the wealth and profusion of nature when, from such common and even trivial phenomena, you derive facts which are not only striking and new but which are often far removed from what we would have imagined. Thousands of times I have observed vibrations especially in churches where lamps, suspended by long cords, had been inadvertently set into motion; but the most which I could infer from these observations was that the view of those who think that such vibrations are maintained by the medium is highly improbable: for, in that case, the air needs to have considerable judgment and little else to do but kill time by pushing to and fro a pendent weight with perfect regularity. But I never dreamed of learning that one and the same body, when suspended from a string a hundred cubits long and pulled aside through an arc of 90° or even 1° or 1/2°, would employ the same time in passing through the least as through the largest of these arcs; and, indeed, it still strikes me as somewhat unlikely. Now I am waiting to hear how these same simple phenomena can furnish solutions for those acoustical problems—solutions which will be at least partly satisfactory.

[104] It is worth noting that, had Galileo known the formula that quantitatively links the oscillation period T of the pendulum to the length of its thread L, and to the gravitational acceleration g

$$T = 2\pi\sqrt{\frac{L}{g}}$$

in order to calculate L, he would not have needed a second pendulum for comparative purposes. The use of two pendulums is necessary if one knows only, as Galileo said, that 'the lengths are to each other as the squares of the times', namely

$$L_1/L_2 = T_1^2/T_2^2$$

Once T_1 and T_2 are measured and the length L_1 of the reference pendulum is known, the only unknown quantity is the length L_2 of the second pendulum. The relationships used above are deduced in the Mathematical Note at the end of this chapter.

Resonance

SALV: First of all one must observe that each pendulum has its own time of vibration so definite and determinate that it is not possible to make it move with any other period than that which nature has given it. For let any one take in his hand the cord to which the weight is attached and try, as much as he pleases, to increase or diminish the frequency of its vibrations; it will be time wasted. On the other hand, one can confirm motion upon even a heavy pendulum which is at rest by simply blowing against it; by repeating these blasts with a frequency which is the same as that of the pendulum one can impart considerable motion. Suppose that by the first puff we have displaced the pendulum from the vertical by, say, half an inch; then if, after the pendulum has returned and is about to begin the second vibration, we add a second puff, we shall impart additional motion; and so on with other blasts provided they are applied at the right instant, and not when the pendulum is coming toward us since in this case the blast would impede rather than aid the motion. Continuing thus with many impulses we impart to the pendulum such momentum that a greater impulse than that of a single blast will be needed to stop it.

SAGR: Even as a boy, I observed that one man alone by giving these impulses at the right instant was able to ring a bell so large that when four, or even six, men seized the rope and tried to stop it they were lifted from the ground, all of them together being unable to counterbalance the momentum which a single man, by properly-timed pulls, had given it.

SALV: Your illustration makes my meaning clear and is quite as well fitted, as what I have just said, to explain the wonderful phenomenon of the strings of the cittern or of the spinet, namely, the fact that a vibrating string will set another string in motion and cause it to sound not only when the latter is in unison but even when it differs from the former by an octave or a fifth.[105] *A string*

[105] *but . . . fifth*—the note which is higher by an interval of an octave, or of a twelfth (octave plus fifth). This happens because, as is well-known, a complex sound—such as that emitted by a string instrument—in addition to the fundamental resonance frequency itself, also contains its harmonics, whose frequencies are double, triple, etc. The first of these higher frequencies corresponds to the same note (in the upper octave, however), the second one to the fifth of the upper octave, and so on.

which has been struck begins to vibrate and continues the motion as long as one hears the sound; these vibrations cause the immediately surrounding air to vibrate and quiver; then these ripples in the air expand far into space[106] *and strike not only all the strings of the same instrument but even those of neighboring instruments. Since that string which is tuned to unison with the one plucked is capable of vibrating with the same frequency, it acquires, at the first impulse, a slight oscillation; after receiving two, three, twenty, or more impulses, delivered at proper intervals, it finally accumulates a vibratory motion equal to that of the plucked string, as is clearly shown by equality of amplitude in their vibrations. This undulation expands through the air and sets into vibration not only strings, but also any other body which happens to have the same period as that of the plucked string. Accordingly if we attach to the side of an instrument small pieces of bristle or other flexible bodies, we shall observe that, when a spinet is sounded, only those pieces respond that have the same period as the string which has been struck; the remaining pieces do not vibrate in response to this string, nor do the former pieces respond to any other tone.*

If one bows the base string on a viola rather smartly and brings near it a goblet of fine, thin glass having the same tone as that of the string, this goblet will vibrate and audibly resound. That the undulations of the medium are widely dispersed about the sounding body is evinced by the fact that a glass of water may be made to emit a tone merely by the friction of the finger-tip upon the rim of the glass; for in this water is

[106] In the *Assayer*, Galileo describes the role of the air in propagating sound as follows (Edizione Nazionale, Vol. VI, p. 349): "sounds are formed and perceived in us when a frequent trembling of the air, rippling in very small waves, moves a cartilage of the drum in our ear. The external sources that are able to produce this rippling of the air . . . are mostly due to simple vibrations of some body". On the role of air trembling, as was the case for other phenomena too, Giovanni Battista Benedetti had been more explicit than Galileo. In *Diversarum speculationum mathematicarum et physicarum liber* (1585) he had written (translation from the original Latin text): "In order for a body to generate sound it is necessary that it trembles or shakes. Nor can sound be produced without air, because air sounds while running to fill the spaces where a vacuum is formed". This view, which had never previously been proposed, was a precursor of the modern view, according to which a vibrating body induces alternate compressions and rarefactions of the air which it is in contact with—and it is these which propagate in the form of pressure waves.

produced a series of regular waves. The same phenomenon is observed to better advantage by fixing the base of the goblet upon the bottom of a rather large vessel of water filled nearly to the edge of the goblet; for if, as before, we sound the glass by friction of the finger, we shall see ripples spreading with the utmost regularity and with high speed to large distances about the glass. I have often remarked, in thus sounding a rather large glass nearly full of water, that at first the waves are spaced with great uniformity, and when, as sometimes happens, the tone of the glass jumps an octave higher I have noted that at this moment each of the aforesaid waves divides into two; a phenomenon which shows clearly that the ratio involved in the octave is two.[107] *[. . .]*

Musical strings

Now comes a discussion about the dependence of the fundamental frequency produced by a string (its "pitch" in musical terminology'[108]) on factors other than its length, that is, tension, diameter, and the material it is made of. Galileo was the first to point out the importance

[107] *the ratio . . . is two*—that is, the value of the octave corresponds to a doubling [of the frequency]. Concerning this conclusion, we report a comment by Settle, who has examined and discussed the problem of the 'singing glasses' in detail (see T.B. Settle, "La rete degli esperimenti galileiani", in *Galileo e la scienza sperimentale*, edited by M. Baldo Ceolin, published by the Physics Department "Galileo Galilei" of the University of Padua, 1995, p. 40):

Galileo was most probably aware that this reasoning might not turn out to be completely convincing. The example was part of a mass of results of empirical investigations (involving pendulums, chronometers and musical acoustics), that he never succeeded in organizing into a demonstrative science that was internally coherent. He must also have perceived that these results could have a value in their own right, and that it was worth reporting them. . . . What we have not been able to reproduce so far is the doubling of the waves. In fact to understand the pattern of the waves is to say the least difficult. [. . .]When one rubs the rim at a single point, so as to produce the fundamental note, one is faced with four groups of perturbations on the surface of the water [. . .]. Therefore one must suppose, at least for the moment, that Galileo actually referred to clusters of waves rather than to the waves themselves. Is it, then, these clusters that split, when the sound pitch jumps an octave?

[108] From ancient times it had generally been held that the pitch of a note depended on the speed of vibration of the strings, that is, on the number of oscillations in unit time, namely the oscillation frequency.

of this last factor, which means for instance that, all other conditions being equal, a heavy brass string will oscillate more slowly than a catgut string.[109] In fact, since the limit to the rapidity with which a string can move is defined by its inertia, which in turn depends on its mass, the thickness and the density of the string act concurrently. The quantitative dependencies that are proposed are quite correct: pitch is inversely dependent on the length of the string and on the square root of the product of density and cross section; it is directly dependent, instead, on the square root of the tension.

SAGR: [. . .] There are three different ways in which the tone of a string may be sharpened, namely, by shortening it, by stretching it and by making it thinner. If the tension and size of the string remain constant one obtains the octave by shortening it to one-half, i.e., by sounding first the open string and then one-half of it; but if length and size remain constant and one attempts to produce the octave by stretching he will find that it does not suffice to double the stretching weight; it must be quadrupled; so that, if the fundamental note is produced by a weight of one pound, four will be required to bring out the octave.

And finally if the length and tension remain constant, while one changes the size of the string he will find that in order to produce the octave the size must be reduced to 1/4 that which gave the fundamental. And what I have said concerning the octave, namely, that its ratio as derived from the tension and size of the string is the square of that derived from the length applies equally well to all other musical intervals [. . .].

[109] In truth, the importance of the weight of the material had already been pointed out by Galileo's friend Guidubaldo del Monte in the above-cited *Meditatiunculae*. He wrote: "For equal tension applied and equal length of the strings, the lighter one emits a higher tone [. . .]. The reason is that, when both are hit, the lighter string acquires a faster back and forth motion, therefore yielding a higher tone". However, del Monte did not specify the quantitative relationship, which Galileo, instead, succeeded in doing (probably on the basis of a careful experimental study). It is also only fair to point out that, in the years in which Galileo was writing up his findings, the French scientist Marin Mersenne (who was in continuous contact by letter with Galileo) formulated virtually identical laws of acoustics and published them in his famous 1637 treatise, *Harmonie Universelle*. (See F. Flora, *op. cit.*, p. 712–14).

SALV: But now before proceeding any farther I want to call your attention to the fact that, of the three methods for sharpening a tone, the one which you refer to as the fineness of the string should be attributed to its weight. So long as the material of the string is unchanged, the size and weight vary in the same ratio. Thus in the case of gut-strings, we obtain the octave by making one string 4 times as large as the other; so also in the case of brass one wire must have 4 times the size of the other; but if now we wish to obtain the octave of a gut string, by use of brass wire, we must make it, not four times as large, but four times as heavy as the gut-string: as regards size therefore the metal string is not four times as big but four times as heavy. The wire may therefore be even thinner than the gut notwithstanding the fact that the latter gives the higher note. Hence if two spinets are strung, one with gold wire the other with brass, and if the corresponding strings each have the same length, diameter, and tension it follows that the instrument strung with gold will have a pitch about one-fifth lower than the other because gold has a density almost twice that of brass.[110] *[. . .]*

Returning now to the original subject of discussion, I assert that the ratio of a musical interval is not immediately determined either by the length, size, or tension of the strings but rather by the ratio of their frequencies, that is, by the number of pulses of air waves which strike the tympanum of the ear, causing it also to vibrate with the same frequency.

The secret of consonance

This fact established, we may possibly explain why certain pairs of notes, differing in pitch produce a pleasing sensation, others a less pleasant effect, and still others a disagreeable sensation. Such an explanation would be tantamount to an explanation of the more or less perfect consonances and of dissonances. The unpleasant sensation produced by the latter arises, I think, from the discordant vibrations of two different tones which strike the ear out of time. Especially harsh is

110 Since the frequency of the tone decreases with the square root of the specific weight of the string, an increase in the specific weight by a factor of 2 corresponds to a lowering of the frequency by $\sqrt{2} = 1.41$, which is not far from the frequency ratio 3/2, between G and C, that is, a fifth below.

the dissonance between notes whose frequencies are incommensurable [. . .]. Agreeable consonances are pairs of tones which strike the ear with a certain regularity; this regularity consists in the fact that the pulses delivered by the two tones, in the same interval of time, shall be commensurable in number, so as not to keep the ear drum in perpetual torment, bending in two different directions in order to yield to the ever-discordant impulses.

The first and most pleasing consonance is, therefore, the octave since, for every pulse given to the tympanum by the lower string, the sharp string delivers two; accordingly at every other vibration of the upper string both pulses are delivered simultaneously[111] *so that one-half of the entire number of pulses are delivered in unison. But when two strings are in unison their vibrations always coincide and the effect is that of a single string; hence we do not refer to it as consonance. The fifth is also a pleasing interval since for every two vibrations of the lower string the upper one gives three,*[112] *so that considering the entire number of pulses from the upper string one-third of them will strike in unison, that is, between each pair of concordant vibrations*[113] *there intervene two single vibrations; and when the interval is a fourth, three single vibrations intervene. In case the interval is a second where the ratio is 9/8 it is only every ninth vibration of the upper string which reaches the ear simultaneously with one of the lower; all the others are discordant and produce a harsh effect upon the recipient ear which interprets them as dissonances.*

[111] The meaning is: the two tones hit the eardrum together every second oscillation of the higher tone.

[112] Let us recall that the ratio between the frequencies of two tones in an interval of fifth (C–G) is 3 to 2.

[113] The reasoning becomes clearer if we lay out schematically the temporal sequence of the "pulses" on the eardrum (in modern terminology, *peaks of the sound pressure*). The diagram shows how the synchronization, in the interval of fifth, occurs every second period of C and third of G. In-between the two instants when synchronization occurs, G exhibits two peaks, which do not coincide with the one peak of C.

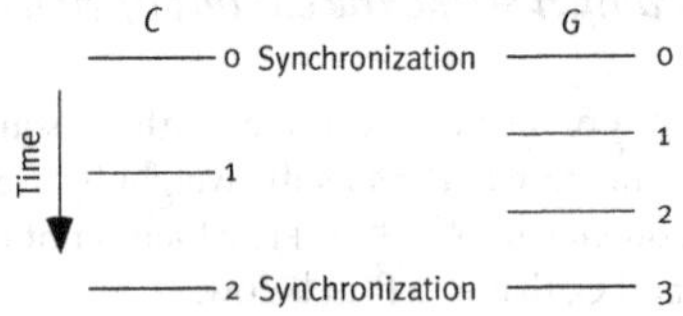

C G
0 Synchronization 0
1
Time
1
2
2 Synchronization 3

At this point Salviati gives some geometrical illustrations (omitted here) of what he has just explained using words. In the end, Sagredo can contain his enthusiasm no longer and reveals the deep emotion that Galileo himself no doubt felt in the face of the wonderful correspondence that exists between musical consonance, physical mechanisms of sound, and their mathematical representation. The same enthusiasm that, 2000 years earlier, had excited Pythagoras. However, whereas Pythagoras had been fascinated by the perfect correspondence that exists between harmony and simple mathematical proportions, Galileo had become fully aware that these mathematical proportions are nothing but the translation into numbers of the physical properties of acoustic sources. And he had even had the insight that these properties find a physical corollary in our auditory system. The proof is the visual analogy of consonance/dissonance obtained by means of a group of pendulums, with which Salviati elegantly closes the discussion ("I must show you a method by which the eye may enjoy the same game as the ear").

On the other hand, there is no reference here to the role played by the psyche. But this—it is worth repeating—is too modern a subject for Galileo to have been aware of its importance. In fact, his conception was strictly materialistic, in that he limited the requirements for appreciating harmony to a definable degree of mechanical perfection, without any need for animism to be called into play. Now, an approach such as this stands in contradiction to those who claim to have found a clear-cut distinction in Galileo's thinking between the physical and the spiritual domains,[114] doing an injustice to his resolutely logical and pragmatic mentality.

It is worth recalling here that Galileo's view on musical harmony was the exact opposite of that expressed by Kepler, who believed it could not be attributed to physiological causes, but rather to the mind. For him, musical harmony consisted in the perception by the mind of its conformity to the geometrical perfection of the Harmony of the Universe. This inclination of Kepler's to resort, at times, to metaphysical or magical explanations was the reason why Galileo,

[114] See, for example, D. Galati, *Galileo, primario matematico e filosofo*, Pagoda Editrice, Rome, 1991.

though holding him in high esteem, had some reservations in his regard.

Finally, Sagredo's description of the perfect fifth interval ("its softness is modified with sprightliness, giving at the same moment the impression of a gentle kiss and of a bite") deserves highlighting: it gives a measure of how deeply Galileo—an accomplished lute player, as mentioned earlier—was involved with the phenomenon of music.

SAGR: I can no longer remain silent; for I must express to you the great pleasure I have in hearing such a complete explanation of phenomena with regard to which I have so long been in darkness. Now I understand why unison does not differ from a single tone; I understand why the octave is the principal harmony, but so like unison as often to be mistaken for it and also why it occurs with the other harmonies. It resembles unison because the pulsations of strings in unison always occur simultaneously, and those of the lower string of the octave are always accompanied by those of the upper string; and among the latter is interposed a solitary pulse at equal intervals and in such a manner as to produce no disturbance[115]*; the result is that such a harmony is rather too much softened and lacks fire. But the fifth is characterized by its displaced beats and by the interposition of two solitary beats of the upper string and one solitary beat of the lower string between each pair of simultaneous pulses; these three solitary pulses are separated by intervals of time equal to half the interval which separates each pair of simultaneous beats from the solitary beats of the upper string. Thus the effect of the fifth is to produce a tickling of the ear drum such that its softness is modified with sprightliness, giving at the same moment the impression of a gentle kiss and of a bite.*

SALV: Seeing that you have derived so much pleasure from these novelties, I must show you a method by which the eye may enjoy

[115] The concept is illustrated clearly by the same type of layout drawn for the interval of fifth (C′ indicates C one octave higher).

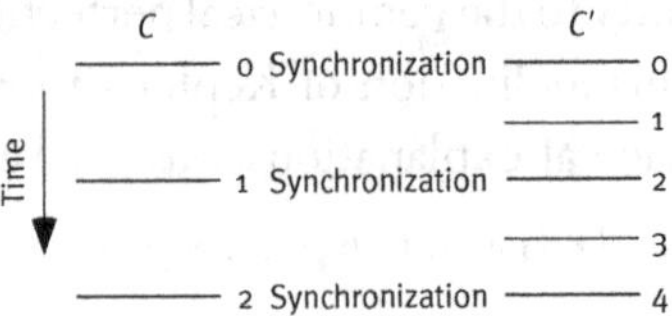

the same game as the ear. Suspend three balls of lead, or other heavy material, by means of strings of different length such that while the longest makes two vibrations the shortest will make four and the medium three;[116] *this will take place when the longest string measures 16, either in hand breadths or in any other unit, the medium 9 and the shortest 4, all measured in the same unit.*

Now pull all these pendulums aside from the perpendicular and release them at the same instant; you will see a curious interplay of the threads passing each other in various manners but such that at the completion of every fourth vibration of the longest pendulum,[117] *all three will arrive simultaneously at the same terminus, whence they start over again to repeat the same cycle. This combination of vibrations, when produced on strings is precisely that which yields the interval of the octave and the intermediate fifth.*[118] *If we employ the same disposition of apparatus but change the lengths of the threads, always however in such a way that their vibrations correspond to those of agreeable musical intervals, we shall see a different crossing of these threads but always such that, after a definite interval of time and after a definite number of vibrations, all the threads, whether three or four, will reach the same terminus at the same instant, and then begin a repetition of the cycle.*

If however the vibrations of two or more strings are incommensurable so that they never complete a definite number of vibrations at the same instant, or if commensurable they return only after a long interval of time and after a large number of vibrations, then the eye is confused by the disorderly succession of crossed threads. In like manner the ear is pained by an irregular sequence of air waves which strike the tympanum without any fixed order. [. . .]

We have pointed out several times that Galileo always felt it necessary to bring together the diverse ways we perceive the various aspects of a phenomenon into an internally consistent, coherent, and unitary whole.[119] Yet further demonstration of this is his idea that, if the effect

[116] The oscillation frequencies are in the same ratio to each other as are those of C, C′, and G.

[117] In reality it is the fourth of the shortest pendulum, which is the fastest.

[118] For instance C–G–C′.

[119] This, says Italian physicist Giorgio Salvini, was symbolic “of the unity of physics, of our mind, of our inspiration”.

of a number of strings vibrating in a coordinated manner (the "commensurability" of the frequencies emitted) is pleasant to the ear, a special effect must also be found for the eye; a behavior which is, let us say, more elegant and more gratifying to the observer than a completely uncoordinated motion. Since this is not observable in strings, which in the interval of acoustic and musical interest vibrate at frequencies too high to be followed by the eye,[120] Galileo resorted topendulums, because they can be made to oscillate much more slowly. The physics was, for him, exactly the same. And he was not mistaken, because both phenomena, as was to be demonstrated years later, are described by one and the same mathematical formula, that relating the deviation from the equilibrium position to a sinusoidal function of time (see Mathematical Note). Both are harmonic motions, as we would say using modern terminology.

The problem of the phase

It is interesting to note that, in Salviati's opinion, in order for the motion of the pendulums to be in harmony ("all the threads . . . will reach the same terminus at the same instant") it is essential that, at the outset, their respective individual motions are in phase coincidence and that this condition is maintained through time. However, this condition is not stated so explicitly for musical consonance, concerning those "pairs of tones which strike the ear with a certain regularity; this regularity consists in the fact that the pulses delivered by the two tones, in the same interval of time, shall be commensurable in number". Modern psychoacoustical research shows that phase coincidence among the tones of a chord is not essential. If the frequencies of the fundamental tones are in the appropriate ratios, vibrations can be out of phase without this influencing either the timbre of individual tones, or the degree of consonance. In fact, owing to the interplay of resonance and decay mechanisms which is typical of musical instruments using the string/harmonic-case system, phase variations

[120] The range of frequencies audible to the human ear is, at best, between 20 and 20,000 Hz (oscillations per second).

during a prolonged emission on these instruments—such as the development of a piano tone—are quite normal.

Finally, we wish to present one more reading, this time taken from the *Dialogue*.[121] In this excerpt suggestions are given as to how the relationship between the oscillation period of the pendulum and the length of the thread, as well as the isochronism of the oscillations of different amplitude, can be verified in a quick and simple way.

Wonderful facts

SALV: [...] Thus I say that one true, natural and even necessary thing is that a single movable body made to rotate by a single motive force will take a longer time to complete its circuit along a greater circle that a lesser circle. This is a truth accepted by all, and in agreement with experiment, of which we may adduce a few.

In order to regulate the time in wheel clocks, especially large ones, the builders fit them with a certain stick which is free to swing horizontally: At its ends they hang leaden weights, and when the clock goes too slowly, they can render each vibration more frequent merely by moving these weights somewhat toward the center of the stick.[122] *On the other hand, in order to retard the vibration, it suffices to draw these same weights out toward the ends, since the oscillations are thus made more slowly and in consequence the hour intervals are prolonged. Here the motive force is constant—the counterpoise—and the moving bodies are the same weights; but their vibrations are more frequent when they are closer to the center; that is, when they are moving along smaller circles.*

[121] *Dialogue Concerning the Two Chief World Systems*, translated by S. Drake, The Modern Library, New York, 2001, p. 521. In Italian: *Dialogo sopra i due massimi sistemi del mondo*, Edizione Nazionale, Vol. VII, p. 474.

[122] A simple diagram illustrates the concept clearly:

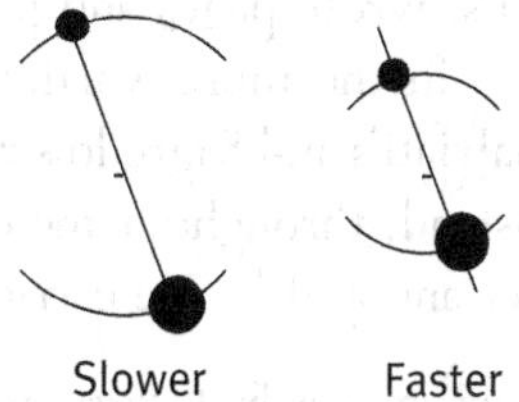

Let equal weights be suspended from unequal cords, removed from the perpendicular, and set free. We shall see the weights on the shorter cords make their vibrations in shorter times, being things that move in lesser circles. Again, attach such a weight to a cord passed through a staple fastened to the ceiling, and hold the other end of the cord in your hand. Having started the hanging weight moving, pull the end of the cord which you have in your hand so that the weight rises while it is making its oscillations. You will see the frequency of its vibrations increase as it rises, since it is going continually along smaller circles.

And here I want you to notice two details which deserve attention. One is that the vibrations of such a pendulum are made so rigorously according to different times, that is quite impossible to make them adopt other periods except by lengthening or shortening the cord. Of this you may readily make sure by experiment, tying a rock to a string and holding the end in your hand. No matter how you try you can never succeed in making it go back and forth except in one definite time, unless you lengthen or shorten the string; you will see that it is absolutely impossible.

The other particular is truly remarkable; it is that the same pendulum makes its oscillations with same frequency, or very little different—almost imperceptibly—whether these are made through large arcs or very small ones along a given circumference.

In all the foregoing excerpts, the reader will have noticed the repeated use of words like "striking, wonderful, remarkable" and also the declarations of amazement, almost incredulity, at the magical properties of the pendulum and other oscillators discussed. Galileo was always emotionally involved in the wonders of Nature, but he appears here to be particularly fascinated by the properties of oscillating systems. Since he was a scientist who was also a good musician there is nothing at all strange in this. However, it may also be a sign of his Aristotelian cultural background: Aristotle, in fact, though never interested in pendulums, wrote pages and pages on sound and on musical consonance. It is furthermore worth noting that the excitement and wonder in Salviati's and Sagredo's remarks are not found in Simplicio's, who instead, throughout the conversation, does not seem pleased, but rather annoyed.[123] He carries Aristotle's grammar

[123] This comment is from an essay by P. Bozzi, which offers an interesting psychological analysis of Galileo's position in regard to oscillating systems (see

crystallized in his mind, but is clearly unable to grasp the universality of his teachings, or the implicit dynamics of learning.

MATHEMATICAL NOTE

In this note we propose a short treatment of the pendulum in the limit of small oscillations, where the motion is known as simple harmonic. From the equation of motion we will deduce an expression for the oscillation period and the effect of isochronism of oscillations. As regards strings, we shall not demonstrate that the deviation from the equilibrium positions exhibits similar behaviour (again, for small oscillations), nor shall we derive the relationship between the frequency and the various parameters of the string. Readers interested in exploring these matters further are advised to consult more specialized textbooks. We shall then give a qualitative account of the phenomenon of musical consonance, as it is described by modern psychoacoustics, in terms of sequences of excitation points on the *basilar membrane* inside the cochlea.

The law of the pendulum

Let us write Newton's equation for an oscillating body of mass m, hanging from a long thread L (of negligible mass). Let us ignore for the moment any friction effect. Looking at the diagram, where θ indicates the tilt angle and g the value of the gravitational acceleration, the acting force is given by the component of the

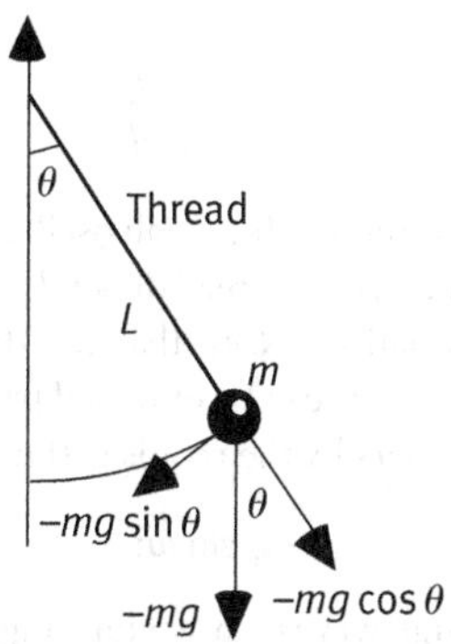

P. Bozzi, "Le ragioni di Simplicio, ossia la base percettiva del moto pendolare e della discesa lungo piani inclinati" ("Simplicio's reasoning, or the basis of the perception of pendulum motion and descent along inclined planes"), edited by M. Baldo Ceolin, published by the Physics Department "Galileo Galilei" of the University of Padua, 1995, p. 118).

weight directed along the tangent to the circumference described by the oscillating mass, namely $-mg\sin\theta$ (the component of the weight directed normal to the trajectory, $-mg\cos\theta$, is completely neutralized by the tension of the wire). If friction forces are negligible, one has

$$-mg\sin\theta = ma$$

where a is the acceleration, equal to the second derivative[124] with respect to time t of the distance traveled (along the circumference arc) and is therefore equal to $L(d^2\theta/dt^2)$. In the case of small oscillations one can approximate $\sin\theta \approx \theta$, and Newton's equation reduces to

$$-g\theta = L\frac{d^2\theta}{dt^2}$$

having also eliminated the mass. The equation of motion, therefore, *does not depend on mass*. A possible solution is given by the function

$$\theta = \theta_0 \sin\omega t \qquad (6.1)$$

where θ_0 is the angle corresponding to the oscillation amplitude and ω is the *pulsation* of the pendulum. By substituting θ into the equation, it is immediately verified that the equation is satisfied if

$$\omega = \sqrt{\frac{g}{L}}$$

Since the period T is linked to the pulsation by the relationship $T = 2\pi/\omega$, we have

$$T = 2\pi\sqrt{\frac{L}{g}} \qquad (6.2)$$

which confirms Galileo's prediction for small oscillations: the period does not depend on the oscillating mass, but instead depends on the length of the thread by way of a square-root relation. Note that the distance x traveled by the mass along the circumference arc can be obtained from solution (6.1) through multiplication of both members by the radius L, thus obtaining

$$x = x_0 \sin\omega t \qquad (6.3)$$

If we look for the standing waves on a vibrating string, it is possible to show that equation (6.3) is equally valid for the transversal deviation of

[124] Instantaneous velocity and acceleration, as defined in terms of derivatives, are briefly discussed in the Mathematical Note of Chapter 12.

the points of a string with respect to their equilibrium positions. It can also be shown that the fundamental oscillation frequency for a string with both ends fixed is given by[125]

$$f = \frac{1}{2L}\sqrt{\frac{T}{\mu}}$$

where T is the tension of the string and μ is its linear density (i.e. the mass per unit length). This result is (in every respect) exactly the same as that empirically deduced by Galileo (indeed, the linear density is given by the volume density, which is proportional to the specific weight, multiplied by the string cross-section).

The effect of damping

Galileo stated that a lead and a cork pendulum have equal oscillation periods. We have shown that in the case of small oscillations and no friction this is definitely the case. In real life, however, the cork pendulum oscillates just a little more slowly because, being lighter, it suffers more from the effects of air friction. If v is the velocity, the resisting force can be introduced into the equation of motion in the form $F_A = -bv$, where b is a friction factor which depends on the viscosity of the medium and on the geometrical shape of the oscillating mass (a very similar problem is dealt with in the Mathematical Note of Chapter 3):

$$-mg \sin\theta - bv = ma$$

Skipping intermediate steps, which can be found in any physics textbook, let us say that the pulsation ω_A as a function of ω in the absence of friction is now given (still in the limit of small oscillations) by:

$$\omega_A = \sqrt{\omega^2 - \left(\frac{b}{2m}\right)^2}$$

that is, it is the smaller the greater is the ratio between the friction factor b and the mass m of the pendulum. Let us take an example. For a cork sphere with an ideal oscillation period of 1 s, as a result of which $\omega = 2\pi$ rad/s, let $b/2m = 0.60$ rad/s: it can be readily calculated that $\omega_A = 0.99\ \omega$. For a lead sphere of equal diameter, the ratio $b/2m$ becomes so small that the pulsation ω_A is virtually identical to ω.

125 For this see any basic physics textbook.

Musical consonance

And now let us examine a model for consonance as proposed by psychoacoustics. We will see that Galileo had some idea of the psycho-physiological aspects which lie at the basis of this model. We start by explaining how the frequencies present in a complex sound act on the *basilar membrane*. This is the organ inside the cochlea which, upon receiving a sound wave through the *oval window* via the ossicles chain, begins to vibrate and conveys nerve impulses into the auditory nerve, which in its turn transfers them to the brain. This process is made possible by the bending of tiny *hair cells*, nerve fibers lined up along the entire length of the basilar membrane. A major characteristic of the basilar membrane is that its point of maximum deformation differs according to the frequency of the wave received, as shown in the diagram, which depicts the situation regarding pure, that is, monochromatic, tones. The diagram illustrates the amplitude of the sound wave along the basilar membrane for a few given values of the frequency, corresponding to a series of C's: the higher the frequency, the closer to the oval window is the maximum in membrane deformation. Low-frequency sounds, instead, excite the membrane near the other end, the *helicotrema*. This behavior gives rise to a spatial discrimination of tones according to their frequencies.

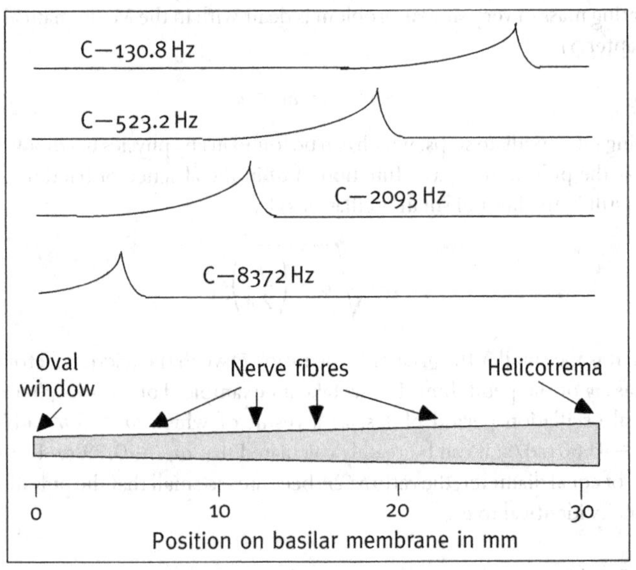

A real sound contains the fundamental frequency plus the frequencies of the higher harmonics, which are multiples of the fundamental frequency. This real sound thus produces on the basilar membrane its own specific pattern of excitation points, which we shall call the *harmonic segment*. The brain recognizes this segment on the basis of the particular nerve fibers which are responsible for conveying the complex signal. If, instead of a C, a G had been played, the harmonic segment would have been different. It is believed that the brain, in receiving a chord of different tones, "feels more at ease" if the respective harmonic segments have some points in common, because a simpler neuron network is activated. Such an easier task would then lead to a gratifying effect—the one which we call "a consonance".

In this model, the phase relations among the various tones in a chord are totally irrelevant. Experimental tests confirm that they are of very small importance even within the different harmonics of a complex tone, whose timbre is affected by them only slightly, if at all.

The excitation pattern of the basilar membrane for a perfect fifth chord is shown schematically in the diagram below, where numbers give the order of the harmonics. Since the fundamental frequencies of G and C are in a ratio of 3 to 2, the second harmonic of G coincides with the third of C (as mentioned at the beginning of this chapter) and consequently they stimulate one and the same nerve fiber; the same is true for the fourth harmonic of G and the sixth of C, and so on.[126]

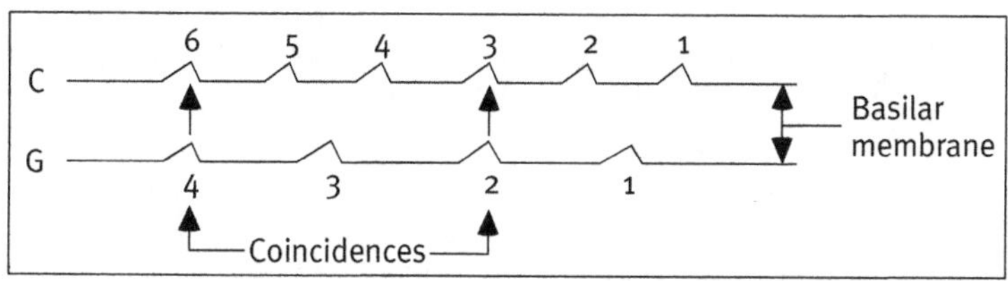

Recent studies in the psychology of perception[127] suggest that infants have a natural preference for chords that present interconnections of the kind just described, as opposed to chords that do not (e.g. the tritone or augmented

[126] To avoid confusing the reader, it needs to be pointed out that, for practical purposes, in this diagram the deformation maxima are represented as a function of their position on the basilar membrane, that is, of the frequency. In the diagrams that appear in the footnotes to the main text, instead, the excitation maxima are represented as a function of time.

[127] See, for instance, M.R. Zentner and J. Kagan, "Perception of music by infants", *Nature*, Vol. 383, 1996, p. 29.

fourth C–F#, historically called *diabolus in musica*). Our perceptive system is, however, capable of learning—and learning a lot. It is not difficult for a trained ear to lose this ontogenetic preference for consonant chords almost entirely—a fact that is cause for optimism about the future of music. Naturally, there was no way Galileo could possibly have imagined this. He knew no music other than that based strictly on the principles of classical harmony, as established by, and handed down from, the Greeks.

PART IV

THE STARRY SKY ABOVE US

7

The Crystal Moon

The Wonders of the Telescope*

Bellissima cosa e assai attraente alla vista
è rimirare il corpo lunare

It is most beautiful and pleasing
to the eye to look upon the lunar body

Galileo discovers the mountains of the moon. Aristotelian philosophers, refusing to accept that a celestial body can be other than spherical and smooth (canons of perfection), counter that the moon is like a transparent sphere of crystal within whose interior there are regions of higher and regions of lower density. After taking a look at the opening pages of *Sidereus Nuncius*, where the wonders of the telescope are described, this chapter will go on to illustrate some of the observations Galileo made using the telescope and, in closing, will present his arguments against the "crystal moon".

Ask the most learned persons you know, perhaps even those involved in science, if they have ever read *Sidereus Nuncius* (or *The Sidereal Messenger*). Their replies will almost invariably be in the negative. And yet this small book is of extraordinary interest. It is no less exciting than a science-fiction story and, although it was originally written in Latin, is available in various translations.[128] The main reason for its

* From *Sidereus Nuncius*.

[128] G. Galilei, *Sidereus Nuncius*, notes by A. Battistini, translated into Italian by M. Timpanaro Cardini, published by Marsilio, Venice, 1993. In English: *Sidereus Nuncius* or *The Sidereal Messenger*, by A. Van Helden, University of Chicago Press, Chicago, IL, 1989; also: *The Starry Messenger*, by S. Drake, in *Discoveries and Opinions of Galileo*, Doubleday & Co., New York, 1957.

very limited readership is that it is never encountered in school curricula, where knowledge is rigidly "boxed". It was written in Latin, but its author lived in the seventeenth century. In Italy literature students do not find it in their bibliographies, partly because of its Latin language, but also because it deals with science—while science students are not expected to devote time to reading an "obsolete" book.

Humans and the universe

At the dawn of a new century, for the first time in human history, a man scans the skies with the help of an instrument—the telescope—which throws open to his eyes wonders never before observed. He does not confine himself, however, merely to enjoying the beauty that until then had been hidden. He badly wants to understand. He is convinced that aesthetic pleasure becomes stronger if it is accompanied by understanding. He wants to investigate the nature of celestial bodies, to unveil the laws that rule their motions. Traditional doctrine can be of no help to him—quite the opposite, it represents an obstacle, removing which will be arduous, due to the blind or prejudiced mistrust of the majority of his contemporaries.

Moving into the third millennium, the thrill of curiosity we all experience at the clear pictures beamed into our homes from Mars must be far inferior to the excited amazement Galileo experienced, all alone, exploring the sky during the cold winter nights of 1609–10. The impetuous progress of science and technology in recent years has dulled our capacity for surprise. What we see is what we expect to see. From the exploration of (what is, for us) the nearby universe, we hardly expect discoveries capable of radically altering our knowledge—let alone our way of thinking.

In pointing his telescope at the sky Galileo was performing a revolutionary act. For years he had been convinced that Copernicus' heliocentric theory was sound (this is clear from a 1597 letter to Kepler[129]), even if he was still obliged to teach the Ptolemaic system. The telescope could provide him with the proof he needed. The act,

[129] Edizione Nazionale, Vol. X, p. 67.

therefore, was that of a man who, like a Ulysses reborn, wished to be the protagonist in expanding the boundaries of knowledge—and rejected knowledge that was "handed-down" from the past, unsupported by logical and experimental evidence.

The dedication to the Grand Duke

The text of *Sidereus Nuncius* is preceded by a dedication (so flattering it is embarrassing, but appropriate to its time and to the purpose) to the young Grand Duke of Florence, Cosimo II De' Medici, to whom Galileo had given mathematics lessons over some summer holidays. Galileo wanted, by now, to leave Padua and its heavy teaching load, private lessons, and low salary. He aspired to having more time to devote to research and to the great work he was already planning on writing about the systems of the world. In Tuscany, he believed, he would find the ambience more encouraging and open.

The solemn *incipit* of the dedication, which was almost certainly later to prove a source of inspiration to Ugo Foscolo[130] in writing his *Sepolcri*, addresses the theme of the survival of the memory of great men. Their name, it asserts, is better linked to stars than to marble or bronze monuments, or to the words of poets, all of which are corruptible in time. Let us read it.[131]

MOST SERENE
COSIMO II DE' MEDICI
FOURTH GRAND DUKE OF TUSCANY

A most excellent and kind service has been performed by those who defend from envy the great deeds of excellent men and have taken it upon themselves to preserve from oblivion and ruin names deserving of immortality. Because of this, images sculpted in marble or cast in bronze are passed down for the memory of posterity; because of this, statues, pedestrian as well as equestrian, are erected; because of this,

[130] Ugo Foscolo was one of the greatest Italian poets of the nineteenth century.

[131] *Sidereus Nuncius*, or *The Sidereal Messenger*, translated from Latin by A. Van Helden, University of Chicago Press, Chicago, IL, 1989, p. 29. In Latin: *Sidereus Nuncius*, Edizione Nazionale, Vol. III, p. 55.

too, the cost of columns and pyramids, as the poet says,[132] *rises to the stars; and because of this, finally, cities are built distinguished by the names of those who grateful posterity thought should be commended to eternity. For such is the condition of the human mind that unless continuously struck by images of things rushing into it from the outside, all memories easily escape from it.*

Others, however, looking to more permanent and long-lasting things, have entrusted the eternal celebration of the greatest men not to marbles and metals but rather to the care of the Muses and to incorruptible monuments of letters. But why do I mention these things as though human ingenuity, content with these [earthly] realms, has not dared to proceed beyond them? Indeed, looking far ahead, and knowing full well that all human monuments perish in the end through violence, weather, or old age, this human ingenuity contrived more incorruptible symbols against which voracious time and envious old age can lay no claim. And thus, moving to the heavens, it assigned to the familiar and eternal orbs of the most brilliant stars the names of those who, because of their illustrious and almost divine exploits, were judged worthy to enjoy with the stars an eternal life. As a result, the fame of Jupiter, Mars, Mercury, Hercules, and other heroes by whose names the stars are addressed will not be obscured before the splendor of the stars themselves is extinguished.

We will now proceed by taking a look at a selection of significant passages from *Sidereus Nuncius*. First, the opening pages of the book,[133] where Galileo announces the extraordinary astronomical discoveries that were becoming possible thanks to the advent of the telescope, the instrument he had improved and which is described here. To illustrate this, we include several excerpts concerning the observation of the moon's face, which Galileo found to be covered with mountains and valleys. This was in direct contradiction to the beliefs of the time, which had it that the moon, as a celestial body, should be incorruptible,

[132] Galileo is referring to Propertius: "*Pyramidum sumptus ad sidera ducti*".

[133] *Sidereus Nuncius* or *The Sidereal Messenger*, translated from Latin by A. Van Helden, University of Chicago Press, Chicago, IL, 1989, p. 35. In Latin: *Sidereus Nuncius*, Edizione Nazionale, Vol. III, p. 59.

spherical, and perfectly smooth. He likens the surface of the moon to that of the earth, and incautiously pushes the analogy so far as to postulate the existence of a gaseous envelope around it.

The method Galileo uses to determine the height of the mountains of the moon is ingenious, yet at the same time simple. He then moves on to his other great discovery—the four "Medicean stars" around Jupiter and to the meticulous observation of their motions, by means of which he was able to establish that they were satellites of the planet. In between these two grand topics, Galileo inserts a few considerations regarding the Milky Way and other nebulae, which he correctly describes as clusters of small stars indiscernible to the naked eye.

The essence of the work

In this short treatise I propose great things for inspection and contemplation by every explorer of Nature. Great, I say, because of the excellence of the things themselves, because of their newness, unheard of through the ages, and also because of the instrument with the benefit of which they make themselves manifest to our sight.

Certainly it is a great thing to add to the countless multitude of fixed stars visible hitherto by natural means and expose to our eyes innumerable others never seen before, which exceed tenfold the number of old and known ones. It is most beautiful and pleasing to the eye to look upon the lunar body, distant from us about sixty terrestrial diameters, from so near as if it were distant by only two of these measures, so that the diameter of the same Moon appears as if it were thirty times, the surface nine-hundred times, and the solid body about twenty-seven thousand times larger than when observed only with the naked eye. Anyone will then understand with the certainty of the senses that the Moon is by no means endowed with a smooth and polished surface, but is rough and uneven and, just as the face of the Earth itself, crowded everywhere with vast prominences, deep chasms, and convolutions.

Moreover, it seems of no small importance to have put an end to the debate about the Galaxy or Milky Way and to have made manifest its essence to the senses as well as the intellect; and it will be pleasing and most glorious to demonstrate clearly that the substance of those stars called nebulous up to now by all astronomers is very different from what has hitherto been thought.

But what greatly exceeds all admiration, and what especially impelled us to give notice to all astronomers and philosophers, is this, that we have discovered four wandering stars, known or observed by no one before us. These, like Venus and Mercury around the Sun have their periods around a certain star notable among the number of known ones, and now precede, now follow, him, never digressing from him beyond certain limits. All these things were discovered and observed a few days ago by means of a glass contrived by me after I had been inspired by divine grace.

Perhaps more excellent things will be discovered in time, either by me or by others, with the help of a similar instrument, the form and construction of which, and the occasion of whose invention, I shall first mention briefly, and then I shall review the history of the observations made by me.

About 10 months ago a rumor came to our ears that a spyglass had been made by a certain Dutchman by means of which visible objects, although far removed from the eye of the observer, were distinctly perceived as though nearby. About this truly wonderful effect some accounts were spread abroad, to which some gave credence while others denied them. The rumor was confirmed to me a few days later by a letter from Paris from the noble Frenchman Jacques Badovere. This finally caused me to apply myself totally to investigating the principles and figuring out the means by which I might arrive at the invention of a similar instrument, which I achieved shortly afterward on the basis of the science of refraction.

Galileo thus gave the credit for having constructed the first telescope to a Fleming. In fact, information exists that rudimentary equipment of the kind was already being built in Italy towards the end of the 1500s.[134] Even Leonardo da Vinci spoke of "glasses for seeing the moon large". Galileo's merit was to have perceived the extraordinary

[134] Reports exist of an Italian telescope bearing the date 1590, of which Dutchmen Hans Lippershey and Zacharias Janssen possibly made a copy in 1608 or perhaps even earlier.

potential of the instrument and to have built prototypes of superior quality, suitable for scientific use. Indeed, it was only after Galileo's announcement that scientists began treating the telescope as a serious research instrument, rather than as a mere curiosity. In 1610, at the time of the publication of *Sidereus Nuncius*, Kepler was still talking in derogatory terms about the "two-lensed tube" (yet the following year he was hurrying to propose his own model with two converging lenses).

Galileo now goes on to give technical and construction details for his improved telescope, which was capable of a linear magnification of about 30. Following this, he proposes a technique for measuring the distances between objects that are being observed, using a scheme of geometrical optics; this, however, does not have the clarity which usually characterizes Galilean descriptions (the reader can find detailed explanations in the Mathematical Note at the end of this chapter). This, coupled with the fact that his competence in optics was limited, suggests that his statement "on the basis of the science of refraction" was not entirely truthful. It is likely that he achieved the optimization of the telescope through a systematic series of adjustments and improvements, rather than on the basis of theoretical principles.

The telescope

And first I prepared a lead tube in whose ends I fitted two glasses, both plane on one side while the other side of one was spherically convex and of the other concave.[135] *Then, applying my eye to the*

[135] Regarding the use of a convex lens and a concave one, in *The Assayer* Galileo writes (S. Drake and O'Malley, *The Assayer*, in *Controversy on the Comets of 1618*, Philadelphia, PA, 1960 (out of print, available in photostatic form, p. 213; in Italian: *Il Saggiatore*, Edizione Nazionale, Vol. VI, p. 259):

The device needs either a single glass or more than one. It cannot consist of one alone, because the shape of that one would have to be either convex [. . .], or concave [. . .], or contained between parallel surfaces. But the last named does not alter visible objects in any way, either by enlarging or reducing them; the concave diminishes them; and the convex, while it does indeed increase them, shows them very indistinctly and confusedly. Therefore a single glass is not sufficient to produce the effect. Passing next to two, and knowing as before that a glass with parallel faces *concave glass, I saw objects satisfactorily large and close. Indeed, they appeared three times closer and nine times larger than when observed with natural vision only. Afterward I made another more perfect one for myself that showed objects more than sixty times larger. Finally, sparing no labor or expense, I progressed so far that I constructed for myself an instrument so excellent that things seen through it appear about a thousand times larger and more than thirty times closer than when observed with the natural faculty only. It would be entirely superfluous to enumerate how many and how great the advantages of this instrument are on land and at sea. But having dismissed earthly things, I applied myself to explorations of the heavens. And first I looked at the Moon from so close that it was scarcely two terrestrial diameters distant. Next, with incredible delight I frequently observed the stars, fixed as well as wandering, and as I saw their huge number I began to think of, and at last discovered, a method whereby I could measure the distances between them. In this matter, it behooves all those who wish to make such observations to be forewarned. For it is necessary first that they prepare a most accurate glass that shows objects brightly, distinctly, and not veiled by any obscurity, and second that it multiply them at least four hundred times and show them twenty times closer. For if it is not an instrument such as that, one will try in vain to see all the things observed in the heavens by us and enumerated below. Indeed, in order that anyone may, with little trouble, make himself more certain about the magnification of the instrument, let him draw two circles or two squares on paper, one of which is four hundred times larger than the other, which will be the case when the larger diameter is twenty times the length of the other diameter. He will then observe from afar both sheets fixed to the same wall, the smaller*

alters nothing, I concluded that the effect would still not be achieved by combining such a one with either of the other two. Hence I was restricted to try to discover what would be done by a combination of the convex and the concave, and you see how this gave me what I sought.

In his *Dioptrice* of 1611, Kepler describes the advantages of a telescope with two convex (converging) lenses and this is, in fact, the dioptric telescope most widely used in astronomy today. These two kinds of telescope, Galilean and Keplerian, are compared in the Mathematical Note at the end of this chapter.

one with one eye applied to the glass and the larger one with the other, naked eye. This can easily be done with both eyes open at the same time. Both figures will then appear of the same size if the instrument multiplies objects according to the desired proportion. After such an instrument has been prepared, the method of measuring distances is to be investigated, which is achieved by the following procedure. For the sake of easy comprehension, let ABCD be the tube and E the eye of the observer. When there are no glasses in the tube, the rays proceed to the object FG along the straight lines ECF and EDG, but with the glasses put in they proceed along the refracted lines ECH and EDI. They are indeed squeezed together and where before, free, they were directed to the object FG, now they only grasp the part HI. Then, having found the ratio of the distance EH to the line HI, the size of the angle subtended at the eye by the object HI is found from the table of sines, and we will find this angle to contain only some minutes, and if over the glass CD we fit plates perforated some with larger and some with smaller holes, putting now this plate and now that one over it as needed, we form at will angles subtending more or fewer minutes. By this means we can conveniently measure the spaces between stars separated from each other by several minutes with an error of less than one or two minutes. Let it suffice for the present, however, to have touched on this so lightly and to have, so to speak, tasted it only with our lips, for on another occasion we shall publish a complete theory of this instrument. Now let us review the observations made by us during the past 2 months, inviting all lovers of true philosophy to the start of truly great contemplation.

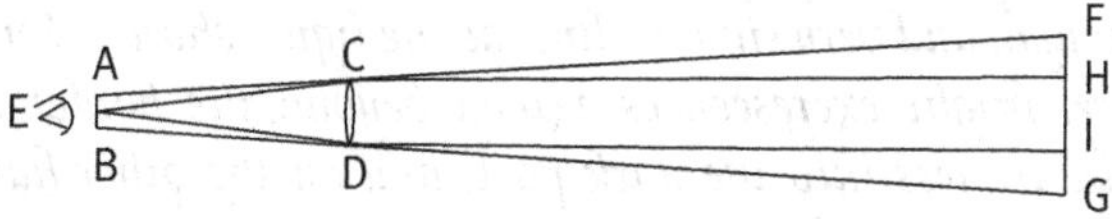

It is here that the description of the lunar surface begins, accompanied by careful illustrations and analyses of the boundary that divides the sun-lit zone from the zone in shadow. Take a careful look at Galileo's drawing of the moon, one of many that he produced—it is so close to reality! Imagine, too, the surprise and mistrust of his contemporaries, accustomed as they were to considering the satellite to be a shiny, polished sphere.

The face of the moon

Let us speak first about the face of the Moon that is turned toward our sight, which, for the sake of easy understanding, I divide into two parts, namely a brighter one and a darker one. The brighter part appears to surround and pervade the entire hemisphere, but the darker part, like some cloud, stains its very face and renders it spotted. Indeed, these darkish and rather large spots are obvious to everyone, and every age has seen them. For this reason we shall call them the large or ancient spots, in contrast with other spots, smaller in size and occurring with such frequency that they besprinkle the entire lunar surface, but especially the brighter part. These were, in fact, observed by no one before us. By oft-repeated observations of them we have been led to the conclusion that we certainly see the surface of the Moon to be not smooth, even, and perfectly spherical, as the great crowd of philosophers have believed about this and other heavenly bodies, but, on the contrary, to be uneven, rough, and crowded with depressions and bulges. And it is like the face of the Earth itself, which is marked here and there with chains of mountains and depths of valleys. The observations from which this is inferred are as follows.

On the fourth or fifth day after conjunction, when the Moon displays herself to us with brilliant horns, the boundary dividing the bright from the dark part does not form a uniformly oval line, as would happen in a perfectly spherical solid, but is marked by an uneven, rough, and very sinuous line, as the figure shows. For several, as it were, bright excrescences extend beyond the border between light and darkness into the dark part, and on the other hand little dark parts enter into the light. Indeed, a great number of small darkish spots, entirely separated from the dark part, are distributed everywhere over almost the entire region already bathed by the light of the Sun, except, at any rate, for that part affected by the large and ancient spots. We noticed, moreover, that all these small spots just mentioned always agree in this, that they have a dark part on the side toward the Sun while on the side opposite the Sun they are crowned with brighter borders like shining ridges. And we have an almost entirely similar sight on Earth, around sunrise, when the valleys are not yet bathed in light but the surrounding mountains facing the Sun are already seen shining with light. And just as the shadows of the earthly valleys are diminished as the Sun climbs higher, so those lunar spots lose their darkness as the luminous part grows.

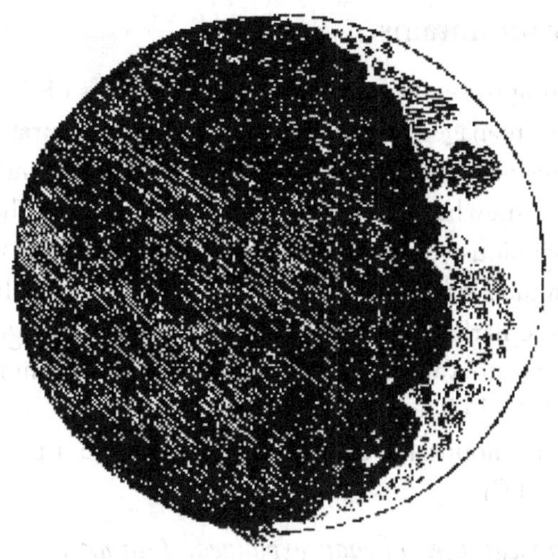

Not only are the boundaries between light and dark on the Moon perceived to be uneven and sinuous, but, what causes even greater wonder, is that very many bright points appear within the dark part of the Moon, entirely separated and removed from the illuminated region and located no small distance from it. Gradually, after a small period of time, these are increased in size and brightness. Indeed, after 2 or 3 hours they are joined with the rest of the bright part, which has now become larger. In the meantime, more and more bright points light up, as if they are sprouting, in the dark part, grow, and are connected at length with that bright surface as it extends farther in this direction. An example of this is shown in the same figure. Now, on Earth, before sunrise, aren't the peaks of the highest mountains illuminated by the Sun's rays while shadows still cover the plain? Doesn't light grow, after a little while, until the middle and larger parts of the same mountains are Illuminated,

and finally, when the Sun has risen, aren't the illuminations of plains and hills joined together? These differences between prominences and depressions in the Moon, however, seem to exceed the terrestrial roughness greatly, as we shall demonstrate below. [. . .]

The lunar mountains

It is fascinating to see how Galileo, in the sections of the discourse which follow, manages to make detailed and very accurate estimates of the dimensions of the objects he has observed. We include here, despite the somewhat tedious geometrical content of the explanations, his ingenious yet simple procedure for calculating the heights of the mountains of the moon (heights which he slightly overestimated). It is remarkable that he was able to achieve such good results despite having no valid data on the heights of the mountains on earth, which he believed to be far lower than their actual heights ("on Earth no mountains exist that reach even to a perpendicular height of 1 mile").

From the appearances already explained, I think it is sufficiently clear that the brighter surface of the Moon is sprinkled all over with prominences and depressions. It remains for us to speak of their magnitudes, demonstrating that the terrestrial roughnesses are far smaller than the lunar ones. I say smaller, speaking absolutely, not merely in proportion to the sizes of their globes. This is clearly shown in the following manner.

As has often been observed by me, with the Moon in various aspects to the Sun, some peaks within the dark part of the Moon appear drenched in light, although very far from the boundary line of the light. Comparing their distance from that boundary line to the entire lunar diameter, I found that this interval sometimes exceeds the twentieth part of the diameter. Assuming this, imagine the lunar globe, whose great circle[136] *is CAF, whose center is E,*

[136] "Great circle" refers to half the circumference.

and whose diameter is CF, which is to the Earth's diameter as 2 to 7.[137] *And since according to the most exact observations the terrestrial diameter contains 7000 Italian miles,*[138] *CF will be 2000 miles, CE 1000, and the twentieth part of the whole of CF will be 100 miles. Now let CF be the diameter of the great circle dividing the luminous from the dark part of the Moon (because of the very great distance of the Sun from the Moon this circle does not differ sensibly from a great circle*[139]*), and let A be distant from point C one-twentieth part of it.*[140] *Draw the semidiameter EA, which, when extended, intersects the tangent GCD (which represents a ray of light) at D. The arc CA or the straight line CD*[141] *will therefore be 100 parts*[142] *of the 1000 represented by CE, and the sum of the squares of CD and CE is 1,010,000, which is equal to the square of ED.*[143] *The whole of ED will therefore be more than 1004,*[144] *and AD more than 4 parts of the 1000 represented by CE. Therefore the height AD on the Moon, which represents some peak reaching all the way up to the Sun's rays GCD and removed from the boundary line C by the distance CD, is higher than 4 Italian miles.*[145] *But on Earth no mountains exist that reach even to a perpendicular*

137 Galileo's estimate is remarkably accurate; since the diameter of the moon is 3476 km and that of the earth is 12,720 km, the exact ratio is 2 to 7.3.

138 Since the Italian mile was equal to 1.851 km, the diameter Galileo used was 12,957 km, a little larger, that is, than the actual diameter.

139 In other words, the moon has almost exactly half of its surface in the light and half in the dark.

140 That is, of CF. Note that, to make the representation clearer, Galileo does not respect the proportions he is assuming. It is worth stressing that segment AD represents the total height of a mountain located in the dark part, except for its summit D, which is just "grazed" by a sun's ray.

141 In a first approximation, arc CA is taken as equal to the tangent segment CD.

142 *Parts*—Italian miles.

143 ED is derived by applying Pythagoras theorem to right-angled triangle CDE. The height of the mountain, AD, is then obtained as the difference between segment ED and radius AE.

144 The square root of 1,010,000 is 1,004.987, that is, almost exactly 1005: it is quite surprising that Galileo, here, chose to round down.

145 On the basis of the preceding note, the altitude AD would in fact be 5 miles, that is 9232 m, as opposed to the 7404 m resulting from Galileo's rounding down.

height of 1 mile.[146] *It is evidence, therefore, that the lunar prominences are loftier than the terrestrial ones. [. . .]*

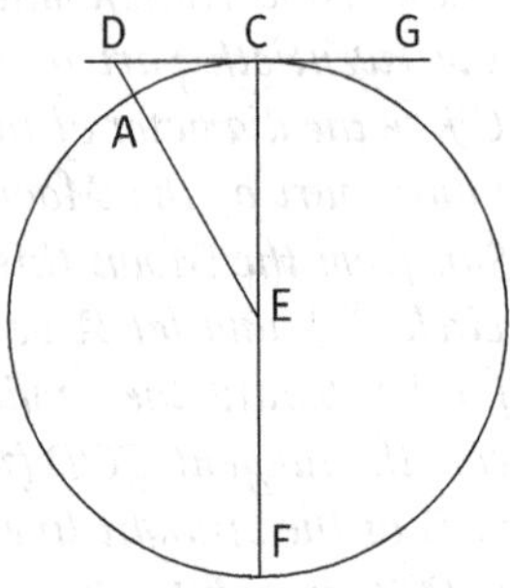

The structure of galaxies

Following this come a few pages regarding the differences between stars and planets and the study of the Milky Way, which Galileo describes (using terms that are close to those employed today) as a "conglomeration of star clusters". Just the part about the Milky Way and the nebulae is presented here.[147]

What was observed by us in the third place is the nature or matter of the Milky Way itself, which, with the aid of the spyglass, may be observed so well that all the disputes that for so many generations have vexed philosophers are destroyed by visible certainty, and we are liberated from wordy arguments. For the Galaxy is nothing else than a congeries of innumerable stars distributed in clusters. To whatever region of it you direct your spyglass, an immense number of stars immediately offer themselves to view, of which very many appear rather large and very conspicuous but the multitude of small ones is truly unfathomable.

And since that milky luster, like whitish clouds, is seen not only in the Milky Way, but dispersed through the ether, many similarly coloured patches shine weakly; if you direct a glass to any of them, you will meet with a dense crowd of stars. Moreover—and what is even more

[146] It is worth underlining that in Galileo's time people had no idea about the real height of mountains on earth.

[147] *Sidereus Nuncius* or *The Sidereal Messenger*, translated from Latin by A. Van Helden, University of Chicago Press, Chicago, IL, 1989, p. 62. In Latin: *Sidereus Nuncius*, Edizione Nazionale, Vol. III, p. 78.

remarkable—the stars that have been called 'nebulous' by every single astronomer up to this day are swarms of small stars placed exceedingly closely together. While each individual one escapes our sight because of its smallness or its very great distance from us, from the commingling of their rays arises that brightness ascribed up to now to a denser part of the heavens capable of reflecting the rays of the stars or Sun. We have observed some of these, and we wanted to reproduce the asterisms of two of them.

In the first there is the nebula called Orion's Head, in which we have counted twenty-one stars.

The second figure contains the nebula called Praesepe, which is not a single star but a mass of more than forty little stars. In addition to the ass-colts we have marked down thirty-six stars, arranged as follows:

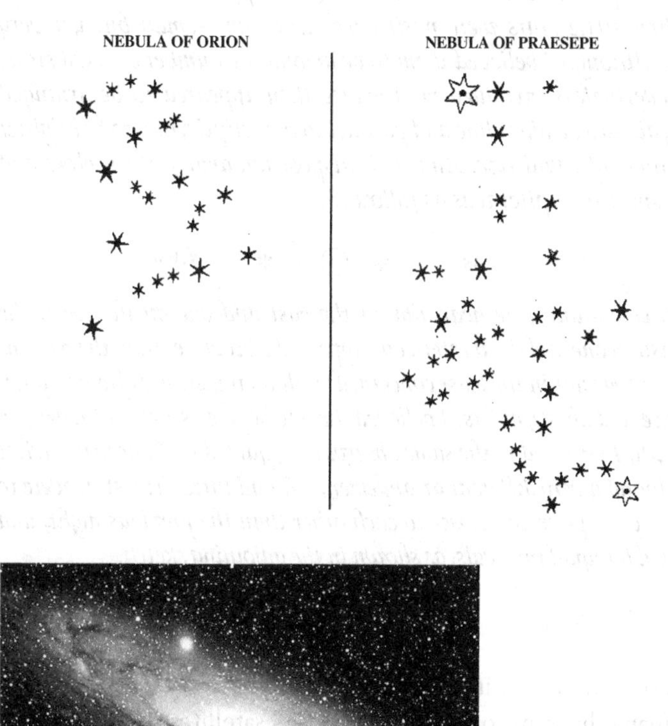

A contemporary photograph: the Andromeda nebula (The Electronic Universe Project)

Jupiter's satellites

Finally, here are some excerpts describing the discovery of the Medicean Planets, that is, the four satellites of Jupiter, the most important astronomical discovery that Galileo was to make.[148] The passages illustrate the scrupulousness of Galileo's approach and the agile and modern style with which he noted and presented the features observed. In the diagrams, Jupiter is depicted as a small circle, its satellites as asterisks.

[...] Accordingly, on the seventh day of January of the present year 1610 at the first hour of the night, when I inspected the celestial constellations through a spyglass, Jupiter presented himself. And since I had prepared for myself a superlative instrument, I saw (which earlier had not happened because of the weakness of the other instruments) that three little stars were positioned near him—small but yet very bright. Although I believed them to be among the number of fixed stars, they nevertheless intrigued me because they appeared to be arranged exactly along a straight line and parallel to the ecliptic, and to be brighter than others of equal size. And their disposition among themselves and with respect to Jupiter was as follows:

East * * ○ * West

that is, two stars were near him on the east and one on the west; the more eastern one and the western one appeared a bit larger than the remaining one. I was not in the least concerned with their distances from Jupiter, for, as we said above, at first I believed them to be fixed stars. But when, on the eighth, I returned to the same observation, guided by I know not what fate, I found a very different arrangement. For all three little stars were to the west of Jupiter and closer to each other than the previous night, and separated by equal intervals, as shown in the adjoining sketch.

East ○ * * * West

The observations continued over the following nights, resulting in a great many diagrams of the planet and its satellites, in which their

[148] *Sidereus Nuncius* or *The Sidereal Messenger*, translated from Latin by A. Van Helden, University of Chicago Press, Chicago, IL, 1989, p. 64. In Latin: *Sidereus Nuncius*, Edizione Nazionale, Vol. III, p. 79.

positions, number, and dimensions constantly change. We resume the story on January 19.

[. . .] On the nineteenth, at the second hour of the night, the formation was like this. There were three stars exactly on a straight line through Jupiter, one to the

East * ○ * * West

east, 6 minutes distant; between Jupiter and the first western one was an interval of 5 minutes, while this star was 4 minutes from the more western one. At this time I was uncertain whether between the eastern star end Jupiter there was a little star, very close to Jupiter, so that it almost touched him. And at the fifth hour, I clearly saw this little star now occupying a place precisely in the middle between Jupiter and the eastern star, so that the formation was as follows:

East * * ○ * * West

Further, the newly perceived star was very small; yet by the sixth hour it was almost equal in magnitude to the others. [. . .]

The investigation went on, with Galileo expressing doubts about, and formulating hypotheses on, the configurations observed. He then began to make quantitative estimates of the distances between Jupiter and its satellites, in terms of the angles between the sight lines by which the various bodies are viewed. His diary of February contains the following notes.

On the eighth, at the first hour, three stars were present, all to the east, as in the figure. The small star closest to Jupiter was 1 minute, 20 seconds distant from

East * * * ○ West

him; the middle star was 4 minutes from this one and rather large; and the very small easternmost star was 20 seconds from that one. I was of two minds whether the one closest to Jupiter was only one, or two little stars, for it seemed now and then that there was another star near it, toward the east, extremely small, and separated from it by only 10 seconds. They were all extended on the same straight line along the zodiac. But at the third hour the star closest to Jupiter nearly touched him. It was only 10 seconds from him, while the others had moved farther from Jupiter, for the middle one was 6 minutes away from Jupiter. Finally, at the fourth hour, the one that before was closest to Jupiter, united with him, was seen no longer.

During the final nights of observation, in order to show the relative motions inside the Jupiter-satellites system, Galileo also drew in the position of a fixed star. The final report is dated March 2, 1610.

On the second, at 0 hours and 40 minutes, three planets were present, two to the east and one to the west, in this configuration.

East ✻∗ ○ ✻ West

✻ fixa

The easternmost planet was 7 minutes from Jupiter, while this one was 30 seconds from the next planet. The western planet was 2 minutes removed from Jupiter. And the outermost planets were brighter and larger than the other one, which appeared very small. The easternmost planet appeared somewhat elevated toward the north above the straight line drawn through Jupiter and the other ones. The fixed star already noted was 8 minutes distant from the western planet along the line drawn to that planet perpendicular to the straight line extended through all the planets, as the figure shows.

I decided to add these comparisons of Jupiter and his adjacent planets with the fixed star so that from them anyone could see that the progress of these planets, in longitude as well as latitude, agrees exactly with the motions that are derived from the tables.

Jupiter and the Medicean satellites in a composite of photographs taken in 1966 by space probe Galileo. From right: Io, Europe, Ganimedes, Callistus

The effect of vapors

These are the observations of the four Medicean planets recently, and for the first time, discovered by me. From them, although it is not yet possible to calculate their periods, something worthy of notice may at least be said. And first, since they sometimes follow and at other times precede Jupiter by similar intervals, and are removed from him toward the east as well as the west by only very narrow limits, and accompany him equally in retrograde and direct motion,[149] *no one can doubt that they complete their revolutions about him while, in the meantime, all together they complete a 12-year period*[150] *about the center of the world. Moreover, they whirl around in unequal circles, which is clearly deduced from the fact that at the greatest separations from Jupiter two planets could never be seen united while, on the other hand, near Jupiter two, three, and occasionally all four planets are found crowded together at the same time. It is further seen that the revolutions of the planets describing smaller circles around Jupiter*[151] *are faster. For the stars closer to Jupiter are often seen to the east when the previous day they appeared to the west, and vice versa, while from a careful examination of its previously accurately noted returns, the planet traversing the largest orb appears to have a semimonthly period. We have moreover an excellent and splendid argument for taking away the scruples of those who, while tolerating with equanimity the revolution of the planets around the Sun in the Copernican system, are so disturbed by the attendance of one Moon around the Earth while the two together complete the annual orb around the Sun that they conclude that this constitution of the universe must be overthrown as impossible. For here we have only one planet revolving around another while both run through a great circle around the*

[149] These statements summarize the detailed and methodical descriptions which Galileo made, night after night, of the Jupiter-satellites system and which he had reported in the preceding pages of his book, accompanied by diagrams of the positions of and distances between the various celestial bodies.

[150] This is the time needed for Jupiter to complete its orbit around the sun (to be precise, the period is 11 years and 315 days).

[151] Kepler's third law, established some years later (1619), states that the squares of the orbital periods are proportional to the cubes of the orbital radii (presumed to be circular). Galileo here shows that he had a qualitative grasp of this relationship.

Sun: but our vision offers us four stars wandering around Jupiter like the Moon around the Earth while all together with Jupiter traverse a great circle around the Sun in the space of 12 years. Finally, we must not neglect the reason why it happens that the Medicean stars, while completing their very small revolutions around Jupiter, are themselves now and then seen twice as large. We can in no way seek the cause in terrestrial vapors, for the stars appear larger and smaller when the sizes of Jupiter and nearby fixed stars are seen completely unchanged. It seems inconceivable, moreover, that they approach and recede from the Earth by such a degree around the perigees and apogees of their orbits as to cause such large changes. For smaller circular motions can in no way be responsible, while an oval motion (which in this case would have to be almost straight) appears to be both inconceivable and by no account harmonious with the appearances. I gladly offer what occurs to me in this matter and submit it to the judgment and censure of right-thinking men.

The statements which follow, when not simply wrong, are entirely questionable. For instance, the apparent growth in diameter of the sun and of the moon when they approach the horizon is today known to be due primarily to a psychological illusion, brought about because there are terrestrial reference objects between us and the horizon, which are absent in the open sky. At dawn and at sunset there is, indeed, a bending of solar rays during their long passage through the atmosphere, because of the gradual variation with altitude of the air refractive index (density, humidity, and temperature are changing). But the only effect of this is a flattening of the solar disk. As for the lunar atmosphere, no comment is needed; Galileo never mentioned it again in any of his subsequent writings.

It is well known that because of the interposition of terrestrial vapors the Sun and Moon appear larger but the fixed stars and planets smaller. For this reason, near the horizon the luminaries appear larger but the stars [and planets] smaller and generally inconspicuous, and they are diminished even more if the same vapors are perfused by light. For that reason the stars [and planets] appear very small by day and during twilight, but not the Moon, as we have already stated above. From what we have said above as well as from those things that will be discussed more amply in our system, it is moreover certain that not only the Earth but also the Moon has its surrounding vaporous orb.[152] *And we can accordingly make the same judgment about the remaining planets, so that it does not appear inconceivable to put around Jupiter an orb denser than the rest of the ether around which the Medicean planets are led like the Moon around the sphere of the elements.*[153] *And at apogee, by the interposition of this orb, they are smaller, but when at perigee, because of the absence or attenuation of this orb, they appear larger. Lack of time prevents me from proceeding further. The fair reader may expect more about these matters soon.*

Hymn to the earth

In Galileo's discourse, earth and moon are considered identical in nature—mountainous, corruptible, enveloped in an atmosphere, lit in equal measure by sunlight, and illuminating each other ("In an equal and grateful exchange, the Earth pays back the Moon with light equal to that which she receives from the Moon almost all the time in the deepest darkness of the night").[154] Between them there exists a true "relationship and likeness". Just when he seems to be detracting from the nobility of the moon, revealing its imperfections and roughness, Galileo feels it necessary to raise the earth to the rank of celestial bodies; he announces that the argument will be developed in a more far-reaching work, the *Dialogue Concerning the Two Chief*

[152] Galileo seems to be convinced, here, that the moon possesses an atmosphere. However, elsewhere, commenting upon his observations of the lunar mountains through the telescope, he talks about "very black shadows, much better defined and sharper than our shadows", which of course may not be the case in the presence of gaseous matter, because of possible effects of optical alteration of images.

[153] *The sphere of the elements*—the terrestrial globe, in Aristotelian terms (the earth being home to the four elements: earth, water, air, fire).

[154] On the subject of the "cinereous light" (namely, the light diffused from the earth which impinges upon the dark side of the moon) see Chapter 8.

World Systems, with these beautiful words:[155]

> *We will say more in our System of the World, where with very many arguments and experiments a very strong reflection of solar light from the Earth is demonstrated to those who claim that the Earth is not to be excluded from the dance of the stars, especially because she is devoid of motion and light. For we will demonstrate that she is movable and surpasses the Moon in brightness, and that she is not the dump heap of the filth and dregs of the universe . . .*

In his *Dialogue* Galileo was to raise a true hymn to his beloved earth, "most noble and admirable" precisely because of the never-ceasing mutations that take place in it.[156]

> *SAGR: [. . .] If, not being subject to any changes, it were a vast desert of sand or a mountain of jasper, or if at the time of the flood the waters which covered it had frozen, and it had remained an enormous globe of ice where nothing was ever born or ever altered or changed, I should deem it a useless lump in the universe, devoid of activity and, in a word, superfluous and essentially nonexistent. This is exactly the difference between a living animal and a dead one [. . .]. What greater stupidity can be imagined than that of calling jewels, silver, and gold "precious", and earth and soil "base"? People who do this ought to remember that if there were as great a scarcity of soil as of jewels or precious metals, there would not be a prince who would not spend a bushel of diamonds and rubies and a cartload of gold just to have enough earth to plant a jasmine in a little pot, or to sow an orange seed and watch it sprout, grow, and produce its handsome leaves, its fragrant flowers, and fine fruit.*

HISTORICAL NOTE

The impact *Sidereus Nuncius* had on contemporary scholars was exceptional and it made Galileo famous throughout Europe. At the

[155] *Sidereus Nuncius* or *The Sidereal Messenger*, translated from Latin by A. Van Helden, University of Chicago Press, Chicago, IL, 1989, p. 57. In Latin: *Sidereus Nuncius*, Edizione Nazionale, Vol. III, p. 72.

[156] *Dialogue Concerning the Two Chief World Systems*, translated by S. Drake, The Modern Library, New York, 2001, p. 67. In Italian: *Dialogo sopra i due massimi sistemi del mondo*, Edizione Nazionale, Vol. VII, p. 83.

same time among traditionalists it aroused grave suspicions about him. To give an idea of the tenor of the counter-arguments that were raised, it suffices to say that one critic[157] dismissed the possibility of Jupiter having satellites on the grounds that the overall number of planets would then have exceeded seven, the number considered perfect and consecrated by tradition.

In 1611, Cardinal Bellarmine,[158] shocked (and perhaps somewhat fascinated) by his reading of the recently published *Sidereus Nuncius*, asked the Jesuits of the Roman College—the most prestigious scientific institution in Europe—to answer the following questions about the validity of the discoveries Galileo had made using the telescope:[159]

I am aware that Your Reverences have news of the recent celestial observations of a worthy mathematician by means of an instrument called *cannone* or *ocular*; and I myself have seen, by means of the same instrument, some very marvelous things about the moon and Venus. Therefore I wish you to do me the favor of telling me sincerely your opinion about the following items: first, if you share the view of a multitude of fixed stars, not visible with the naked eye, and in particular that the Milky Way and the nebulae are conglomerations of very minute stars; second, that Saturn is not a simple star, but three stars joined together; third, that the star of Venus changes appearance, waxing and waning like the moon;[160] fourth, that the moon has a rough and uneven surface; fifth, that four mobile stars revolve about planet Jupiter, their motions being different from each other and very rapid. This I wish to know, because I hear various opinions expressed about the matter; and since Your Reverences are experts in the mathematical sciences, I trust you will easily be able to tell me whether these new inventions are well founded or whether they are merely illusory and not real.

In view of the way events were subsequently to develop, it is interesting to note that the Roman College confirmed each of the discoveries,

[157] Francesco Sizzi in *Dianoia astronomica, optica, fisica*, 1611.

[158] Cardinal Robert Bellarmine was the leading theologian of the Order of Jesuits. "Inquisitor" in the trial of Giordano Bruno, he was made a saint in 1930.

[159] Edizione Nazionale, Vol. XI, p. 87.

[160] The observation of phases similar to those of the moon (i.e. full, half, crescent, new) in the planet Venus, was decisive in convincing Galileo of the validity of the heliocentric system.

except for some reservations about the mountains of the moon voiced by the most reputable Jesuit astronomer, Father Christopher Clavius, who it appears was the only one to express an opinion on the matter. It should be remembered that the perfect sphericity of all celestial bodies was an idea which Aristotelians maintained more doggedly even than geocentricism. In respect of this item, the College replied:[161]

> one cannot deny the great inequality of the moon; however, it seems to Father Clavius more probable that the surface is not uneven but rather that the lunar body is not uniformly dense, having parts that are more dense and parts that are more rarefied, in the same manner as those ordinary spots [on the moon] which can be seen with the naked eye. Others think that the surface is indeed uneven: but up to now we are not certain enough about this that we can state it without doubts.

Notwithstanding this authoritative and substantial confirmation, five years later Cardinal Bellarmine was to command Galileo to stop making his ideas public. It seems reasonable to assume that Bellarmine arrived at this decision *ex-officio*, rather than out of personal conviction—a view supported by the fact that he limited himself to warning Galileo against publicizing his ideas, he did not prohibit him from formulating them.

The sphere of crystal

Returning to the answers given by the Roman College, it is worth pointing out a curious misunderstanding that several science historians have fallen prey to, including Ludovico Geymonat.[162] A strange conjecture, aimed at preserving the perfectly spherical shape of the moon as a celestial body, is attributed to Father Clavius. He is supposed, drawing upon the ideas of some German scholars, to have advanced the notion that the moon is enclosed in a spherical shell similar to crystal, so transparent as to render visible eventual irregularities within its interior, but perfectly smooth and polished on the surface. However, Father Clavius was too good a scientist to

[161] Edizione Nazionale, Vol. XI, p. 92.

[162] L. Geymonat, *Galileo Galilei*, Einaudi, Turin, 1969, p. 69. English translation by S. Drake, MacGraw-Hill Book Co., New York, 1965.

have come up with such grossness. It was instead Lodovico delle Colombe,[163] in a letter he sent to Father Clavius in appreciation of the judgement issued by the Roman College, who suggested a parallel between the appearance of the moon from the viewpoint of a distant observer—its interior being made of zones of variable density—and a sphere of crystal enclosing white-enameled figures.

To enable readers to form their own opinion on this matter, here is an excerpt from the letter in question:[164]

Most Reverend Father,

I have seen the reply that your Fathers gave to the Most Illustrious Cardinal Bellarmine and I appreciate that you in particular do not approve the view that the moon's surface is uneven and mountainous, as *Signor* Galileo believes and wishes to persuade others is so. Those mountains that can be seen in the moon may, indeed, exist, because from their shadows and lighted parts and the changes in these, they do appear to be real and bodily[165] and not to be only surface features as if they were painted. But the main difference between my view and that of Galileo is that he deems that these mountains are on the surface, just like on earth which is surrounded by air, while I believe they are inside that body and not on the surface. In fact, they are the denser parts, whereas the remainder of the body is more rarefied, and in this way it is all one body with a smooth surface and nowhere is it uneven or indented. However, since our senses are deceived by so great a distance, it being impossible for us to see those less dense parts because the sun rays are not reflected by them, that body appears unequal and not polished and spherical [. . .]. This is exactly what would happen with a large crystal ball in which were contained many kinds of figures made of white enamel and placed high up, far away from our eyes: it would not look round because we could not see the clear parts of that crystal [. . .].

Lodovico delle Colombe's letter was passed on to Galileo by a certain Gallanzone Gallanzoni, to whom the scientist sent a reply

[163] Lodovico delle Colombe was an amateur philosopher and author. In 1611, he wrote a poor-quality tract opposing the idea of the motion of the earth. In Galileo's circle, delle Colombe and his friends were dubbed the League of the Pidgeon, to indicate their bird brains. [Translator's note: "colombo" in Italian means "pidgeon"].

[164] Letter of 27 May, 1611, Edizione Nazionale, Vol. XI, p. 118.

[165] *To be bodily*—that is, to be three-dimensional, to have a volume.

which was to become historically renowned:[166]

[. . .] I will try to answer what is in my understanding about the content of the letter written to the Most Rev. Father Clavius by Signor Lodovico delle Colombe [. . .] and I do this all the more willingly as I see this to be the last refuge of those philosophers who would wish to make the works of nature fit within their inveterate opinions. This new proposal of a very transparent environment around the lunar body to fill and level off its visible cavities and prominences was written to me many months ago by the Most Illustrious Signor Mark Welser of Aachen,[167] as being a conjecture of some philosophers of that land. I answered him, and perhaps succeeded in satisfying and persuading those philosophers, since no objection has reached me. I do not know what will happen to me in Rome, where this same idea finds, as you have written to me, many people who share it.

[. . .] There is no need for me to recount here the observations from which I deduce my demonstrations, because I have already written about them elsewhere and spoken about them very many times. Moreover, the opponents we are presently dealing with do not dispute them, nor do they dispute the manifest lunar inequalities. But what they are saying, in substance, is that the moon is not simply that globe which sensibly with our eyes we see and have, until now, always seen. Instead, in addition to what is seen by men, there is all around it a certain completely transparent environment, similar to crystal or diamond, which is totally imperceptible to our senses. Filling all cavities and reaching the highest lunar summits, this environment surrounds that original and visible globe, terminating in a smooth and perfectly neat spherical surface. And it does not prevent the sun's rays from penetrating. They can be reflected off the submerged mountains and cause shadows to be projected on their opposite sides, making the old moon perceptible to our senses. This image is truly beautiful, except that it has neither been demonstrated, nor is it demonstrable. Who can fail to see that this is a pure and arbitrary fiction which establishes nothing, but just proposes a mere non repugnance? If our mind's

166 Letter of July 1611, Edizione Nazionale, Vol. XI, p. 141.

167 Marc Welser was a banker and an amateur scientist, who was in correspondence with Galileo.

dreams were able to have an influence upon the choices made by nature, I could with equal authority say that the surface of the earth is perfectly spherical, polished and smooth; meaning by earth, not just this opaque body where the solar rays terminate, but together with it that part of the transparent environment which fills all the valleys and surrounds the globe to levels reaching the peaks of the highest mountains.

Viewed through the Galilean telescope, Saturn appeared to be "three stars joined together". In this photograph, taken by space probe VOYAGER 2, the moons of the planet can be seen

MATHEMATICAL NOTE

The Keplerian telescope

The refraction telescope, also called the dioptric telescope, is used to bring objects closer to the observer's eye, so that, because of the increased viewing angle, they appear enlarged. The Keplerian telescope, first described in 1611 and the type most used in astronomy, consists of two converging lenses having different focal lengths. The first lens (*objective*), with a long focal length f_{ob}, forms an upside-down, real image of the object; the second lens (*ocular*), with a short focal length f_{oc}, acts as a magnifier enabling the image to be observed. The two lenses are aligned on the same axis and separated by a distance such that the first focus F_{oc} of the second lens coincides with the second focus F_{ob} of the first lens (see the diagram). When the observed object A is very far away, the image B produced by the objective falls approximately at its second focus (the height of A is readily found by drawing the ray that, originating at the top of the object, passes through the center of the lens, as this particular ray is not deviated). Since this image also falls at the first focus of the second lens, the ray that passes through the center of the ocular defines the angle at which the eye sees it.

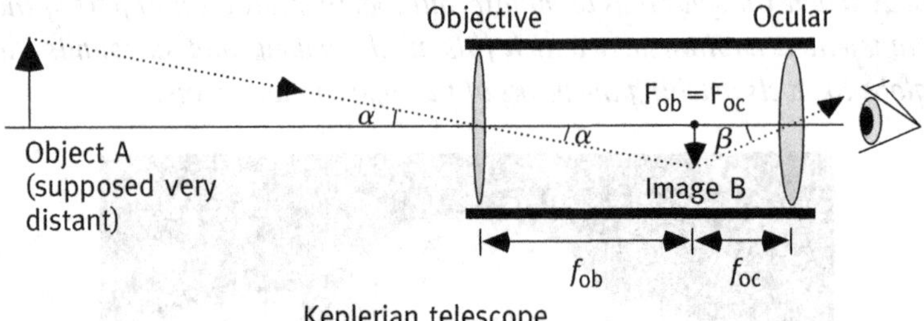

Keplerian telescope

The visual enlargement is given by the ratio of angles β and α. The image, as seen by the eye, is upside-down.

The visual enlargement is given, in absolute value, by the ratio of angles β and α. Considering the two right-angle triangles inside the telescope, it can be seen that the proportion $\mathrm{tg}\,\beta/\mathrm{tg}\,\alpha = f_{ob}/f_{oc}$ holds. If the angles are sufficiently small, one can approximate $\mathrm{tg}\,\beta/\mathrm{tg}\,\alpha \approx \beta/\alpha$, finally obtaining

$$G = \frac{-f_{ob}}{f_{oc}} < 0$$

where the minus sign has been used to indicate that the *image is upside-down. The visual enlargement is therefore the greater, the greater the ratio of the focal distances is.*

In a telescope, it is not only the magnifying power that matters but also the luminosity, which increases with the diameter of the objective. The biggest telescope of this kind (Keplerian) is located at the Yerkes Observatory of the University of Chicago. It has an objective of 102 cm diameter and focal lengths for the objective and ocular of, respectively, 19.5 m and 10 cm; thus, the visual enlargement is calculated as $G = -195$. The construction and putting into operation of lenses as big as this presents technological and physical problems (e.g., strong chromatic aberrations are present) and, for this reason, reflection telescopes, in which mirrors replace the lenses, are now preferred.

It should be noted that the ocular, instead, can be tiny, with a diameter precisely $|G|$ times smaller than that of the objective, a feature that can prove extremely convenient. This is immediately clear if one considers that the ocular has to collect nothing more than the light beam which emerges from the objective, whose boundary rays are illustrated in the following diagram.

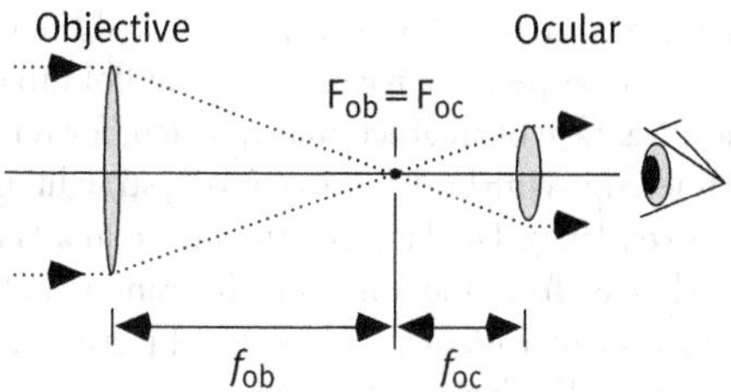

The dimensions of the two lenses in a Keplerian (astronomical) telescope. The high-low inversion of the two boundary rays clearly shows that the image in the eye appears upside-down

Finally, it is worth highlighting an advantage that the Keplerian telescope has with respect to the Galilean one: due to the fact that image B, acting as a source for the second lens, is real, it is possible to place in its plane a micrometric-scale grid, thus enabling precise measurements to be made.

The Galilean telescope

In Galileo' refraction telescope, built in 1609, the second lens is double-concave and diverging. It is again placed in a position such as to make the second focal plane of the objective coincide with the first one of the ocular. However, since in a diverging lens the positions of the focal points are interchanged (which results in negative focal distances), the arrangement of the telescope is as shown in the following diagram.

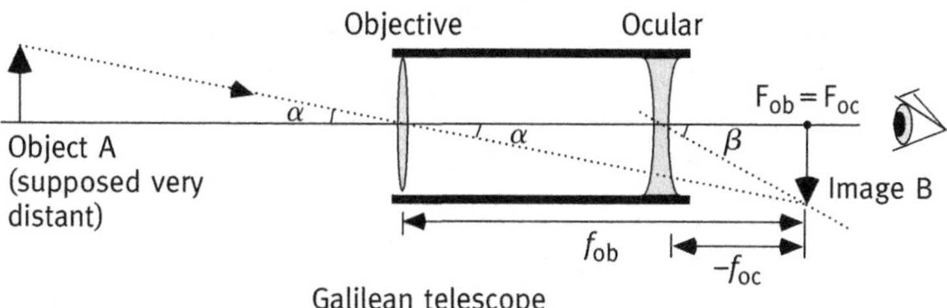

The visual enlargement is given by the ratio of angles β and α. Thanks to the diverging lens, the image seen by the eye is erect

Reasoning exactly as in the case of Keplerian telescope, one finds that the visual enlargement is again given by

$$G = \frac{-f_{ob}}{f_{oc}} > 0$$

with the difference that, since $f_{oc} < 0$, G is positive and the eye sees an erect image. The Galilean telescope has a narrower visual field than the Keplerian one, but offers the advantage of an erect image, which, for terrestrial use, is to be preferred (in a telescope with both lenses convex, straightening of the image requires a third convex lens). The fact that the image in a Galilean telescope is erect can be clearly seen from the following diagram, which also shows that the ocular may have a smaller diameter than the objective, and by a factor of precisely G, just as for the Keplerian telescope.

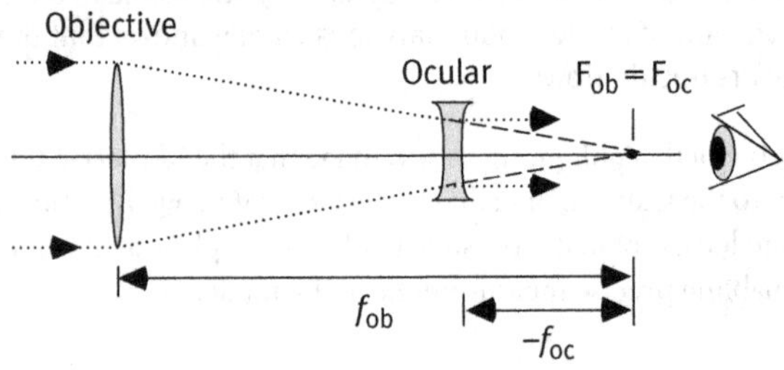

Dimensions of the two lenses in the Galilean telescope. The absence of high–low inversion of the rays results in an erect image for the eye

8

In The Moonlight

The Origin of the Moon's Silvery Whiteness*

dove adoprereste voi colori più oscuri,
nel dipignere il muro o pur nel dipigner lo specchio?

if you had to paint a picture of that wall with the mirror hanging on it,
where would you use the darkest colors?

In contrast to what was commonly believed at the time, Galileo demonstrates that the extreme brightness of the part of the lunar surface lit by the sun at any moment is proof that it is irregular and rugged, rather than perfectly spherical and mirror-like, as Aristotelians would have had it. The faint "cinereous light" of the part of the surface not lit by the sun he attributes to light diffusion by the earth. Galileo dares come to the almost sacrilegious conclusion that the earth and the moon are identical in nature. The arguments used provide him with the opportunity to express his great admiration for the human mind and his confidence in its powers.

The telescope enabled Galileo to provide direct experimental evidence for the existence of mountains on the moon and also to make good estimates of their highest altitudes. However, the most obstinate of those who claimed that the moon had to be perfect (and therefore smooth and spherical like all other celestial bodies) initially even doubted that Galileo's telescope could show things the way they

* From *Dialogue*.

really are, rather than provide a picture that was distorted and entirely open to interpretation.

Evidence that the moon is not, as Aristotelians would have had it, polished like a mirror can, however, also be derived by applying simple optics arguments to observations made with the naked eye. Galileo became certain of this only after he had established, beyond all doubt that the moon's surface is rugged and uneven using the telescope. He considered his argument so persuasive that he has Sagredo say: "If I were on the moon itself I do not believe that I could touch the roughness of its surface with my hand more definitely than I now perceive it by understanding your argument".

Photographs of moon craters Copernicus and Eratosthenes taken by space probe Apollo 17 (NASA)

In this chapter we present excerpts of the discussion which takes place among Salviati, Sagredo, and Simplicio about the silvery-whiteness of the moon. At the start they talk about the so-called *cinereous light*, that faint shining of the part of the moon in shadow which enables its presence to be perceived. This cinereous light is not generated by the moon itself, as Simplicio maintained. Rather it is light coming from the earth, which is merely diffusing the light it receives from the sun. Such an idea was almost sacrilegious, because it implied that a heavenly body can be subject to the action of an imperfect body—one which, like the earth, is corrupted and corruptible.

Specular and diffused reflection

We have used the verb *to diffuse*—an equivalent term would be "to reflect in a diffuse manner"—but not *to reflect* as is often said, incorrectly. This distinction, it turns out, is the key point of the discourse. The surface of the earth is not mirror-like, but rough, wrinkled, and composed of opaque matter. This is why Simplicio refuses to believe that it can behave as a good reflector, giving rise to the cinereous light on the moon. Indeed, for Aristotelians the brightness of the sunlit part of the moon was a sure sign that our satellite cannot be similar to the earth, but must be polished to a shine like a mirror. This conviction was based on confused ideas about the two basic mechanisms by which a body, struck by light rays, can return them to its surroundings. These two mechanisms are reflection and diffusion, or—in more specific terms—specular reflection and diffused reflection (for more details see the Mathematical Note at the end of this chapter).

Salviati does not waste time on verbal descriptions, he limits himself to giving a proof that is beyond question—an experimental one. He leads his friends outdoors and invites them to judge which is the brighter between a mirror hanging on a white wall and the white wall itself ("if you had to paint a picture of that wall with the mirror hanging on it, where would you use the darkest colors?"). Simplicio cannot deny that the wall always wins out in brightness over the mirror, and by a long way. Except, that is, for just one precisely defined direction, corresponding to the angle which equals the angle of incidence of the sun's rays (law of specular reflection). In Salviati's words: "You see how the reflection that comes from the wall diffuses itself over all the points opposite to it, while that from the mirror goes to a single place no larger than the mirror itself". Owing to its roughness, protrusions, and cavities, the wall scatters light in all directions and thus appears brighter from any observation point, even if under no circumstances can it match the glare coming from the mirror when this is viewed from the specular reflection angle. If the moon were smooth it would mainly appear to us as dark—in fact, we would only be able to see it on those very rare occasions when the angle was appropriate, and then it would shine almost as brightly as the sun. The moon, therefore, "diffuses", it does not "reflect": hence its surface must be rough.

Oblique illumination

With equally lucid reasoning and probative verifications, Salviati demonstrates that the same conclusions can be drawn if a comparison is made with a spherical mirror, such as the moon would be under the Aristotelian conception. The discourse then enters into explanations of why the asperities of the moon's surface cannot be slight, but must consist of mountains and valleys of considerable dimensions. Salviati makes use of another practical demonstration: he folds a sheet of white paper to form an angle and shows how the more perpendicular the folded surface is to the sun rays striking it, the brighter it appears. High mountains situated near the edges of the lunar disk have extended slopes whose inclination is optimal for conveying the sun's light towards us (Salviati: "Do you not see that their peaks and ridges, being elevated above the convexity of a perfectly spherical surface, are exposed to the sun and accommodated to receive the rays much less obliquely, and therefore to look as much lighted as the rest?"). If this were not the case, the full moon would not appear to us to shine with even brightness across all points on its surface (i.e. at the periphery as well as the center).

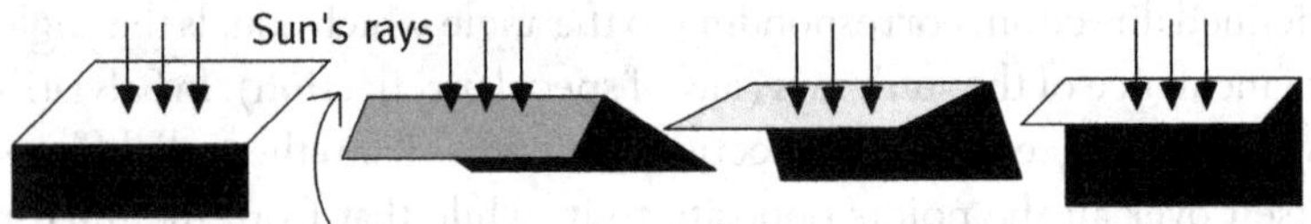

Simplicio persists in his refusal to believe that the earth, seen from the moon, would be even brighter than the full moon seen from the earth. The argument Salviati puts forward is simple and brilliant. In order to compare the part of the earth which is lit with the part of the moon which is lit, it would be necessary to observe it against a dark sky, that is, by night—a condition which is obviously impossible for us terrestrials. Hence comparison can only be made with the moon as seen in full daylight, when it appears to us no brighter than a small cloud in the blue sky. Had space flight existed in Galileo's time it would have been enough to show Simplicio photographs of the earth taken by orbiting astronauts. The sky on these is black indeed, because in the absence of atmosphere there is no light diffusion and the earth stands out brightly like a giant moon.

The text

The reader will note, in the pages which follow, how particularly lively the dialogue is, the natural way the interlocutors take their turns and their deep and untiring satisfaction at finding themselves capable of progressing in knowledge ("Stop there, Salviati, and allow me the pleasure of showing you"). But above all there is the action, almost theatrical, which unfolds before our eyes. Here Salviati and Sagredo are not designing future experiments, or describing ones already carried out. Instead they actually perform three—reflection by a plane mirror and by a spherical mirror, and diffuse reflection by a folded sheet of paper—in "real time", as it were. The utterances of the spoken dialogue accompany the actions required in carrying out the experiment step by step, and Galileo shows masterly skill in accomplishing this synchronism. Listen to Salviati, for example: "Now please take that mirror which is hanging on the wall and let us go out into that court; come with us, Sagredo. Hang the mirror on that wall, there, where the sun strikes it. Now let us withdraw into the shade. Now, there you see two surfaces struck by the sun, the wall and the mirror. . . ."

The three friends move across the scene in front of us in a relaxed and natural way, revealing their personalities, the sharpness of their minds, their curiosity, diffidence, craftiness, irony, courage, and hesitancy. And yet Galileo is never indulging in psychological speculation or dramatic sketches: his clear aim is constantly to achieve understanding, which may in part also derive from the dialectics of the different temperaments of the interlocutors.

Nevertheless, here more than elsewhere, it is obvious that Salviati and Sagredo are the ones who really have Galileo's sympathies and that, basically, he lays the blame for the mistakes Simplicio makes on his character, which is of a type all-too-commonly found among people. While Salviati and Sagredo are driven by the desire to make progress, Simplicio is driven by that of blindly defending tradition *per se*. This is why, rather than rejoicing at a new discovery, he gets cocky every time he thinks he has caught his interlocutors at fault ("Very clever, my dear sir; and is this the best experiment you have to offer? You have placed us where the reflection from the mirror does not strike. But come with me a bit this way; no, come along"). He gets upset by difficulties, and does not hesitate to admit it, coming out with vivid metaphors (impossible to preserve in the English translation, unfortunately) which reveal his uneasiness. Or he tries to slow down the pace of the discussion, with modesty that sounds falsely self-ironic ("Please explain further for me, since I am not that quick-witted"). Naturally, providing a range of different levels, paces, and ways of understanding was an expedient Galileo used to render the dialogue more enjoyable and realistic. But above all it also offered to readers of different cultural backgrounds and age, or who were simply more or less "quick-witted", a variety of ways of accessing the ideas.

Simplicio's difficulties, on the other hand, do not arise so much from the quest for new solutions, as from the arduous task of defending Aristotle. And this is what irritates his interlocutors, who do not spare him their pungent irony, striking all the deeper precisely where he shows himself to be pedantic and unbending, draped in Latin and biblical quotations. Some of the witty exchanges bring to mind Calandrino—a character in Boccaccio's *Decameron*—who, without realizing he is totally in Bruno's and Buffalmacco's power, thinks he can take them in.

Finally, it is worth underlining how modern the figures of Salviati and Sagredo are. In many respects we perceive them as men of our time, not only because of their approach, which is rationalistic, but also because of their use of language, which is straightforward and functional. Mental convolutions and verbal ceremonies ("I give you my word as a gentleman") make Simplicio, instead, a thoroughly seventeenth century man.

We now pass the word to Galileo's characters.[168]

Cinereous light

SALV: [...] just as the moon supplies us with the light we lack from the sun a great part of the time, and by reflection of its rays makes the nights fairly bright, so the earth repays it by reflecting the solar rays when the moon most needs them, giving a very strong illumination—as much greater than what the moon gives us, it would seem to me, as the surface of the earth is greater than that of the moon.

SAGR: Stop there, Salviati, and allow me the pleasure of showing you how from just this first hint I have seen through the cause of an event which I have thought about a thousand times without ever getting to the bottom of it.

You mean that a certain baffling lights which is seen on the moon, especially when it is horned, comes from the reflection of the sun's light from the surface of the earth and the sea; and this light is seen most clearly when the horns are the thinnest. For at that time the luminous part of the earth that is seen from the moon is greatest, in accordance with your conclusion a little while ago that the luminous part of the earth shown to the moon is always as great as the dark part of the moon which is turned toward the earth. Hence when the moon is thinly horned and consequently in large part shadowy, the illuminated part of the earth seen from the moon is large, and so much the more powerful is its reflection of light.

SALV: That is exactly what I meant. Really, it is a great pleasure to talk with discriminating and perceptive persons, especially when people

[168] *Dialogue Concerning the Two Chief World Systems*, First Day, translated by S. Drake, The Modern Library, New York, 2001, p. 76 and following. In Italian: *Dialogo sopra i due massimi sistemi del mondo*, Giornata prima, Edizione Nazionale, Vol. VII, p. 91 and following. Elsewhere in the same book (p. 297), by way of Sagredo, Galileo offers us this delightful image of the moon:

> an event . . . from which . . . one may learn how easily anyone may be deceived by simple appearances, or let us say by the impressions of one's senses. This event is the appearance to those who travel along a street by night of being followed by the moon, with steps equal to theirs, when they see it go gliding along the eaves of the roofs. There it looks to them just as would a cat really running along the tiles and putting them behind it; an appearance which, if reason did not intervene, would only too obviously deceive the senses.

are progressing and reasoning from one truth to another. For my part I more often encounter heads so thick that when I have repeated a thousand times what you have just seen immediately for yourself, I never manage to get it through them.

SIMPL: [...] I reply that I consider the moon more solid than the earth, not for the reason you already gave, of the roughness and ruggedness of its surface, but on the contrary from its being suited to receive a polish and a lustre superior to that of the smoothest mirror, as observed in the hardest stones on earth. For thus must be its surface in order to make such a vivid reflection of the sun's rays. The appearances you speak of, the mountains, rocks, ridges, valleys, etc., are all illusions. I have heard it strongly maintained in public debates against these innovators that such appearances belong merely to the unevenly dark and light parts of which the moon is composed inside and out. We see the same thing occur in crystal, amber, and many perfectly polished precious stones, where, from the opacity of some parts and the transparency of others,[169] *various concavities and prominences appear to be present. [...] I think it most false that the moon can receive light from the earth, which is completely dark, opaque, and unfit to reflect sunlight as the moon reflects it so well to us. And as I have said, I consider the light which is seen over the rest of the face of the moon (outside the horns brightly illuminated by the sun) to be the moon's own proper and natural light, and it would be quite a feat to make me think otherwise. [...]*

The mirror and the wall

SALV: We are inquiring, Simplicio, whether in order to produce a reflection of light similar to that which comes to us from the moon, it is necessary that the surface from which the reflection comes shall be as smooth and polished as a mirror, or whether a rough and ill-polished surface, neither smooth nor shiny, may not be better suited. Now if two reflections should come to us, one brighter than the other, from two surfaces situated opposite to us, I ask you which of the two surfaces you believe would look the lighter to our eyes, and which the darker?

[169] This was an opinion expressed by Peripatetic philosophers and was drawn upon by Father Christopher Clavius—Jesuit and foremost astronomer of the Roman College—in giving his views on Galileo's discoveries (see Chapter 7).

SIMPL: I think without any doubt that the surface which reflected the light more brilliantly would look lighter to me, and the other darker.
SALV: Now please take that mirror which is hanging on the wall and let us go out into that court; come with us, Sagredo. Hang the mirror on that wall, there, where the sun strikes it. Now let us withdraw into the shade. Now, there you see two surfaces struck by the sun, the wall and the mirror. Which looks brighter to you: the wall, or the mirror? What, no answer?
SAGR: I am going to let Simplicio answer; he is the one who is experiencing the difficulty. For my part, from this small beginning of an experiment I am persuaded that the moon must indeed have a very badly polished surface.
SALV: Tell me, Simplicio; if you had to paint a picture of that wall with the mirror hanging on it, where would you use the darkest colors? In depicting the wall or the mirror?
SIMPL: Much darker in depicting the mirror.
SALV: Now if the most powerful reflection of light comes from the surface that looks brightest, the wall here would be reflecting the rays of the sun more vividly than the mirror.
SIMPL: Very clever, my dear sir; and is this the best experiment you have to offer? You have placed us where the reflection from the mirror does not strike. But come with me a bit this way; no, come along.
SAGR: Perhaps you are looking for the place where the mirror throws its reflection?
SIMPL: Yes, sir!
SAGR: Well, just look at it—there on the opposite wall, exactly as large as the mirror, and little less bright than it would be if the sun shone there directly.
SIMPL: Come along, then, and look at the surface of the mirror from there, and then tell me whether I should say it is darker than that of the wall.
SAGR: Look at it yourself; I am not anxious to be blinded, and I know perfectly well without looking that it looks as bright and vivid as the sun itself, or little less so.
SIMPL: Well, then, what do you say? Is the reflection from a mirror less powerful than that from a wall? I notice that on this opposite wall, which receives the reflection from the illuminated wall along with that of the mirror, the reflection from the mirror is much the brighter. And I see likewise that from here the mirror itself looks very much brighter to me than the wall.

SALV: You have got ahead of me by your perspicacity, for this was the very observation which I needed for explaining the rest. You see the difference, then, between the reflections made by the surface of the wall and that of the mirror, which are struck in exactly the same way by the sun's rays. You see how the reflection that comes from the wall diffuses itself over all the points opposite to it, while that from the mirror goes to a single place no larger than the mirror itself. You see likewise how the surface of the wall always looks equally light in itself, no matter from what place you observe it, and somewhat lighter than that of the mirror from every place except that small area where the reflection from the mirror strikes; from there, the mirror appears very much brighter than the wall. From this sensible and palpable experiment it seems to me that you can very readily decide whether the reflection which comes here from the moon comes like that from a mirror, or like that from a wall; that is, whether from a smooth or a rough surface.

SAGR: If I were on the moon itself I do not believe that I could touch the roughness of its surface with my hand more definitely than I now perceive it by understanding your argument. The moon, seen in any position with respect to the sun and to us, always shows the surface exposed to the sun equally bright. This effect corresponds precisely with that of the wall, which seen from any place appears equally bright; it conflicts with that of the mirror, which from one place alone looks luminous and from all others dark. Besides, the light that comes to me from the reflection of the wall is weak and tolerable in comparison with that from the mirror, which is extremely strong and little less offensive to the eyes than the primary and direct rays of the sun. It is in just such a way that we can calmly contemplate the face of the moon. If that were like a mirror, appearing as large as the sun because of its closeness, it would be of an absolutely intolerable brilliance, and would seem to us almost as if we were looking at another sun.

And now comes the experiment with the spherical mirror:

The spherical mirror

SALV: Please, Sagredo, do not attribute to my demonstration more than belongs to it. I am about to confront you with a fact that I think you

will find not so easy to explain. You take it as a great difference between the moon and the mirror that the former yields its reflections equally in all directions, as the wall does, while the mirror sends its reflection to one definite place alone. From this, you conclude that the moon is like the wall and not like the mirror. But I tell you that this mirror sends its reflection to one place alone because its surface is flat, and since reflected rays must leave at equal angles with incident rays, they have to leave a plane surface as a unit toward one place. But the surface of the moon is not flat, it is spherical; and the rays incident upon such a surface are found to be reflected in all directions at angles equal to those of incidence, because of the infinity of slopes which make up a spherical surface. Therefore the moon can send its reflections everywhere and need not send them all to a single place like those of a plane mirror.
SIMPL: This is exactly one of the objections which I wanted to make.
[...]
SAGR: [...] But I return to the first point raised by Salviati, and I tell you that in order to make an object appear luminous, it is not sufficient for the rays of the illuminating body to fall upon it; it is also necessary for the reflected rays to get to our eyes. This is to be clearly seen in the case of the mirror, upon which no doubt the rays of the sun are falling, but which nevertheless does not appear to be bright and illuminated unless we put our eyes in the particular place where the reflection is going.

Let us consider this in regard to what would happen if the mirror had a spherical surface. Unquestionably we should find that of the whole reflection made by the illuminated surface, only a small part would reach the eyes of a particular observer, there being only the very least possible part of the entire surface which would have the correct slope to reflect the rays to the particular location of his eyes. Hence only the least part of the spherical surface would shine for his eyes, all the rest looking dark. If then the moon were smooth as a mirror, only a very small part would show itself to the eyes of a particular person as illuminated by the sun, although an entire hemisphere would be exposed to the sun's rays. The rest would remain, to this observer's eyes, unilluminated and therefore invisible. To conclude, the whole moon would be invisible, since that particle which gave the reflection would be lost by reason of its smallness and great distance. And just as the moon would remain invisible to the eyes, so its illumination would remain nil; for it is indeed impossible that a luminous body should by its splendor take away our darkness, and we be unable to see it.

SALV: Wait a minute, Sagredo, for I see certain signs in Simplicio's face and actions which indicate to me that he is neither convinced nor satisfied by what you, with the best evidence and with perfect truth, have said. And now it occurs to me how to remove all doubt by another experiment. I have seen in a room upstairs a large spherical mirror; have it brought here. And while it is on its way, Simplicio, consider carefully the amount of light which comes from the reflection of the flat mirror to this wall here under the balcony.

SIMPL: I see that it is little less lighted than if the sun were striking it directly.

SALV: So it is. Now tell me; if, taking away that little flat mirror, we were to put the large spherical one in its place, what result do you think that would have upon the reflection on this same wall?

SIMPL: I think it would produce a much greater and broader light.

SALV: But what would you say if the illumination should be nil, or so small that you could hardly perceive it?

SIMPL: When I have seen the effect, I shall think up a reply.

SALV: Here is the mirror, which I wish to have placed beside the other. But let us first go over there, near the reflection from the flat mirror, and note carefully its brightness. You see how bright it is here where it strikes, and how you can distinctly make out these details of the wall.

SIMPL: I have looked and observed very closely; now place the other mirror beside the first.

SALV: That is where it is. It was placed there as soon as you began to look at the detail, and you did not perceive it because the increase of light over the rest of the wall was just as great. Now take away the flat mirror. See there, all the reflection is taken away, although the large convex mirror remains. Remove that also, and then replace it as you please; you will see no change whatever in the light upon the whole wall. Thus you see it shown to your senses how the reflection of the sun made from a spherical convex mirror does not noticeably illuminate the surrounding places. Now what have you to say to this experiment?

SIMPL: I am afraid you have introduced some trickery. Yet I see, in looking at that mirror, that it gives out a dazzling light that almost blinds me; and what is more significant, I see it all the time, wherever I go, changing place on the surface of the mirror according as I look at it from this place or that; a conclusive proof that the light is reflected very vividly on all sides, and consequently upon the entire wall as upon my eyes.

SALV: Now you see how carefully and with what reserve one must proceed in giving assent to what is shown by argument alone. There is no doubt that what you say is plausible enough, and yet you can see that sensible experience refutes it.

[...]

Diffuse reflection

SIMPL: I am more perplexed than ever; I must bring up the other difficulty. How can it be that the wall, being of so dark a material and so rough a surface, is able to reflect light more powerfully and vividly than a smooth and well-polished mirror?

SALV: Not more vividly, but more diffusely. As to vividness, you see that the reflection of that little flat mirror, where it is thrown there under the balcony, shines strongly; and the rest of the wall, which receives a reflection from the wall to which the mirror is attached, is not lighted up to any great extent (as is the small part struck by the reflection from the mirror). If you wish to understand the whole matter, consider how the surface of this rough wall is composed of countless very small surfaces placed in an innumerable diversity of slopes, among which of necessity many happen to be arranged so as to send the rays they reflect to one place, and many others to another. In short, there is no place whatever which does not receive a multitude of rays reflected from very many little surfaces dispersed over the whole surface of the rough body upon which the luminous rays fall. From all this it necessarily follows that reflected rays fall upon every part of any surface opposite that which receives the primary incident rays, and it is accordingly illuminated.

It also follows that the same body on which the illuminating rays fall shows itself lighted and bright all over when looked at from any place. Therefore the moon, by being a rough surface rather than smooth, sends

the sun's light in all directions, and looks equally light to all observers. If the surface, being spherical, were as smooth as a mirror, it would be entirely invisible, seeing that that very small part of it which can reflect the image of the sun to the eyes of any individual would remain invisible because of the great distance, as we have already remarked.
[...]

Today it is easy to check the validity of Salviati's conclusions by examining the photographs of the earth taken from artificial satellites. Land masses, even if they reflect less light than the oceans, always appear brighter.[170] Let us now take a look at the evidence for the existence of mountains on the moon. This is based on the fact that, when it is full, the moon appears evenly bright, and not resplendent at its centre and dimmer towards its rim. Salviati's main argument is based on the observation that the more perpendicularly the sun's rays strike a surface, the brighter this surface appears.

Obliquity of the sun's rays

SAGR: Then if these doctors of philosophy were content to grant that the moon, Venus, and the other planets had surfaces not as smooth and bright as a mirror, but were something short of that, like a silver plate merely bleached but not burnished, would this be sufficient to make it visible and fit to reflect the sun's light for us?
SALV: Partly, but it would not make a light as powerful as is made by its being mountainous and full of great prominences and cavities. However, these doctors of philosophy never do concede it to be less polished than a mirror; they want it more so, if that can be imagined, for they deem that only perfect shapes suit perfect bodies. Hence the sphericity of the heavenly globes must be absolute. Otherwise, if they were to concede me any inequality, even the slightest, I would grasp without scruple for some other, a little greater; for since such perfection consists in indivisibles, a hair spoils it as badly as a mountain.

[170] Galileo, naturally, declared himself certain that this was the case: "I never doubted that, looking from a distance at the terrestrial globe lit by the sun, the solid earth surface would appear lighter, the aqueous darker" (*The Starry Messenger*, translation from Latin).

SAGR: This gives rise to two questions on my part. One is to understand why the greater irregularity of the surface makes the reflection of the light more powerful, and the other is why these Peripatetic gentlemen want such a precise shape.
SALV: I shall answer the first and let Simplicio worry about replying to the second. You must know, then, that a given surface receives more or less illumination from the same light according as the rays of light fall upon it less or more obliquely; the greatest illumination occurs where the rays are perpendicular. And here I shall show you this by means of your senses. First I fold this sheet of paper so that one part makes an angle with the other, and now I expose it to the light reflected from that wall opposite to us. You see how this part that receives the rays obliquely is less light than this other where the rays fall at right angles Note how the illumination becomes weaker as I make it receive them more and more obliquely.
SAGR: I see the effect, but I do not understand the cause.
SALV: If you thought about it a minute you would find it, but so as to save time, here is a sort of proof in this figure.
SAGR: Just seeing the diagram has cleared the whole matter up, so go on.
SIMPL: Please explain further for me, since I am not that quick-witted.
SALV: Imagine that all the parallel lines which you see leaving from

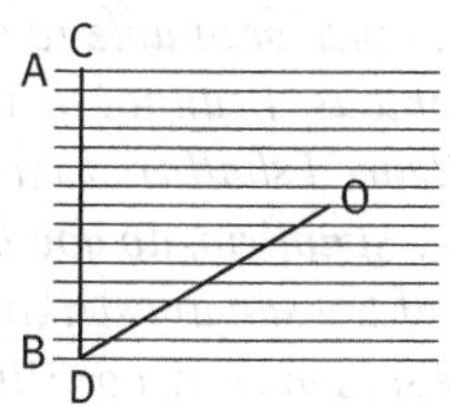

between the points A and B are rays that strike the line CD at right angles. Now tilt CD so that it leans like DO. Do you not see that many of the rays which struck CD pass by without touching DO? And if DO is illuminated with less rays, it is surely reasonable that the light it receives from them is weaker.

Now let us get back to the moon, which, being spherical in shape would, if its surface were as smooth as this paper, receive much less light near the edges of its lighted hemisphere than upon the central parts; for the rays would fall upon the former quite obliquely, and upon the latter at right angles. For that reason at full moon, when we see nearly all the hemisphere illuminated, the central parts ought to look brighter than those near the edges; but that is not what is seen. Now imagine the face of the moon to be full of high mountains. Do you not see that their peaks and ridges, being elevated above the convexity of a perfectly spherical surface, are exposed to the sun and accommodated to receive the rays much less obliquely, and therefore to look as much lighted as the rest?

SAGR: All right, but even if there are such mountains and it is true that the sun strikes them much straighter than it would the slopes of a smooth surface, still it is also true that the valleys among these mountains would remain dark because of the great shadows which the mountains would cast at such a time; whereas the central parts, though full of mountains and valleys, would remain without shadows through the elevation of the sun. Therefore they would be much lighter than the parts at the edge, those being spotted with shadows no less than with light. Yet no such difference is observed.

SIMPL: I was turning over in my mind a like difficulty.

SALV: How much quicker Simplicio is to perceive difficulties that strengthen Aristotle's position than he is to see their solutions! But I suspect that sometimes he deliberately keeps those to himself; and having in the present instance been able to see the objection, which incidentally is quite ingenious, I cannot believe that he has not also discovered the answer. Hence I shall try to worm it out of him, as the saying goes. Now tell me, Simplicio, do you believe that there can be shadows where the rays of the sun are striking?

SIMPL: I do not believe so; I am sure not. The sun being the strongest light, which scatters darkness with its rays, it is impossible that darkness could remain where it arrived. Besides, we know by definition that tenebrae sunt privatio luminis.[171]

SALV: Then the sun, looking at the earth or moon or any other opaque body, never sees any of its shady parts, having no other eyes to see with than its light-bearing rays. Consequently anyone who was located on the sun would never see anything shady, because the rays of his vision would always travel in company with the illuminating sunshine.

171 *tenebrae sunt privatio luminis*—darkness is absence of light.

SIMPL: That is very true; it is beyond contradiction.
SALV: Now when the moon is in opposition to the sun,[172] *what difference is there between the path which the rays of your vision take and the way the rays of the sun go?*
SIMPL: Oh, now I understand you. You mean that since the rays of vision and those of the sun are going along the same lines, we can never see any of the shaded valleys of the moon. But please give up your opinion that I am a hypocrite or a dissembler; I give you my word as a gentleman that I did not perceive this reply, and I might never have discovered it without your help or without long study.

Salviati strengthens his arguments with a great many details and even succeeds in persuading Simplicio—or this, at least, is what it seems. However, further on in the text Simplicio replies to Sagredo's question concerning the shape of heavenly bodies, only to then immediately shift the discussion to a second argument he cannot make out. This is the fact that the earth, which is so dark and such a poor reflector, is capable of diffusing light to the moon ("cinereous light"). In a moment we will resume our reading from this point.

This section of the dialogue is an admirable example of maieutic art. With subtle skill Salviati leads Simplicio, little by little, to accept an idea which initially he had rejected ("such an effect strikes me as quite impossible"). He demolishes a prejudice, showing how it is possible to arrive at a truth using exclusively items of information which one already has in one's possession. These items, however, since they had been accepted passively, had remained unconnected to each other and, as a consequence, had hitherto been of no significance. In these passages the contrast between the old approach and the new is illustrated by thematic words, such as *believe* in the first part and *doubt* in the second; just as the insistence on *know* indicates that Simplicio has finally attained knowledge by way of reason.

The moon in daytime

SIMPL: . . .] for the present we can [. . .] pass on to the point that comes next, setting forth the reasons you have for believing that the

[172] *Opposition*—that is, when the earth is between the sun and the moon.

earth can reflect light of the sun no less strongly than the moon can. For to me it seems so dark and opaque that such an effect strikes me as quite impossible.

SALV: What you think is a cause making the earth unfit for illuminations, Simplicio, is really not one at all. Would it not be interesting if I should see into your reasoning better than yourself?

SIMPL: Whether I reason well or badly, you might indeed know better than I do; but whether I reason well or badly I shall never believe that you can see into my reasoning better than I.

SALV: Even that I shall make you believe in due course. Tell me, when the moon is nearly full, so that it can be seen by day and also in the middle of the night, does it appear more brilliant in the daytime or at night?

SIMPL: Incomparably more at night. It seems to me that the moon resembles those pillars of cloud and fire which guided the children of Israel; for in the presence of the sun it shows itself like a little cloud, but then at night it is most splendid. Thus I have observed the moon by day sometimes among small clouds, and it looked like a little bleached one; but on the following night it shone very splendidly.

SALV: So that if you had never happened to see the moon except by day, you would not have judged it brighter than one of those little clouds?

SIMPL: I do believe you are right.

SALV: Now tell me, do you believe that the moon is really brighter at night than by day, or just that by some accident it looks that way?

SIMPL: I believe that it shines intrinsically as much by day as by night, but that its light looks greater at night because we see it in the dark field of the sky. In the daytime, because everything around it is very bright, by its small addition of light it appears much less bright.

SALV: Now tell me, have you ever seen the terrestrial globe lit up by the sun in the middle of the night?

SIMPL: That seems to me to be a question that is not asked except in sport, or only of some person notorious for his lack of wit.

SALV: No, no; I take you for a very sensible man, and ask the question in earnest. So answer just the same, and then if it seems to you that I am talking nonsense, I shall be taken for the brainless one; for

he is a greater fool who asks a silly question than he to whom the question is put.
SIMPL: Then, if you do not take me for a complete simpleton, pretend that I have answered you by saying that it is impossible for anyone who is on the earth, as we are, to see by night that part of the earth where it is day; that is to say, the part which is struck by the sun.
SALV: So you have never chanced to see the earth illuminated except by day, but you see the moon shining in the sky on the darkest night as well. And that, Simplicio, is the reason for your believing that the earth does not shine like the moon; for if you could see the earth illuminated while you were in a place as dark as night, it would look to you more splendid than the moon. Now if you want to proceed properly with the comparison, we must draw our parallel between the earth's light and that of the moon as seen in daytime; not the nocturnal moon, because there is no chance of our seeing the earth illuminated except by day. Is that satisfactory?
SIMPL: So it must be.
SALV: Now you yourself have already admitted having seen the moon by day among little whitish clouds, and similar in appearance to one of them. This amounts to granting at the outset that these little clouds, though made of elemental matter, are just as fit to receive light as the moon is. More so, if you will recall in memory having seen some very large clouds at times, white as snow. It cannot be doubted that if such a one could remain equally luminous on the darkest night, it would light up the surrounding regions more than a hundred moons.

If we were sure, then, that the earth is as much lighted by the sun as one of these clouds, no question would remain about its being no less brilliant than the moon. Now all doubt upon this point ceases when we see those same clouds, in the absence of the sun, remaining as dark as the earth all night long. And what is more, there is not one of us who has not seen such a cloud low and far off, and wondered whether it was a cloud or a mountain; a clear indication that mountains are no less luminous than those clouds.
[. . .]
SALV: If you are satisfied now, Simplicio, you can see how you yourself really knew that the earth shone no less than the moon, and that not my instructions but merely the recollection of certain things already

known to you have made you sure of it. For I have not shown you that the moon shines more brilliantly by night than by day; you already knew it, as you also knew that a little cloud is brighter than the moon. Likewise you knew that the illumination of the earth is not seen at night, and in short you knew everything in question without being aware that you knew it. Hence there should be no reason that it should be hard for you to grant that reflection from the earth can illuminate the dark part of the moon with no less a light than that with which the moon lights up the darkness of the night. More, because the earth is forty times[173] *the size of the moon.*
[...]

The earth is like the moon

SALV: [...] From all this we conclude that the reflection from the earth is very powerful on the moon. What is more important is that there follows from this another beautiful resemblance, which is that if it is true that the planets act reciprocally upon the earth by their motion and by their light, perhaps the earth is no less potent in acting upon them by its own light—and possibly by its motion, too. [...]
SIMPL: No philosopher is to be found who ever said that these inferior bodies act upon celestial ones, while Aristotle said clearly the opposite.
SALV: Aristotle and the others who did not know that the earth and moon reciprocally illuminate each other deserve to be excused. But they would equally deserve to be reprehended if, while wanting us to give in to them and believe that the moon acts on the earth by light, they should insist on denying us the action of the earth on the moon when we had demonstrated to them that the earth lights up the moon.
SIMPL: All in all, I find in my heart a great reluctance to grant this companionship between the earth and moon of which you want to persuade me, placing the earth in the host of the stars, so to speak. For even if there were nothing else, the immense separation and distance

[173] Actually, what matters in diffusion is not volume, but surface area, and the surface area of the earth is only about thirteen times greater than that of the moon.

between the earth and the heavenly bodies seems to me to imply necessarily a great dissimilarity.

SALV: See what an inveterate affection and a deeply rooted opinion can do, Simplicio. It is so strong that you make the very things seem to favor your opinion which you yourself adduce against it. If separation and distance are valid facts for arguing a great difference in natures, it is necessary on the other hand that closeness and contiguity should mean similarity; and how much closer is the moon to the earth than it is to the other heavenly bodies! Confess then, by your own admission (and you will have plenty of other philosophers for company), the great affinity between the earth and the moon. [...]

With this conclusion Salviati completes his argument on the similarity between the earth and its satellite, denying to the moon those noble virtues that, according to Aristotelians, were its due as a celestial body. And thus he destroys poor Simplicio, who at the start of the dialogue, provoking a fierce objection from Sagredo, had stated with touching boldness:

SIMPL: Oh, there is no doubt whatever that the earth is more perfect the way it is, being alterable, changeable, etc., than it would be if it were a mass of stone or even a solid diamond, and extremely hard and invariant. But to the extent that these conditions bring nobility to the earth, they would render less perfect the celestial bodies, in which they would be superfluous. For the celestial bodies—that is, the sun, the moon, and the other stars, which are ordained to have no other use than that of service to the earth—need nothing more than motion and light to achieve their end.

SAGR: Has nature, then, produced and directed all these enormous, perfect, and most noble celestial bodies, invariant, eternal, and divine, for no other purpose than to serve the changeable, transitory, and mortal earth? To serve that which you call the dregs of the universe, the sink of all uncleanness? Now to what purpose would the celestial bodies be made eternal, etc. in order to serve something transitory, etc.?[174]

[174] See Chapter 7 for the hymn Galileo raises to the earth precisely because it is corruptible and lacking perfection.

The Hubble space telescope orbits at a distance of 615 km from the earth's surface

In praise of the human mind

We now come to Salviati's final considerations, where, even though he pays due tribute to the incommensurable power of God, he expresses all his admiration for the human mind, supreme proof in itself of divine magnificence. Sagredo, as elsewhere in the text, echoes his words and illustrates humanity's great achievements in the arts and crafts (the miracle of writing: "speaking to those who are not yet born and will not be born for a thousand or ten thousand years; and with what facility, by the different arrangements of twenty characters upon a page!"). Thus, in glory, ends the first day of the *Dialogue*.

SALV: [...] Now these advances, which our intellect makes laboriously and step by step, run through the Divine mind like light in an instant; which is the same as saying that everything is always present to it.

I conclude from this that our understanding, as well in the manner as in the number of things understood, is infinitely surpassed by the Divine; but I do not thereby abase it so much as to consider it absolutely null. No, when I consider what marvelous things and how many of them men have understood, inquired into, and contrived, I recognize and understand only too clearly that the human mind is a work of God's, and one of the most excellent.

SAGR: I myself have many times considered in the same vein what you are now saying, and how great may be the acuteness of the human mind. And when I run over the many and marvelous inventions men have discovered in the arts as in letters, and then reflect upon my own knowledge, I count myself little better than miserable. I am so far from being able to promise myself, not indeed the finding out of anything new, but even the learning of what has already been discovered, that I feel stupid and confused, and am goaded by despair. If I look at some excellent statue, I say within my heart: "When will you be able to remove the excess from a block of marble and reveal so lovely a figure hidden therein? When will you know how to mix different colors and spread them over a canvas or a wall and represent all visible objects by their means, like a Michelangelo, a Raphael, or a Titian?" Looking at what men have found out about arranging the musical intervals and forming precepts and rules in order to control them for the wonderful delight of the ear, when shall I be able to cease my amazement? What shall I say of so many and such diverse instruments? With what admiration the reading of excellent poets fills anyone who attentively studies the invention and interpretation of concepts! And what shall I say of architecture? What of the art of navigation?

But surpassing all stupendous inventions, what sublimity of mind was his who dreamed of finding means to communicate his deepest thoughts to any other person, though distant by mighty intervals of place and time! Of talking with those who are in India; of speaking to those who are not yet born and will not be born for a thousand or ten thousand years; and with what facility, by the different arrangements of twenty characters upon a page!

Let this be the seal of all the admirable inventions of mankind and the close of our discussions for this day. The hottest hours now being past, I think that Salviati might like to enjoy our cool ones in a gondola; and tomorrow I shall expect you both so that we may continue the discussions now begun.

MATHEMATICAL NOTE

Light diffusion

We have said that the lunar surface gives rise to a light diffusion mechanism and as a consequence is silvery-white in appearance and uniformly bright. The origin of the silvery-whiteness lies in the fact that the lunar surface scatters light in an essentially achromatic way, effectively retaining the spectral distribution characteristic of the sun, that is, it has the same color content.[175] In fact, while the sun's spectrum, according to black-body theory,[176] corresponds to an incandescent body of about 6000°C (temperature at the surface, known as the *color temperature* of the sun), the moon's spectrum is characteristic of a black-body temperature of about 5000°. Why? Because the moon is not an incandescent body, but a mere diffuser of light from other sources. If we acknowledge, for instance, that the diffusion mechanism is not equally efficient for all wavelengths of radiation, but instead is just a bit weaker on the violet side, the resulting spectrum can correspond to a lower color temperature.

Color temperature, in fact, decreases when the chromatic components of the light in the blue-violet region—that is, those of short wavelength—become weaker; it increases if the components on the red side become weaker. Typical examples: on the one hand, the sun at sunset, which appears red due to the loss, during the long passage of the sun's rays through the atmosphere, of the short-wavelength components, as a result of light diffusion by gas molecules; on the other hand, the sky, which, due to the same mechanism, appears blue. In these examples the diffusion is anything but achromatic, as it greatly favors blue-violet components: thus sunlight, when it comes to us by direct transmission through the atmosphere, retains the red hues, whereas when it comes to us through diffusion by the molecules of the atmosphere, is dominated by blue hues.

Then how is it that the moon, instead, gives rise to virtually achromatic diffusion? It is a question of the size of the diffusing bodies as compared to the wavelength of the radiation. This latter, for the visible range, falls between 0.45 μm (blue-violet) and 0.75 μm (deep red). When we are dealing with molecules, that is, very small objects, so-called "Rayleigh diffusion" occurs, which strongly privileges blue-violet. When, instead, we are dealing with large-scale prominences, such as the mountains and valleys of a planet, diffusion occurs via a mechanism which does not present chromatic effects. It is as if the reflecting surface were composed of a myriad of randomly oriented mirrors. Let us recall Galileo's astute description: "consider how the surface

[175] Well-known examples of substances upon whose surfaces the diffusion is almost perfectly achromatic are chalk, snow, flour, talcum powder, and letter paper. When, instead, diffusion privileges certain chromatic components over others, the surface appears colored. This is the case with non-white paper and chalk, paints, dyed fabrics, etc. A surface that neither diffuses, nor reflects, any component in the light spectrum looks black. The same applies to the sky seen by astronauts in the vacuum of space.

[176] See any general physics textbook. See also the Mathematical Note of Chapter 18.

of this rough wall is composed of countless very small surfaces placed in an innumerable diversity of slopes. . . .”

To conclude, we provide a diagrammatic illustration of how, in comparison to specular reflection, diffuse reflection gives rise to a more even luminosity of the observed surface.

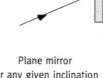

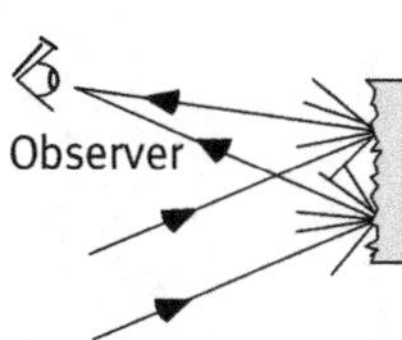

Diffuser

Whatever the observation position, the light reaches it from all points of the diffusing surface

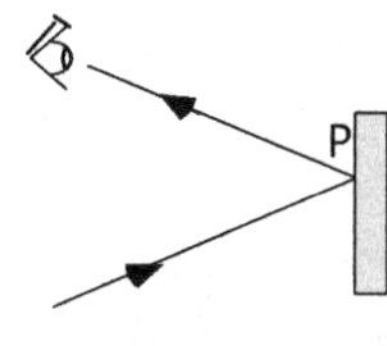

Plane mirror

For any given inclination of the light rays striking the mirror, there is only one observation position which satisfies the condition for specular reflection

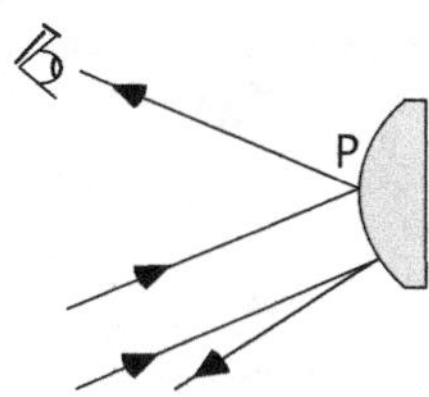

Spherical mirror

For any given observation position, there is only one single point on the mirror surface which reflects the light rays in the appropriate direction

PART V

THE SPINNING EARTH

9

An Ingenious Wrong Theory

The Ebb and Flow of the Tide*

Figuriamoci dunque una tal barca venirsene
con mediocre velocità per la Laguna,
portando placidamente l'acqua della quale ella sia piena

Let us imagine to ourselves such a barge
coming along the lagoon with moderate speed,
placidly carrying the water with which it is filled

Galileo believed he could provide incontrovertible proof that the earth moves using a model in which the rise and fall of the waters of the seas is caused by the accelerations and decelerations that points on the earth's surface are subjected to, as a consequence of the planet's combined motions of rotation and revolution about the sun. The model is as ingenious as it is flawed. Other phenomena capable of demonstrating the earth's rotation were right there in front of Galileo, but, as far as we know, either he did not notice them, or he deemed them not worth considering. Simplicio's comment in closing the dialogue offended the sensibilities of Pope Urban VIII.

Galileo would doubtless have been prepared to give up years of his life, if only he could have found irrefutable proof that the earth rotates on its axis and moves in orbit around the sun. Following on from his main source of inspiration, Giovanni Battista Benedetti, who had confuted the Aristotelian thesis in support of the earth's immobility, Galileo had devised enough valid arguments to remove all doubt that

* From *Dialogue*.

any phenomenon observable on earth would be capable of *disproving* that it moved. These arguments had led him to formulate the principle of relativity (see Chapters 4 and 5). However, he still had no proof capable of demonstrating such motion, and this was a weak point in his world system that he could not tolerate.

He believed he had finally found the decisive proof he had been looking for in the phenomenon of tides, for which he considered the existing explanations to be totally lacking in foundation. Among the many hypotheses there was one, advanced by Bishop Marcantonio de Dominis, which related peaks in sea level to the attraction exerted by the moon upon water masses. This idea was not particularly new, since already in Plinius and then later in Dante, references can be found to the role of the moon in the tidal effect ("*E come 'l volger del ciel de la luna cuopre e discuopre i liti sanza posa* . . .": "And the same way as the turning of the moon in the sky restlessly covers and uncovers the shores . . ."). Nevertheless, the idea came very close to the explanation for the phenomenon which is accepted today. To Galileo, however, it was far too similar to many others not worthy of attention and he consigned it to the "bunch of old ridiculous [ideas]".

One feature of tides, in particular, was a mystery to him: the moon passes over any given point on the earth's surface every 24 h and 50 min, while high-tides follow each other at intervals of approximately 12 h and 25 min, that is, they occur, rather counter-intuitively, both when the moon is directly above a given point and when it is directly above the antipodes of that point[177] ("the moon not only retains this

[177] The earth completes a full rotation on its axis, from west to east, in 24 h, while the moon orbits around it, also from west to east, in about 27 days and a half. The moon, therefore, passes above any given meridian approximately every 24 h and 50 min, which, divided by 2, gives 12 h, 25 min. The diagram below should help make this clear.

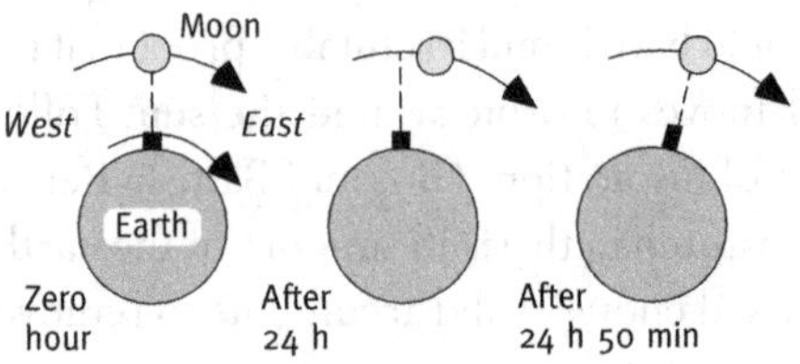

faculty naturally in itself, but in this case has also the power to confer it upon the opposite sign of the zodiac"). Aspects of the explanation proposed by de Dominis, however, were almost magical in character and so incapable of convincing Galileo, contrary as he was to anything that smacked of scanty reasoning.

Galileo's idea

He thus came up with an idea which, however ingenious it may appear, was nevertheless a long way from explaining the observed phenomenon, and soon afterwards was to be revealed as altogether wrong. This was not the only occasion upon which Galileo allowed himself to get carried away by the goal he had in mind, and thereby lost sight of the essentials of the problem. As a starting point he took the example of the movements of a cargo of drinking water loaded into a barge. The level of this water rises at the bow and falls at the stern whenever the boat decelerates (as a result of a collision against, say, a sand bank) and vice versa in the case of acceleration. The waters of the seas, he said, should behave in the same fashion, provided a good reason could be found for points on the earth's surface being subjected to acceleration. Galileo's argument was that, if it is accepted that the earth moves in orbit around the sun

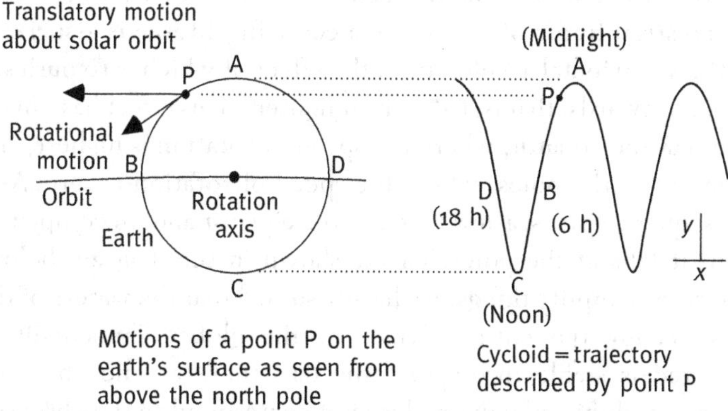

Motions of a point P on the earth's surface as seen from above the north pole

Cycloid = trajectory described by point P

("the wide orbit"), while at the same time spinning on its own axis, then every point on its surface will describe a trajectory similar to that shown on the right-hand side of the above diagram (to make things clearer, the velocity due to the earth's rotation has been exaggerated with respect to that due to the orbital motion). Such a trajectory, known as "cycloid", is the one described, for example, by the air valve of a rotating bicycle wheel. It is thus easy to see that the component of the velocity directed along the orbit around the sun reaches a maximum at point A (*midnight*), where the velocities of the two motions (translational and rotational) add to each other—and a minimum at point C (*noon*), where the two velocities subtract from each other. Consequently, the velocity is not constant and it is this which leads to periodic acceleration and deceleration.

Where the error lies

However, had Galileo pondered on this mechanism just a little longer, he would have realized that it is incapable of explaining the observed effect. As in the case of the cargo of water in a barge, the problem immediately becomes clear if one reasons on the basis of acceleration and of the forces involved in it. The largest contribution to the acceleration of points on the earth's surface is that arising from rotation. This is the centripetal acceleration, always directed towards the center of the earth. Now, within a terrestrial frame of reference, a centrifugal force is associated to this centripetal acceleration, the effect of which is to push sea water outwards, that is, to "make it lighter". This effect has a maximum at the equator, where the speed of rotation is highest, and is absent at the poles, where the speed of rotation is zero. As a consequence, the sea-level rises at the equator and, in compensation, it falls at the poles. This is shown in the diagram below, where, to simplify things, we have assumed that the waters of the seas occupy the entire surface of the globe. The condition depicted is stable throughout the day, there are no ebbs and flows—and this, of course, does not correspond to the observed behaviour of tides.

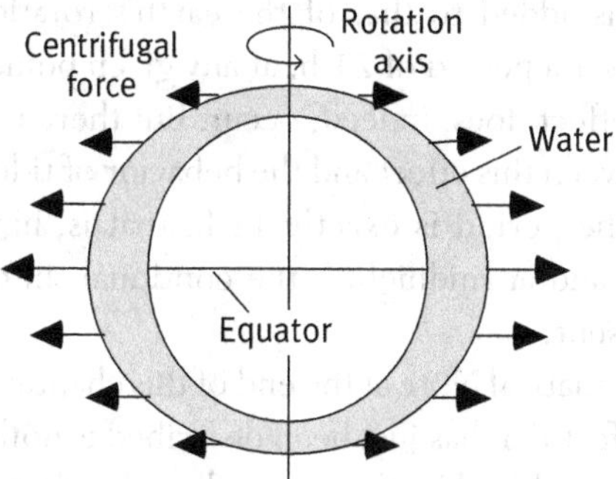

Effect of the centrifugal force due to the earth's rotation

The second contribution to the acceleration comes from the earth's orbital motion around the sun. The center of the earth lies on the path of the orbit, where the attracting gravitational force of the sun and the centrifugal force exactly balance each other out. However, this equilibrium between the two forces is not equally precise at all other points on the earth: for instance, at point A the force of gravity is weaker than at the centre, since the point is more distant from the sun, and this results in the centrifugal force (directed upwards in the diagram) prevailing; at point C the force of gravity is stronger, therefore the contrary holds (centrifugal force directed downwards).

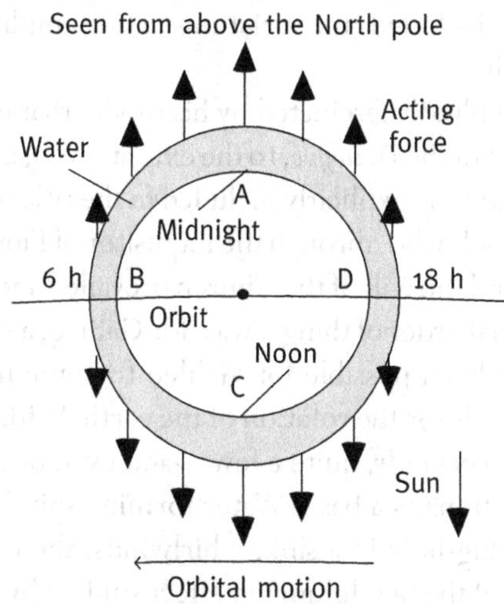

Effect of the overall force associated to the orbital motion about the sun

This effect is added to that of the earth's rotation. It can thus be seen that, over a period of 24 h, at any given point on the earth's surface a tide effect does, indeed, occur. But there is a fundamental difference between this effect and the behavior of tides that we actually observe: the period is exactly 12 h, that is, high tides always occur at noon and at midnight—the continual shifting in the tide timetable is absent.

In the Mathematical Note at the end of this chapter the reader will see how the effect that has just been described is nothing other than the correction produced by the sun to the tide of lunar origin. This, of course, has little to do with what Galileo thought. His description was too qualitative, let us say too intuitive, even, to permit a detailed analysis of the phenomenon. He seemed to realize this himself, as he by-passed the problem of the discrepancy in the timing of tides, asserting that rotation and revolution were responsible neither for the period, nor for the amplitude of the tide, but were, instead, determined locally by the extension and depth of the sea.

The correct explanation for tides in terms of lunar attraction will also be summarized in the Mathematical Note. This explanation, which is reminiscent of that proposed by de Dominis (and so summarily dismissed by Galileo) fully accounts for the two maxima during the course of 24 h and 50 min, as well as (when the effects due to the sun are included) the variations in tide amplitude over the course of a month.

Galileo was so blindly fascinated by his model that he considered it the pivotal point of the *Dialogue*, to the extent, even, that he wanted the phrase *"ebb and flow"* explicitly included in the title of the book. It was the Pope himself who, through the Inquisitor of Florence, vetoed this idea. In the end the will of the Church prevailed and this (besides being in the normal order of things) was, for Galileo, a stroke of luck.

Would it have been possible for Galileo to come up with other experimental proofs for the rotation of the earth? With hindsight we can say, well yes, certainly, quite a few—and even ones which used everyday observations as a basis. Water forming spiral vortices as it runs down the plug-hole in a sink, whirlwinds, the rotation of the oscillation plane of the pendulum (an effect studied by Foucault two centuries later) are all phenomena that occur as a consequence of the earth's rotation. A body moving in a rotating system is, in fact, subjected to lateral deviation, so that the trajectories it describes are curved. This happens as a result of the so-called Coriolis force (described in the Mathematical Note of Chapter 4), the power of which can easily be verified by trying to walk from the centre to the edge of a moving merry-go-round, a classic example of a rotating system.

The text

The reader may find the excerpts which follow rather heavy-going, owing to the detailed, somewhat dry and protracted explanations (particularly evident on Salviati's part) and to the dialogue itself, which is not as lively and functional as elsewhere. Nevertheless, we thought it appropriate to include these passages for two reasons. First, they provide a very good example of the Galilean approach: the scientist explores a natural phenomenon in a systematic way (in this case the behavior of water in a vessel undergoing acceleration) and then extends the conclusions to similar phenomena (here, the large oceanic masses). There can be no discrepancy in the behavior of the two, Galileo asserts, because the laws of Nature cannot contradict one another.

Second, on this occasion we find that it is Galileo who, notwithstanding his frequent warnings that emotion must not influence scientific discourse, falls into the trap himself. Passion blinds him twice over: it makes him brand as "frivolous" a theory which is close to the truth, and it also cramps his imagination. If the laws of Nature cannot contradict one another, then it may be that some seemingly contradictory effects stem from causes that are not immediately apparent. The scientist must be capable of considering all the alternatives and of thinking in a way which is creative, yet still within the logic schemes of the possible. In the pages we are about to examine it seems, almost, that Galileo has forgotten this.

In his defense, however, seeing that he did not know the laws of gravitational attraction, stands his healthy repugnance for astrology at a time when too many things—from human character traits to the virtues of precious gems—were attributed to the influence of the moon and of heavenly bodies. A repugnance that still today, centuries later, a large number of thinking human beings are incapable of nurturing.

The passage which follows is taken from the *Dialogue*.[178]

The tidal effect

SALV: [. . .] it will be good for us to get to the matter in hand without wasting any more words.

Let us see, then, [. . .] how, reciprocally, this ebb and flow itself cooperates in confirming the earth's mobility. Up to this point the indications of that mobility have been taken from celestial phenomena, seeing that nothing which takes place on the earth has been powerful enough to establish the one position any more than the other. This we have already examined at length by showing that all terrestrial events from which it is ordinarily held that the earth stands still and the sun and the fixed stars are moving would necessarily appear just the same to us if the earth moved and the others stood still.[179] *Among all sublunary things it is only in the element of water (as something which is very vast and is not joined and linked with the terrestrial globe as are all its solid parts, but is rather, because of its fluidity, free and separate and a law unto itself) that we may recognize some trace or indication of the earth's behavior in regard to motion and rest. After having many times examined for myself the effects and events, partly seen and partly heard from other people, which are observed in the movements of the water; after, moreover, having read and listened to the great follies which many people have put forth as causes for these events, I have arrived at two conclusions which were not lightly to be drawn and granted. Certain necessary assumptions having been made, these are that if the terrestrial globe were immovable, the ebb and flow of the oceans could not occur naturally;*

[178] *Dialogue Concerning the Two Chief World Systems*, Fourth Day, translated by S. Drake, The Modern Library, New York, 2001, p. 483 and following. In Italian: *Dialogo sopra i due massimi sistemi del mondo*, Giornata quarta, Edizione Nazionale, Vol. VII, p. 442 and following.

[179] The arguments which Salviati refers to are discussed in Chapters 4 and 5.

and that when we confer upon the globe the movements just assigned to it, the seas are necessarily subjected to an ebb and flow agreeing in all respects with what is to be observed in them. [. . .]

I say, then, that three periods are observed in the flow and ebb of the ocean waters. The first and principal one is the great and conspicuous daily tide, in accordance with which the waters rise and fall at intervals of some hours; these intervals in the Mediterranean are for the most part about six hours each—that is, six hours of rising and six more of falling. The second period is monthly, and seems to originate from the motion of the moon; it does not introduce other movements, but merely alters the magnitude of those already mentioned, with a striking difference according as the moon is full, new, or at quadrature[180] *with the sun. The third period is annual, and appears to depend upon the sun; it also merely alters the daily movements by rendering them of different sizes at the solstices*[181] *from those occurring at the equinoxes.*[182]

We shall speak first about the diurnal period, as it is the principal one, and the one upon which the actions of the moon and the sun are exercised secondarily in their monthly and annual alterations. [. . .]

Salviati's words indicate that Galileo was aware of the monthly and annual variations of the tides and, moreover, that he did not seem to be against attributing these to mechanisms depending on the moon and sun. Nevertheless, he was to rule out the possibility that the main effect, the diurnal one, could depend on these celestial bodies, attributing it exclusively to the motion of the earth. Later in the dialogue Salviati proceeds to describe in detail the effects of the tide at different locations in the Mediterranean Sea, stating that he could not account for these if the earth were stationary. Let us resume the discourse from this last point.

Tidal models

SALV: [. . .] Now it seems to me that these actual and known effects alone, even if no others were to be seen, would very probably persuade

180 *Quadrature*—an angle, moon–earth–sun, of 90° or 270°.
181 *Solstices*—the beginning of summer and winter.
182 *Equinoxes*—the beginning of spring and autumn.

anyone of the mobility of the earth who is willing to stay within the bounds of nature; for to hold fast the basin of the Mediterranean and to make the water contained within it behave as it does surpasses my imagination, and perhaps that of anyone else who enters more than superficially into these reflections.

SIMPL: These events, Salviati, did not just commence; they are very ancient, and have been observed by innumerable men, many of whom have contrived to give one reason or another to account for them. Not far from here there is a great Peripatetic who gives for them a cause recently dredged out of one of Aristotle's texts which had not been well understood by his interpreters. From this text, he deduces that the true cause of these movements stems from nothing else but the various depths of the seas. The deepest waters, being more abundant and therefore heavier, expel the waters of lesser depth; these, being raised up, then try to descend, and from this continual strife the tides are derived.

Then there are many who refer the tides to the moon, saying that this has a particular dominion over the waters; lately a certain prelate[183] *has published a little tract wherein he says that the moon, wandering*

[183] This prelate was Bishop Marcantonio de Dominis (1566–1624), author of the book *Euripus, seu de fluxu et refluxu maris sententia*. His explanation approached that which is accepted today. In a macabre rite, the Church prosecuted de Dominis after his death, condemned his writings, and then had them burned together with his disinterred corpse. An event of this kind had not happened since 1415, when John Wyclif (whose cadaver, unlike that of de Dominis, was not present in the courtroom) was convicted posthumously for having asserted that bread and wine in the Eucharist could neither be annihilated nor transubstantiated. It is worth recalling that de Dominis was the editor of Paolo Sarpi's *Istoria del Concilio Tridentino*, a book that met with great success all over Europe and was strongly opposed by the Jesuits. The main doctrinal decision of the Trent Council was the dogma of the Eucharistic transubstantiation:in his book, Sarpi revealed the disputes and doubts among the Council members concerning a dogma which was not to be found in the Gospels and was contrary to reason. In his *De repubblica ecclesiastica*, de Dominis went as far as to propose that, had the Eucharist been allowed to retain its character of obscure mystery, as handed down by the Fathers, major controversies within Christianity would have been avoided. After a solemn abjuration, pronounced in Rome in 1622, de Dominis found great favor, but soon fell into disgrace again. In 1624, just before dying in prison, he abjured once more, or at least this is what his inquisitors reported (for a deeper discussion of this argument, see P. Redondi, *Galileo Heretic*, translated by R. Rosenthal, Princeton University Press, Princeton, 1987). It is interesting to observe how Catholic believers today accept quite readily dogmatic assumptions which centuries ago were the subject of fierce debates. The importance of the events

through the sky, attracts and draws up toward itself a heap of water which goes along following it, so that the high sea is always in that part which lies under the moon. And since when the moon is below the horizon, this rising nevertheless returns, he tells us that he can say nothing to account for this effect except that the moon not only retains this faculty naturally in itself, but in this case has also the power to confer it upon the opposite sign of the zodiac. Others, as I think you know, say that the moon also has power to rarefy the water by its temperate heat, and that thus rarefied, it is lifted up. Nor are those lacking who . . .
SAGR: Please, Simplicio, spare us the rest; I do not think there is any profit in spending the time to recount them, let alone the words to refute them. If you should give assent to any of these or to similar triflings, you would be wronging your own judgment—just when, as we know, it has been much unburdened of error.
SALV: I am a little more easygoing than you, Sagredo, and I shall put in a few words for Simplicio's benefit if he thinks that some probability attaches to the things he has been telling us.

Thus, with a few cutting remarks, Sagredo tosses aside the various hypotheses proffered by Simplicio. And Salviati, though declaring himself more open to considering these hypotheses, in fact immediately goes on—as we are about to see—to reject Marcantonio de Dominis's sound theory in a manner which is not only hasty, but also superficial. He attributes no importance, for example, to the fact that the Mediterranean sea is small and almost completely closed, which means its waters cannot be exchanged with those of the ocean within the space of just a few hours.

In the opening exchanges of the next passage the reader will once again note the delightful contraposition of the characters. Simplicio, haughty as ever, worrying more about form than substance, points out "with that frankness which is permitted here among ourselves" that Salviati's explanation is "fictitious". He asks (*he* asks!) to be presented with reasons "more agreeable to natural phenomena", or else he will have to attribute the tidal effect to a miracle. Salviati,

recounted here will become clearer in Chapter 18, in connection with the atomistic theory of matter. This was expounded by Galileo in *The Assayer* and it gave some of his enemies grounds for accusing him of Eucharistic heresy.

exasperated, just to make his reluctant interlocutor proceed with the discussion, astutely uses Urban VIII's argument[184] ("But do you not believe that the terrestrial globe could be made movable supernaturally, by God's absolute power?"). He chooses, that is, the same weapon Simplicio had often used to get himself out of trouble. But the way Galileo has Salviati phrase the concept, plus the unscrupulous use he puts it to, reveal in what low esteem he holds a viewpoint such as this within a scientific discussion. And the words that follow—"returning to our discussion"—are intended to indicate the resumption of arguments of substance.

SALV: [. . .] You might tell your prelate that the moon travels over the whole Mediterranean every day, but the waters are raised only at its eastern extremity and for us here at Venice.
SIMPL: [. . .] But I must say, with that frankness which is permitted here among ourselves, that to introduce the motion of the earth and make it the cause of the tides seems to me thus far to be a concept no less fictitious than all the rest I have heard. If no reasons more agreeable to natural phenomena were presented to me, I should pass on unhesitatingly to the belief that the tide is a supernatural effect, and accordingly miraculous and inscrutable to the human mind—as are so many others which depend directly upon the omnipotent hand of God.
SALV: You argue very prudently, and also in agreement with Aristotle's doctrine; at the beginning of his Mechanics, as you know, he ascribes to miracles all things whose causes are hidden. [. . .] But do you not believe that the terrestrial globe could be made movable supernaturally, by God's absolute power?
SIMPL: Who can doubt this?
SALV: Then, Simplicio, since we must introduce a miracle to achieve the ebbing and flowing of the oceans, let us make the earth miraculously move with that motion by which the oceans are naturally moved. This operation will indeed be as much simpler and more natural among things

[184] The argument is that God can make natural phenomena occur for causes other than those discovered by the human mind. We shall return to this point later on in the chapter.

miraculous, as it is easier to make a globe turn around (which we see so many of them do) than to make an immense bulk of water go back and forth more rapidly in some places than in others; rise and fall, here more, there less, and in other places not at all, and to make all these variations within the same containing vessel. Besides, these are many miracles, while the other is only one. [. . .]

If it has to be a miracle, let us make it the least miraculous possible: this is, in effect, what Salviati is saying. Then straight away after, as if repenting of this concession (tinged with sarcasm though it is), he hastens to affirm once again his strict deterministic credo. His motto is "from one and uniform cause only one and uniform effect can follow"[185]: it is a pity that sometimes, as in this case, not all the potential causes can be known.

The water in the vessel

SALV: [. . .] Now, returning to our discussion, I reply and reaffirm that it has never previously been known how the waters contained in our Mediterranean basin can make those movements which they are seen to make, so long as this basin and containing vessel rests motionless. [. . .]

As these effects must be consequences of the motions which belong naturally to the earth, it is not only necessary that they encounter no obstacle or impediment, but that they follow easily. Nor must they merely follow easily; they must follow necessarily, in such a way that it would be impossible for them to take place in any other manner. For such is the property and condition of things which are natural and true.

Having established, then, that it is impossible to explain the movements perceived in the waters and at the same time maintain the immovability of the vessel which contains them, let us pass on to considering whether the mobility of the container could produce the required effect in the way in which it is observed to take place.

[185] *Dialogue Concerning the Two Chief World Systems*, translated by S. Drake, The Modern Library, New York, 2001, p. 515. In Italian: *Dialogo sopra i due massimi sistemi del mondo*, Edizione Nazionale, Vol. VII, p. 469.

Canaletto: "La Dogana e il Canale della Giudecca" in Venice (Milan, Mario Crespi collection)

There follows a somewhat verbose explanation of the possible movements of water in a vessel. It is alleviated a little, however, by a brief, but evocative, scene taken from everyday life—one to which Galileo's attention had no doubt been drawn many times. This was of barges continually plowing "placidly" through the lagoon to supply Venice with drinking water. The example of the barges serves to illustrate the physical effect under investigation and confirms once again that Galileo's speculations were never abstract, but were taken from the environments in which he found himself, inexhaustible sources for him of curiosity and intellectual stimulus. In this fleeting image drawn from memory, a faint note of nostalgia may be detected, for a time and a place that had once been dear to Galileo, but were now remote.

SALV: Two sorts of movement may be conferred upon a vessel so that the water contained in it acquires the property of running first toward one end and then toward the other, and rise and sink there. The first would occur when one end is lowered and then the other, for under those conditions the water, running toward the depressed part, rises and sinks alternately at either end. [. . .]

The other sort of motion would occur when the vessel was moved without being tilted, advancing not uniformly but with a changing velocity, being sometimes accelerated and sometimes retarded. From this variation it would follow that the water (being contained within the vessel but not firmly adhering to it as do its solid parts) would because of its fluidity be almost separate and free, and not compelled to follow all the changes of its container. Thus the vessel being retarded, the water would retain a part of the impetus already received, so that it would run toward the forward end, where it would necessarily rise. On the other hand, when the vessel was speeded up, the water would retain a part of its slowness and would fall somewhat behind while becoming accustomed to the new impetus, remaining toward the back end, where it would rise somewhat.

These effects can be very clearly explained and made evident to the senses by means of the example of those barges which are continually arriving from Fusina filled with water for the use of this city. Let us imagine to ourselves such a barge coming along the lagoon with moderate speed, placidly carrying the water with which it is filled, when either by running aground or by striking some obstacle it becomes greatly retarded. Now the water will not thereby lose its previously received impetus equally with the barge; keeping its impetus, it will run forward toward the prow, where it will rise perceptibly, sinking at the stern. But if on the other hand the same barge noticeably increases its speed in the midst of its placid course, then the water which it contains (before getting used to this and while retaining its slowness) will stay back toward the stern, where it will consequently rise, sinking at the prow. This effect is indubitable and clear; it may be tested experimentally at any time, and there are three things about it which I want you to note particularly.

The first is that in order to make the water rise at one extremity of the vessel, there is no need of new water, nor need the water run there from the other end.

The second is that the water near the middle does not rise or sink noticeably [. . .].

The third thing is that whereas the parts around the center make little change as to rising or sinking with respect to the water at the ends, yet they run to and fro a great deal in comparison with the water at the extremities.

Now, gentlemen, what the barge does with regard to the water it contains, and what the water does with respect to the barge containing it, is precisely the same as what the Mediterranean basin does with regard to the water contained within it, and what the water contained does with respect to the Mediterranean basin, its container. The next thing is for us to prove that it is true, and in what manner it is true, that the Mediterranean and all other sea basins (in a word, that all parts of the earth) move with a conspicuously uneven motion, even though nothing but regular and uniform motions may happen to be assigned to the globe itself. [. . .]

As will be demonstrated more fully in a while, this last statement of Salviati's is, to say the least, inaccurate—as is the statement that now follows. In both cases the motions being dealt with are circular and thus, even if the velocity is constant in value, the fact that direction keeps changing leads to centripetal acceleration being present.

Galileo's model

SALV: We have already said that there are two motions attributed to the terrestrial globe[186] *[. . .]. From the composition of these two motions, each of them in itself uniform, I say that there results an uneven motion in the parts of the earth. In order for this to be understood more easily, I shall explain it by drawing a diagram.*

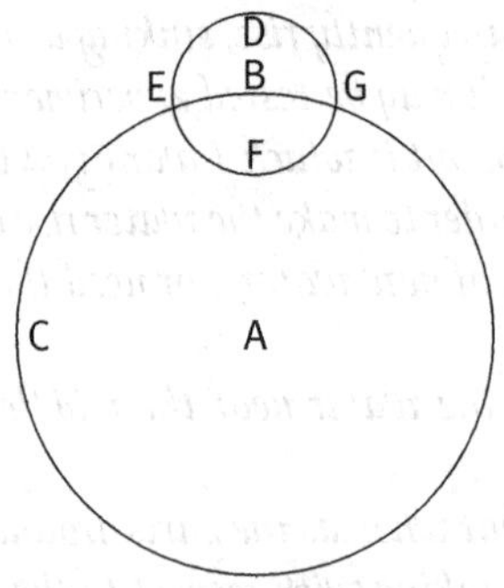

[186] The movement of the earth as it advances along its orbit about the sun ("annual motion") and as it rotates upon its axis ("diurnal motion"). Both motions, in the accompanying diagram, are counterclockwise.

First I shall describe around the center A *the circumference of the earth's orbit* BC, *on which the point* B *is taken; and around this as center, let us describe this smaller circle* DEFG, *representing the terrestrial globe. We shall suppose that its center B runs along the whole circumference of the orbit from west to east; that is, from* B *toward* C. *We shall further suppose the terrestrial globe to turn around its own center* B *from west to east, in the order of the points* D, E, F, G, *during a period of twenty-four hours. Now here we must carefully note that when a circle revolves around its own center, every part of it must move at different times with contrary motions. This is obvious, considering that when the part of the circumference around the point* D *is moving toward the left (toward* E*), the opposite parts, around* F, *go toward the right (toward* G*); so that when the point* D *gets to* F, *its motion will be contrary to what it was originally when it was at* D. *Moreover, in the same time that the point E descends, so to speak, toward* F, G *ascends toward* D. *Since this contrariety exists in the motion of the parts of the terrestrial surface when it is turning around its own center, it must happen that in coupling the diurnal motion with the annual, there results an absolute motion of the parts of the surface which is at one time very much accelerated*[187] *and at another retarded by the same amount.*[188] *[...]*

Around the points E *and* G, *the absolute motion remains equal to the simple annual motion, since the diurnal motion acts upon it little or not at all, tending neither to left nor to right, but downward and upward. From this we conclude that just as it is true that the motion of the whole globe and of each of its parts would be equable and uniform if it were moved with a single motion, whether this happened to be the annual or the diurnal, so is it necessary that upon these two motions being mixed together there results in the parts of the globe this uneven motion, now accelerated and now retarded by the additions and subtractions of the diurnal rotation upon the annual revolution.*

Now if it is true (as is indeed proved by experience) that the acceleration and retardation of motion of a vessel makes the contained water run back and forth along its length, and rise and fall at its extremities, then

[187] Point D is being referred to here. In reality, the effect of the earth's rotation is weak: the orbital motion about the sun takes place at a speed in excess of 100,000 km/h, while the motion due to rotation occurs at a speed slightly over 1500 km/h (for a point on the equator, where it is highest).

[188] Point F is being referred to here.

who will make any trouble about granting that such an effect may—or rather, must—take place in the ocean waters? For their basins are subjected to just such alterations; especially those which extend from west to east, in which direction the movement of these basins is made.

Now this is the most fundamental and effective cause of the tides, without which they would not take place. [. . .]

In the preceding discussion, the phrase: "just as it is true that the motion of the whole globe and of each of its parts would be equable and uniform if it were moved with a single motion . . ." reveals the particular understanding that Galileo had of "equable and uniform" motion. He refers to the absolute value of speed alone, overlooking the fact that both the orbital motion and the rotational motion are characterized by continuous change in direction and, because of this, are accelerated individually (as a matter of fact, a centrifugal force is associated with them). An accelerated motion could never derive from the composition of two *truly uniform* motions, as the very principle of inertia, discovered by Galileo himself, teaches us. This position, together with others related to it, has led some historians (Koyré,[189] for example) to state that Galileo's ideas about inertia were not fully focused. The conception of the earth–sun system Galileo had in mind can probably be likened to one of those fairground roundabouts which have, along the perimeter of the rotating platform, seats which are free to rotate separately. People riding in the seats experience continual variations in acceleration, resulting in thrilling increases and decreases in centrifugal push. Indeed, if it had been described in these terms, Galileo's model would have been closer to the truth, and worthy of greater appreciation.

SIMPL: I do not think it can be denied that your argument goes along very plausibly, the reasoning being ex suppositione, *as we say; that is, assuming that the earth does move in the two motions assigned to it by Copernicus. But if we exclude these movements, all the rest is vain and invalid; and the exclusion of this hypothesis is very clearly pointed out to us by your own reasoning. Under the assumption of the two*

[189] A. Koyré, *Études galiléennes*, Hermann, Paris 1966 (English translation by J. Mepham: *Galileo Studies*, Humanities Press, Atlantic Highlands, NJ, 1978).

terrestrial movements, you give reasons for the ebbing and flowing; and vice versa, arguing circularly, you draw from the ebbing and flowing the sign and confirmation of those same two movements. Passing to a more specific argument, you say that on account of the water being a fluid body and not firmly attached to the earth, it is not rigorously constrained to obey all the earth's movements. From this you deduce its ebbing and flowing.

In your own footsteps, I argue the contrary and say: The air is even more tenuous and fluid than the water, and less affixed to the earth's surface, to which the water adheres (if for no other reason) because of its own weight, which presses it down much more than the very light air. Then so much the less should the air follow the movements of the earth; hence if the earth did move in those ways, we, its inhabitants, carried along at the same velocity, would have to feel a wind from the east perpetually beating against us with intolerable force. That such would necessarily follow, daily experience informs us; for if, in riding post with no more speed than eight or ten miles an hour in still air, we feel in our faces what resembles a wind blowing against us not lightly, just think what our rapid course of eight hundred or a thousand miles per hour[190] *would have to produce against air which was free from such motion! Yet we feel nothing of any such phenomenon.*

Wind and waves on the sea

SALV: To this objection, which seems so persuasive, I reply that it is true that the air is much more tenuous and much lighter than the water, and by its lightness is much less adherent to the earth than heavy and bulky water. But the consequence which you deduce from these conditions is false; that is, that because of its lightness, tenuity, and lesser adherence to the earth it must be freer than water from following the movements of the earth, so that to us who participate completely in those movements its disobedience would be made sensible and evident. In fact, quite the opposite happens. For if you will remember carefully, the cause of the ebbing and flowing of the water assigned by us consisted

[190] This is the velocity that is being attributed to the rotational motion of the earth.

in the water not following the irregularity of motion of its vessel, but retaining the impetus which it had previously received, and not diminishing it or increasing it in the exact amount by which this is increased or diminished in the vessel. Now since disobedience to a new increase or diminution of motion consists in conservation of the original received impetus, that moving body which is best suited for such conservation will also be best fitted for exhibiting the effect that follows as a consequence of this conservation. How strongly water is disposed to preserve a disturbance once received, even after the cause impressing it has ceased to act, is demonstrated to us by the experience of water highly agitated by strong winds. Though the winds may have ceased and the air become tranquil, such waves remain in motion for a long time, as the sacred poet so charmingly sings: Qual l'alto Egeo, etc.[191] *The continuance of the commotion in this way depends upon the weight of the water, for as has been said on other occasions,*[192] *light bodies are indeed much easier to set in motion than heavier ones, but they are also much less able to keep the motion impressed upon them, once the cause of motion stops. The air, being a thing that is in itself very tenuous and extremely light, is most easily movable by the slightest force; but it is also most inept at conserving the motion when the mover ceases acting.*

As to the air that surrounds the terrestrial globe, I shall therefore say that it is carried around by its adherence no less than the water, and especially those parts of it which are contained in vessels, these vessels being plains surrounded by mountains.[193] *[...]*

It has to be acknowledged that, even if its objective was primarily polemical, Simplicio's conjecture that the atmosphere experiences tidal effects was correct. Despite the fact that they are not easy to detect, today we know that these phenomena do actually exist (it could not be otherwise, because the atmosphere, too, is subject to lunar attraction). These effects are called atmospheric tides.

Next we come to a pause in the discussion, almost as if to allow our friends to give their minds a rest. They appear worn out by the intense thinking about phenomena which are getting more and more

[191] Torquato Tasso, *Gerusalemme liberata*, XII, 63.

[192] See Chapter 5.

[193] For a more detailed discussion of this argument see Chapter 10.

complicated and seem to be in contradiction with one another. Into Salviati's speech Galileo introduces elements that are strongly autobiographical, recalling the hard work (physical, too) needed for research; the despair that assails him when he fears he will never succeed in getting to the bottom of a problem; but, above all, the difficulty of ridding himself of the huge weight of the authority of those who have gone before him.

SAGR: I feel myself being gently led by the hand; and although I find no obstacles in the road, yet like the blind I do not see where my guide is leading me, nor have I any means of guessing where such a journey must end.

SALV: There is a vast difference between my slow philosophizing and your rapid insights; yet in this particular with which we are now dealing, I do not wonder that even the perspicacity of your mind is beclouded by the thick dark mists which hide the goal toward which we are traveling. All astonishment ceases when I remember how many hours, how many days, and how many more nights I spent on these reflections; and how often, despairing of ever understanding it, I tried to console myself by being convinced, like the unhappy Orlando, that that could not be true which had been nevertheless brought before my very eyes by the testimony of so many trustworthy men. [. . .]

The tower of knowledge

At this point Salviati turns his attention to explaining secondary effects, in particular the variations in the level of high and low tides, with monthly and annual periods. When he has finished, Sagredo is in awe of the captivating progression of the reasoning and its illuminating graduality. He uses the image of an enormously high tower, the top of which can, step by step, nevertheless still be arrived at, as a splendid metaphor for the way in which only with patience, humility, wisdom, and perseverance, is it possible to progress (without apparent effort and beyond all expectations) to the highest peaks of knowledge.

SAGR: If a very high tower were shown to someone who had no knowledge of any kind of staircase, and he were asked whether he dared to scale such a supreme height, I believe he would surely say no, failing to understand that it could be done in any way except by flying. But being shown a stone no more than half a yard high and asked whether he thought he could climb up on it, he would answer yes, I am sure; nor would he deny that he could easily climb up not once, but ten, twenty, or a hundred times. Hence if he were shown the stairs by which one might just as easily arrive at the place he had adjudged impossible to reach, I believe he would laugh at himself and confess his lack of imagination.

You, Salviati, have guided me step by step so gently that I am astonished to find I have arrived with so little effort at a height which I believed impossible to attain. It is certainly true that the staircase was so dark that I was not aware of my approach to or arrival at the summit, until I had come out into the bright open air and discovered a great sea and a broad plain. And just as climbing step by step is no trouble, so one by one your propositions appeared so clear to me, little or nothing new being added, that I thought little or nothing was being gained. So much the more is my wonder at the unexpected outcome of this argument, which has led me to a comprehension of things I believed inexplicable.
[. . .]

And so we come to the end of the fourth and final day of discussions, with the concluding points of the three interlocutors. Salviati returns to and criticizes the hypothesis that attributes the diurnal effect of tides to lunar influences, which to him seems more astrological in character than scientific (he also takes the opportunity to express strong reservations about Kepler, who is an adherent of the hypothesis). In Salviati's peremptory judgement one can perceive an attempt to exorcize any possibility that his own theory is mistaken. It is quite true that soon afterwards he goes on to describe his theory as an "invention, which may very easily turn out to be a most foolish hallucination and a majestic paradox". However, it is plain that the last to concede that his theory was a "hallucination" would be Salviati himself. In fact, this phrase, coming as it does at the end of the dialogue (i.e. in the conclusion to what has been a systematic and pitiless demolition of ancient knowledge), rings false. It appears to have been uttered solely in order that the other interlocutors, gratified by the compliments and thanks heaped upon them by Salviati, should protest that it was not the case. Indeed, the phrase was included by Galileo for reasons of censorship. The *Dialogue* was supposed to be an impartial, balanced account of the two systems, but in actuality it was a sustained criticism of the Ptolemaic system. The phrase, therefore, must have struck the Inquisition judges as neither genuine, nor sufficient to rebalance the one-sidedness of the arguments.

An angelic doctrine

While Salviati, in this last page of the *Dialogue*, appears prepared to desist from his customary "too heated and opinionated speech", Simplicio remains as much himself as ever, manifesting all the haughtiness and clumsy arrogance of a provincial man of learning. Unfortunately (unfortunately for Galileo, that is, because damaging consequences for him were not to be slow in arriving), Simplicio, in his final intervention, straight after having displayed the mediocrity of his character yet once again, pulls out of the hat, by way of conclusion, "a most solid doctrine [. . .] before which one must fall silent". He is referring to Urban VIII's argument that God can make natural phenomena occur for causes other than those conceivable by the human mind: "From this I forthwith conclude that, this being so, it would be excessive boldness for anyone to limit and restrict the Divine power and wisdom to some particular fancy of his own".

In truth, the insertion of this reference had been imposed on Galileo by the Inquisition when the text of the *Dialogue* had been examined for the concession of the *imprimatur*. But the fact that it had been put in Simplicio's mouth led Urban VIII to suspect that it was none other than he himself who had been portrayed in the character of Simplicio. This annoyed him, and contributed not a little to his decision to send Galileo for trial (strangely enough, on the sole occasion when Simplicio, at the scientific level, had sound reasons for being skeptical).[194] A further aggravation was that Salviati, in his final intervention, accepts Urban VIII's argument hurriedly and without reservation—"An admirable and angelic doctrine . . .", he says—using terms which, for anyone who by this stage knows Salviati–Galileo well, cannot but sound imprudently ironic, almost sarcastic. Elsewhere in his writing, in fact, Galileo states resolutely: "the order of the world is just one, and it has never been else: therefore he who looks for something different from what it is, looks for the false and the impossible", and also: "I must tell you that in natural phenomena human authority is worthless. Like a lawyer, you seem to capitalize on it; but Nature, dear Sir, makes fun of constitutions and decrees of princes, emperors, and monarchs, and at their request it would not change one iota of its laws and statutes".[195] Another peremptory assertion on this theme is to be found in the *Dialogue*, where Salviati, in replying to Simplicio's idea that velocity in accelerated motion increases in steps, affirms[196]: "I did not say, nor dare I, that it was impossible for Nature or for God to confer immediately that velocity which you speak of. I do indeed say that *de facto* Nature does not do so—that the doing of this would be something outside the course of Nature, and therefore miraculous". And Galileo could believe in anything except miracles.

SALV: [. . .] Likewise it is completely idle to say (as is attributed to one of the ancient mathematicians) that the tides are caused by the conflict arising between the motion of the earth and the motion of the lunar sphere, not only because it is neither obvious nor has it been explained how this must follow, but because its glaring falsity is revealed by the rotation of the earth being not contrary to the motion of the moon, but in the same direction. Thus everything that has been previously

[194] Urban VIII was that self same Cardinal Maffeo Barberini who, a few years earlier, in 1620, had written a poem in Latin in praise of Galileo.

[195] *Galileo's reply to Ingoli*, translated by M.A. Finocchiaro, in *The Galileo Affair: A Documentary History*, University of California Press, p. 178. In Italian: Lettera del 1624 a Francesco Ingoli, Edizione Nazionale, Vol. VI, p. 538.

[196] *Dialogue Concerning the Two Chief World Systems*, translated by S. Drake, The Modern Library, New York, 2001, p. 23. In Italian: *Dialogo sopra i due massimi sistemi del mondo*, Edizione Nazionale, Vol. VII, p. 45.

conjectured by others seems to me completely invalid. But among all the great men who have philosophized about this remarkable effect, I am more astonished at Kepler than at any other. Despite his open and acute mind, and though he has at his fingertips the motions attributed to the earth, he has nevertheless lent his ear and his assent to the moon's dominion over the waters, to occult properties, and to such puerilities. [. . .]

The offense against the Pope

SALV: Now, since it is time to put an end to our discourses, it remains for me to beg you that if later, in going over the things that I have brought out, you should meet with any difficulty or any question not completely resolved, you will excuse my deficiency because of the novelty of the concept and the limitations of my abilities; then because of the magnitude of the subject; and finally because I do not claim and have not claimed from others that assent which I myself do not give to this invention, which may very easily turn out to be a most foolish hallucination and a majestic paradox.

To you, Sagredo, though during my arguments you have shown yourself satisfied with some of my ideas and have approved them highly, I say that I take this to have arisen partly from their novelty rather than from their certainty, and even more from your courteous wish to afford me by your assent that pleasure which one naturally feels at the approbation and praise of what is one's own. And as you have obligated me to you by your urbanity, so Simplicio has pleased me by his ingenuity. Indeed, I have become very fond of him for his constancy in Sustaining so forcibly and so undauntedly the doctrines of his master. And I thank you, Sagredo, for your most courteous motivation, just as I ask pardon of Simplicio if I have offended him sometimes with my too heated and opinionated speech. Be sure that in this I have not been moved by any ulterior purpose, but only by that of giving you every opportunity to introduce lofty thoughts, that I might be the better informed.

SIMPL: You need not make any excuses; they are superfluous, and especially so to me, who, being accustomed to public debates, have heard disputants countless times not merely grow angry and get excited at each other, but even break out into insulting speech and sometimes come very close to blows.

As to the discourses we have held, and especially this last one concerning the reasons for the ebbing and flowing of the ocean, I am really not entirely convinced; but from such feeble ideas of the matter as I have formed, I admit that your thoughts seem to me more ingenious than many others I have heard. I do not therefore consider them true and conclusive; indeed, keeping always before my mind's eye a most solid doctrine that I once heard from a most eminent and learned person,[197] *and before which one must fall silent,*[198] *I know that if asked whether God in His infinite power and wisdom could have conferred upon the watery element its observed reciprocating motion using some other means than moving its containing vessels, both of you would reply that He could have, and that He would have known how to do this in many ways which are unthinkable to our minds. From this I forthwith conclude that, this being so, it would be excessive boldness for anyone to limit and restrict the Divine power and wisdom to some particular fancy of his own.*

SALV: An admirable and angelic doctrine, and well in accord with another one, also Divine,[199] *which, while it grants to us the right to argue about the constitution of the universe (perhaps in order that the working of the human mind shall not be curtailed or made lazy) adds that we cannot discover the work of His hands. Let us, then, exercise these activities permitted to us and ordained by God, that we may recognize and thereby so much the more admire His greatness, however much less fit we may find ourselves to penetrate the profound depths of His infinite wisdom.*

SAGR: [...] I shall be waiting impatiently to hear the elements of our Academician's new science[200] *[...]. Meanwhile, according to our*

[197] Here reference is being made to Pope Urban VIII.

[198] Regarding the words "one must fall silent" it is worth recalling the meeting Galileo had had with Cardinal Maffeo Barberini before he was made Pope, during which he had received advance notice of the "most solid doctrine". According to Cardinal Oregio, who witnessed the meeting, in the face of Barberini's argumentations, *quievit vir ille doctissimus*, the great scientist kept silent. Years later Galileo was even obliged to adopt the doctrine, and he obeyed, using, however, words laced with sarcasm and clear superiority. And this was his undoing.

[199] *also Divine*—also the Pope's.

[200] Sagredo is referring to the *Discourses*, which Galileo had plans for and started writing immediately after his conviction and abjuration.

custom, let us go and enjoy an hour of refreshment in the gondola that awaits us.

Sagredo's words bringing the *Dialogue* to a close give advance notice that it is to be resumed, and return us graciously (as on other occasions) to the placid daily-life of a far-off summertime Venice. But to us, the tranquil scene is subtly disquieting, aware as we are of the dark turn events were about to take.

MATHEMATICAL NOTE

The acceleration at the earth's surface

In order to analyze Galileo's model quantitatively, let us go back to the first diagram of this chapter (left side). Let $r = 6360$ km be the earth's radius and $R = 150{,}000{,}000$ km be the radius of its orbit around the sun, assumed to be circular; taking the origin of the co-ordinates at the sun, the x and y co-ordinates of point P (located at the equator), as a function of time, are:

$$x(t) = -r \cos \omega t - R \cos \Omega t$$
$$y(t) = r \sin \omega t + R \sin \Omega t$$

where $\omega = 7.27 \times 10^{-5}$ rad/s is the angular rotation velocity of the earth (corresponding, for a point on the equator, to a peripheral velocity of 1,665 km/h), and $\Omega = 2 \times 10^{-7}$ rad/s is that of the orbital motion (corresponding to a peripheral velocity of 107,000 km/h, that is, 65 times greater). The cycloid-like trajectory is obtained by plotting y versus x, as was done for the right side of the diagram.

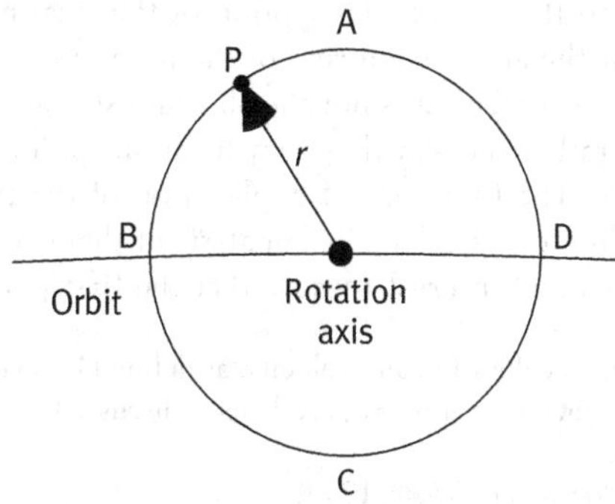

Let us now calculate the acceleration of the generic point P(*x*,*y*), whose components are obtained as the second derivatives of the co-ordinates with respect to time.[201] For the part concerning the rotation, one has:

$$d^2x/dt^2 = \omega^2 r \cos \omega t$$
$$d^2y/dt^2 = -\omega^2 r \sin \omega t$$

which yield, for the acceleration ***a*** expressed in vector form:

$$\boldsymbol{a} = -\omega^2 \boldsymbol{r}$$

where ***r*** is the vector radius of point P, with its origin at the earth's centre. The written expression is just the *centripetal acceleration* associated with terrestrial rotation, for which calculation gives $\omega^2 r = 3.4\ \text{cm/s}^2$ at the equator. This value decreases with increasing latitude and reaches zero at the poles. The effect of this, as discussed in the Mathematical Note of Chapter 4, is a corresponding "lightening" of the waters of the seas. As a consequence, in the ideal case of an earth covered entirely with water, one would expect a rise (constant through time) in sea level at the equator, with respect to the poles—but no other effect whatsoever.

The centripetal acceleration associated with orbital motion, calculated in the same way, is $\Omega^2 R = 0.6\ \text{cm/s}^2$. It is virtually the same for all points on the earth. As already mentioned in the text, the centrifugal force is largely balanced by the force of solar attraction. However, it produces its own tidal effect and this is added to the effect resulting from the earth's rotation, with maxima always at noon and midnight, and minima always at 6.00 and at 18.00 h. This is the "solar" contribution to the tide and below it will be calculated in an approximate way, so as to provide a comparison with the main tidal effect due to the moon.

Calculation of the lunar tide

Let us now turn to the correct description of the tidal mechanism. This takes into account the attractive forces of the moon in a manner which is not entirely obvious. In fact, it is not the absolute strength of these forces that matters, but rather the way they vary from one point to another over the earth's surface. The first correct explanation of the tides was due to Isaac Newton, while the first detailed analysis of them was due to Pierre Laplace.[202] It may be of interest to note that the first person to estimate

[201] Instantaneous acceleration and velocity, as defined in terms of derivatives, together with the concept of integral, are briefly discussed in the Mathematical Note in Chapter 12.

[202] P. Laplace, *Mechanique Célèste*, 1773.

the distribution of the net forces acting upon different points of the earth's surface as a result of the presence of the moon was George H. Darwin, son of the originator of the theory of evolution. He derived the diagram shown below, which clearly implies two "swellings" at points A and C and to two "dips" at points B and D. To give an idea of how this comes about, let us perform the calculation of the tide-generating forces in the particular case of points A and C.

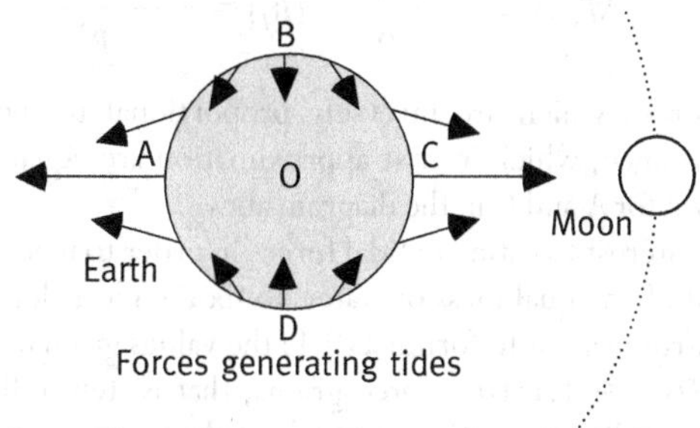

Forces generating tides

Tidal forces have their origin in the fact that, if the earth is considered as a large ensemble of small masses, the lunar attraction each of these experiences differs because each is located at a different distance from the moon. Let us take as a reference the center of the earth, O, where the lunar attractive force and the centrifugal force—resulting from the rotation of the earth–moon system around its center of mass—exactly balance each other out. The tidal forces that act on any particular mass located at the surface of the earth must therefore correspond to the amount by which the gravitational force at that point deviates from the value it assumes in O, that is, positive in C and negative in A. By Newton's law of gravitation, the tidal forces acting on a particle of mass *m*, are given by

$$\text{Point O: } F(\text{O})=\frac{GmM_L}{R^2} \quad \text{Point C:} F(\text{C})=\frac{GmM_L}{(R-R_T)^2} \quad \text{Point A:} F(\text{A})=\frac{GmM_L}{(R+R_T)^2}$$

where G is the universal gravitational constant, M_L is the lunar mass, R is the distance between the centres of the two bodies, and R_T is the earth's radius.[203] Thus, the forces that can produce tides are, respectively:

$$\Delta F(\text{C})=F(\text{C})-F(\text{O}) \quad \Delta F(\text{A}):F(\text{A})-F(\text{O}).$$
(positive) (negative)

[203] $G = 6.67 \cdot 10^{-11}\ \text{Nm}^2/\text{kg}^2$, $M_L = 7.34 \cdot 10^{22}$ kg, $R = 3.84 \cdot 10^8$ m, $R_T = 6.36 \cdot 10^6$ m.

As the earth's radius is very small compared to the earth–moon distance, the above two differences can be approximated in terms of derivatives of F with respect to R. One has

$$\Delta F(\mathrm{C}) \cong \frac{d(GmM_{\mathrm{L}}/R^2)}{dR}(-R_T) = \frac{2GmM_{\mathrm{L}}R_{\mathrm{T}}}{R^3}$$
$$\Delta F(\mathrm{A}) \cong \frac{d(GmM_{\mathrm{L}}/R^2)}{dR}(R_T) = -\frac{2GmM_{\mathrm{L}}R_{\mathrm{T}}}{R^3} \tag{9.1}$$

namely, two forces which are inversely proportional to the cube of the earth–moon distance, which in first approximation are equal and opposite, just as was shown for A and C in the diagram above.

It may be of interest to estimate tidal forces in order to make a comparison with the weight of an equal mass of water. To fix the ideas, let us take a mass equal to 1 g. Introducing into formula (9.1) the values given in footnote 203, one obtains $\Delta F(\mathrm{C}) = 1.1 \cdot 10^{-7}$ force-grams, that is, ten million times less than the corresponding weight! Nevertheless, there are points on the earth where the difference between tidal maxima and minima can be as great as 15 m (e.g. near Fundy Bay in Canada). In Europe, the difference is more than 12.5 m at Mont St Michel in Normandy, but only half a meter in the Mediterranean Sea. This is because the size of tides depends on local morphology. Tides are also small in the open ocean, because they require displacements of huge masses of water over great distances and within times that are very short.

The sun's contribution

Just a few more words on the alteration to tidal effects produced by the gravitational attraction of the sun. If we write formulas (9.1) for the mass of the sun and the sun–earth distance,[204] we see that the force producing tides increases by a factor $27 \cdot 10^6$, due to the ratio of the masses, but decreases by a factor $(390)^3$, due to the ratio of the distances: as a result, the tidal force due to the sun is about 0.46 times that due to the moon. The former combines with the latter to give the variations in maxima and minima which are actually observed from one tide to the next.

More pronounced swelling occurs when the moon is in conjunction with the sun (new moon), or is in opposition to it (full moon), because the tidal forces due to the two bodies add to each other (*sizigial tides*). The swelling is minimal, instead, when the two bodies are at a 90° angle with respect to the earth (*quadrature tides*). It should be noted that Galileo attributed these

[204] Respectively, $M_{\mathrm{S}} = 1.99 \cdot 10^{30}$ kg and $R_{\mathrm{TS}} = 1.49 \cdot 10^{11}$ m.

variations to the moon, though he had deprived it of its principal role in giving rise to the daily oscillations. Nevertheless, he could not avoid perceiving the importance of the position of the moon relative to that of the sun ("The second period is monthly, and seems to originate from the motion of the moon; it does not introduce other movements, but merely alters the magnitude of those already mentioned, with a striking difference according as the moon is full, new, or at quadrature with the sun").

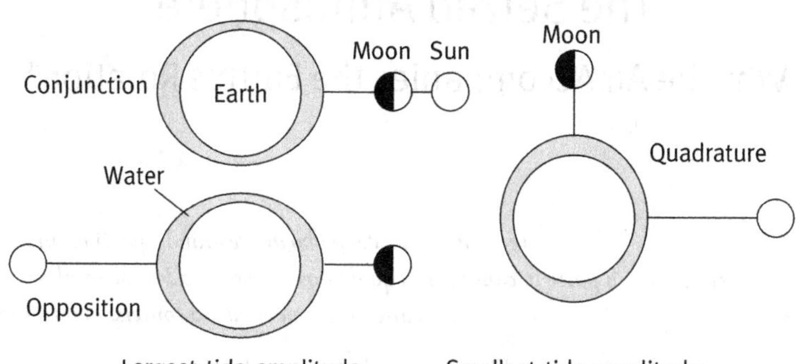

10

The Seized Atmosphere

Why the Air Accompanies the Earth's Rotation*

dove la superficie del globo avesse grandi spazii piani . . .
cesserebbe in parte la causa per la quale l'aria ambiente dovesse totalmente
obbedire al rapimento della conversion terrestre

where the earth's surface has large flat spaces . . .
the reason for the surrounding air to obey entirely the seizure of the
terrestrial rotation would be partly removed

Yet another dialogue on whether or not the earth rotates. In support of the hypothesis that it does, Galileo advances two arguments, one conceptual and one practical: the former holds that it is more *rational* that it should be the earth that moves, as opposed to the unlikely proposition, favored by Aristotelians, that it is all the stars that move collectively; the latter is based on the existence of the trade-winds which, over the vast expanses of the surface of the oceans, blow permanently towards the west, just as might be expected if the atmosphere were not quite keeping pace with the earth's rotation. Galileo's explanation seems simple and straightforward, but, as was the case with his far more elaborate theory concerning tides, it is wrong.

The argument about whether or not the earth rotates was at the center of so many Galilean discourses that it merits revisiting. In Chapter 9, we saw how the feverish search for positive proof that the earth rotates

* From *Dialogue*.

—proof, that is, capable of demonstrating that it actually does rotate, instead of merely not excluding that it rotates (as Aristotelians did)—had so taken hold of Galileo's thinking that it led him to devise a model for tides which, though fascinating, was conceptually mistaken. If, within the corpus of Galileo's work, this model sits at the apex of his failures, the probative passion with which he pursued terrestrial rotation was to cause damage in other respects, too.

The trade-winds and the earth's motion

Against the background of his broad, in-depth analyses of terrestrial phenomena, ones which, according to his relativity principle, would not occur differently whether the earth rotated or was immobile, Galileo believed he had found experimental confirmation that the earth moves in the behavior of the trade-winds. These are light winds which blow more or less constantly in a zone between the equator and a little over 30° of latitude either side of it; from northeast to southwest in the Northern Hemisphere and from southeast to northwest in the Southern Hemisphere. Sagredo says: "I discovered . . . that . . . the voyages from east to west over the Mediterranean were made in proportionately less time than those in the opposite direction, in a ratio of 25 per cent", and he goes on to remark that those who sailed the open seas, especially at low latitudes, experienced even greater differences in east–west/west–east sailing times. For Salviati, this amounted to proof that the earth rotates, carrying the air along with it in a way which is very efficient over dry land, where valleys and mountains cause it to be dragged along (by friction), but a little less efficiently over extensive flat areas, like the oceans.

Please note that the argument that the air is stationary with respect to us inhabitants of the earth was one which Aristotelians advanced most forcibly in demonstration that the planet is immobile. Galileo confuted it again and again. We have already examined several of these confutations in earlier chapters, e.g., in discourses concerning the flight of birds (achieved with equal effort in all directions); cannon balls (which travel in an identical fashion, whether shot towards the east or the west); phenomena aboard ship (where everything occurs just as it would on dry land) etc. Now the first part of the dialogue presented in this chapter offers yet a further argument on the matter. For Galileo, there was no doubt that the air, at least that in the lower atmosphere, fully shares the motion of the earth ("the roughness of the terrestrial surface catches and carries along with it that part of the air which is contiguous to it"). But the temptation was simply too strong: if only we could just feel a whiff of persistent wind blowing from east to west, dreamed Galileo, Aristotelians would be "*belli e serviti*" (i.e. the case against the Aristotelians would be all sewn up). Thus, as had already happened regarding tides, he did not linger long enough on the problem and allowed himself to commit a small infraction in his general picture. To all appearances, trade-winds fit the bill perfectly: as the diagram below shows: they have a component towards the west (i.e. in precisely the right direction), they blow throughout the entire year and, what is more, they are to be found in the central zone of the globe, where the peripheral velocity of the earth is greatest. The mistake, in this case, was certainly more excusable than the mistake in the tide model. As we have said elsewhere (Mathematical Notes in Chapters 4 and 5), a body moving in a rotating system is subjected to the Coriolis force, which makes it deviate. In Galileo's time this force was unknown. It is, in fact, the Coriolis force which is responsible for deviating to the west air moving from the high-pressure subtropical zone towards the low-pressure equatorial zone.

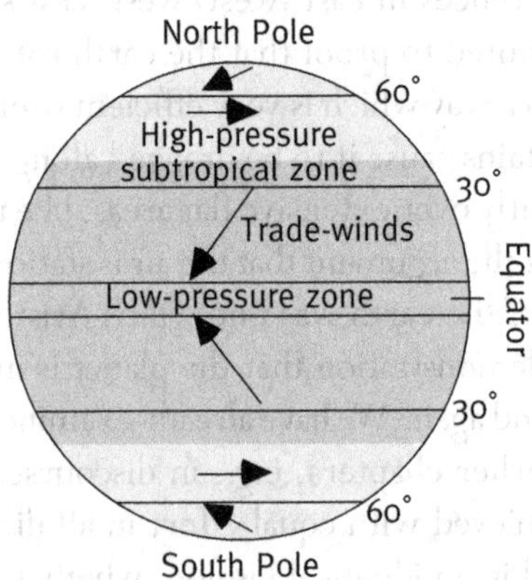

Distribution of constant winds on earth: the dark area is where trade-winds blow

Apart from local effects, such as those linked to the presence of lands or seas, areas of high and low pressure on the globe are determined by differences in temperature. Even without going into details it appears intuitive that at the equator warm air will rise, creating a low pressure zone at sea level. In the Northern Hemisphere winds should therefore blow towards the south. But the Coriolis force deviates them in a southwesterly direction, giving rise to the trade-winds. The average velocity of trade-winds is about 18 km/h, with peaks of 32 km/h over the vast expanses of the oceanic surfaces, where the air meets with fewer obstacles.

To sum up, the oceanic winds Galileo talks about do have an east–west component which is, indeed, caused by the earth's rotation. However, the origin of these winds has nothing to do with the atmosphere "slipping behind" with respect to the surface of the earth. Galileo was right in affirming that the atmosphere above land masses keeps pace with the rotation of the earth, but he was wrong in attributing this behavior to the unevenness of the crust. Later in this chapter we shall give the arguments why the atmosphere travels at the same pace as the terrestrial globe, as if the two were a single body.

The rationality of Nature

So, we begin our reading with yet another discussion about the motion of the earth:[205] this motion, owing to the relativity principle, is imperceptible to anyone who is located on the earth. It can only be revealed by the behavior of the sun, moon, and stars, which do not participate in the motion of the earth. The main argument in this passage concerns the fundamental rationality of the "choices" made by Nature. If a given phenomenon could be accounted for using a simple and "inexpensive" mechanism (the earth rotating about its own axis) then Galileo did not accept that it should be explained using a mechanism that was far more intricate and complex (keeping the

[205] *Dialogue Concerning the Two Chief World Systems*, Second Day, translated by S. Drake, The Modern Library, New York, 2001, p. 132 and following. In Italian: *Dialogo sopra i due massimi sistemi del mondo*, Giornata seconda, Edizione Nazionale, Vol. VII, p. 139 and following.

earth fixed and having the entire firmament go around it synchronously and collectively). *The Copernican system, therefore, was more correct mainly because it was more economical and more rational*. This argument is a great tribute to the rationality of the universe, as if to say that any yearning for the arcane, any search for the mysterious or supernatural, was nothing but inexcusable human weakness.

The reasoning is never abstract but, as usual, is interwoven with observation of the world of experience. Galileo's eye turns to the motion of the stars in "the immense bulk of the starry sphere" with the same admiring curiosity as it did to the motion of the cargo-laden ships departing Venice for Aleppo: all that happens "below deck" is governed by the same laws that rule the universe, and the converse is true, too.

This Galilean man who reads and interprets the "Great Book of Nature", unveiling its rational meanings, was the last and highest product of Renaissance humanism. If the earth was no longer the centre of the universe, man, nonetheless, was; and he appeared all the greater, the smaller, and more lost among galaxies was his observation platform—the earth. From the earth he explored space, nearby and remote, discovering with patient humility, but also with great pride, that using the only instruments he possessed—the senses and reason—he could find his way in it. In the following passages the reader should note the keywords "reasonable", "irrational", "credible", "absurdity", which make explicit what the evaluation and judgement criteria ought to be in the field of knowledge.

In the decades that immediately followed, this rational assuredness was to be replaced by uncertainty and anxiety. Humanity was to end up feeling exiled and neglected on its obscure peripheral planet and was to prefer other comforts to reason, ones which were less demanding and perhaps more rewarding in the short term. But with Galileo humanity lived, even if it was to be for a brief period only, its most exciting season: that of challenge and discovery.

The impossibility of perceiving terrestrial motion

SALV: Then let the beginning of our reflections be the consideration that whatever motion comes to be attributed to the earth must necessarily remain imperceptible to us and as if nonexistent so long as we look only at terrestrial objects; for as inhabitants of the earth, we consequently participate in the same motion. But on the other hand it is indeed just as necessary that it display itself very generally in all other visible bodies and objects which, being separated from the earth, do not take part in this movement. So the true method of investigating whether any motion can be attributed to the earth, and if so what it may be, is to observe and consider whether bodies separated from the earth exhibit some appearance of motion which belongs equally to all. For a motion which is perceived only, for example, in the moon, and which does not affect Venus or Jupiter or the other stars, cannot in any way be the earth's or anything but the moon's.

Now there is one motion which is most general and supreme over all, and it is that by which the sun, moon, and all other planets and fixed stars—in a word, the whole universe, the earth alone excepted—appear to be moved as a unit from east to west in the space of twenty-four hours. This, in so far as first appearances are concerned, may just as logically belong to the earth alone as to the rest of the universe, since the same appearances would prevail as much in the one situation as in the other. [. . .]

[. . .]

SALV: [. . .] I shall set forth, commencing with the most general things, those reasons which seem to favor the earth's motion, so that we may then hear their refutation from Simplicio.

First, let us consider only the immense bulk of the starry sphere in contrast with the smallness of the terrestrial globe, which is contained in the former so many millions of times.[206] *Now if we think of the velocity of motion required to make a complete rotation in a single day and night, I cannot persuade myself that anyone could be found who would think it the more reasonable and credible thing that it was the celestial sphere which did the turning, and the terrestrial globe which remained fixed.*

[. . .]

206 By saying "many million times", Galileo doubtless believed he was providing a majestic image of the "starry sphere". In reality, on the basis of big-bang theory, today the estimated radius of the universe is of the order of 10^{23} km, giving a ratio between its volume and that of the earth greater than 10^{57} (i.e. 1 followed by 57 zeros, or, one thousand billion, billion, billion, billion, billion, billion).

From the top of the cupola

SAGR: If, throughout the whole variety of effects that could exist in nature as dependent upon these motions, all the same consequences followed indifferently to a hairsbreadth from both positions, still my first general impression of them would be this: I should think that anyone who considered it more reasonable for the whole universe to move in order to let the earth remain fixed would be more irrational than one who should climb to the top of your cupola[207] *just to get a view of the city and its environs, and then demand that the whole countryside should revolve around him so that he would not have to take the trouble to turn his head. Doubtless there are many and great advantages to be drawn from the new theory and not from the previous one (which to my mind is comparable with or even surpasses the above in absurdity), making the former more credible than the latter. But perhaps Aristotle, Ptolemy, and Simplicio ought to marshal their advantages against us and set them forth, too, if such there are; otherwise it will be clear to me that there are none and cannot be any.*
SALV: Despite much thinking about it, I have not been able to find any difference, so it seems to me I have found that there can be no difference; hence I think it vain to seek one further. For consider: Motion, in so far as it is and acts as motion, to that extent exists relatively to things that lack it; and among things which all share equally in any motion, it does not act, and is as if it did not exist.[208] *Thus the goods with which a ship is laden leaving Venice, pass by Corfu, by Crete, by Cyprus and go to Aleppo. Venice, Corfu, Crete, etc. stand still and do not move with the ship; but as to the sacks, boxes, and bundles with which the boat is laden and with respect to the ship itself, the motion from Venice to Syria is as nothing, and in no way alters their relation among themselves. This is so because it is common to all of them and all share equally in it. If, from the cargo in the ship, a sack were shifted from a chest one single inch, this alone would be more of a movement for it than the two-thousand-mile journey made by all of them together.*
[. . .]

[207] The *cupola* (dome) of Florence cathedral, of course, Salviati being a Florentine.

[208] This is the Galilean relativity principle, according to which no physical phenomenon occurring inside a given system permits an observer in that same system to tell whether the system is stationary, or is moving with uniform rectilinear motion.

Nature is thrifty

SALV: [. . .] Now if precisely the same effect follows whether the earth is made to move and the rest of the universe stay still, or the earth alone remains fixed while the whole universe shares one motion, who is going to believe that nature (which by general agreement does not act by means of many things when it can do so by means of few) has chosen to make an immense number of extremely large bodies move with inconceivable velocities, to achieve what could have been done by a moderate movement of one single body around its own center?
SIMPL: I do not quite understand how this very great motion is as nothing for the sun, the moon, the other planets, and the innumerable host of the fixed stars. Why do you say it is nothing for the sun to pass from one meridian to the other, rise above this horizon and sink beneath that, causing now the day and now the night; and for the moon, the other planets, and the fixed stars to vary similarly?
SALV: Every one of these variations which you recite to me is nothing except in relation to the earth. To see that this is true, remove the earth; nothing remains in the universe of rising and setting of the sun and moon, nor of horizons and meridians, nor day and night, and in a word from this movement there will never originate any changes in the moon or sun or any stars you please, fixed or moving. All these changes are in relation to the earth, all of them meaning nothing except that the sun shows itself now over China, then to Persia, afterward to Egypt, to Greece, to France, to Spain, to America, etc. And the same holds for the moon and the rest of the heavenly bodies, this effect taking place in exactly the same way if, without embroiling the biggest part of the universe, the terrestrial globe is made to revolve upon itself.

The atmosphere accompanies the earth

And now let us turn to the problem of those winds—the trade-winds—which, in blowing with some constancy towards the west, provided, according to Galileo, proof of the eastward rotation of the earth. Needless to say, incorrigible Simplicio (with touching dedication) says he is sure that, even if the earth were stationary, it could still be that the atmosphere were being dragged around by the moon and the stars, so that Salviati's argument proved nothing. Today we know that the mechanism Galileo suggested does not take place, either in the lower atmosphere, or (please note!) at higher altitudes, that is, well above mountain ridges and thus in absence of obstacles capable of "sweeping" the air along. Before getting into the dialogue, let us try to explain, as simply as possible, why we can safely state that Galileo did not see things the right way.

The fact that the atmosphere strictly follows the rotational motion of the globe is a consequence of the process by which our planet was originally formed (we shall consider this process in a moment). No variations can subsequently have taken place, such as to slow down the motion of the air with respect to the solid earth, because of a fundamental principle of physics, namely, that an isolated system (and the earth system is isolated) cannot change its angular momentum.[209] To take an example, if an airplane, in taking off, were not to carry along with itself the rotational motion it had shared with all other terrestrial objects when it was stationary on the ground, then there would be a variation in the overall angular moment of the earth–air system, and this cannot occur. Of course, this also holds for birds (Galileo dealt with this, as we saw in Chapter 5), for insects—and for every single particle of the atmosphere.

Now let us look at the origins of the atmosphere. The sun and the planets were formed, as an effect of particle aggregation, from a primitive nebula consisting of finely dispersed, incandescent matter. The bodies gradually increased in size as a result of gravitational aggregation (larger aggregates of particles captured smaller ones), until the present dimensions were attained. This aggregation process brought about temperature increases large enough to melt matter

[209] A well-known example of the conservation of angular momentum is provided by the pirouettes of a ballerina, which become faster and faster the more the arms are withdrawn close to the body. It is not the case to go deeply into this problem here: those who wish to know more can consult any general physics textbook or encyclopaedia. We will limit ourselves to saying that the angular momentum for a rotating body is equal to the product of its moment of inertia times its angular velocity. In the example of the ballerina, the withdrawal of the arms implies a reduction in her moment of inertia and a corresponding increase in her angular rotation velocity.

and free imprisoned, or chemically bound, gases (water vapor, for instance). The primitive atmosphere, therefore, had its origin *inside* the system. Even if the composition of this atmosphere was subsequently to undergo modifications, it kept (and keeps) rotating along with the solid part of the earth owing to the conservation of angular momentum. The presence of unevenness on the earth's surface is completely immaterial in respect of its dragging effect.

The highest layers of the atmosphere would not accompany the terrestrial (solid body) rotation only if forces acting from the *outside* were present, for instance friction forces against a hypothetical material "broth" (motionless and different from a vacuum) in which the earth were immersed. The simple diagrams, below, need no explanation (the diagram on the left, viewed upside down, would correspond to the behavior of water-flow in a river, fast at the free surface and almost motionless at the bottom, owing to friction with the river bed). Galileo's guess was that the velocity of air molecules, as dragged along by the earth, corresponds to the situation shown on the left, despite the fact that there is no external material shell capable of slowing down the highest layers of the atmosphere. Therefore his hypothesis is untenable.

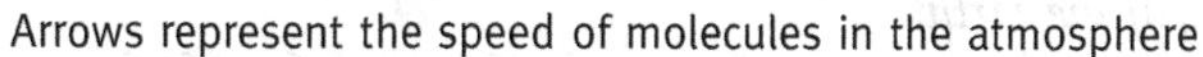

Arrows represent the speed of molecules in the atmosphere

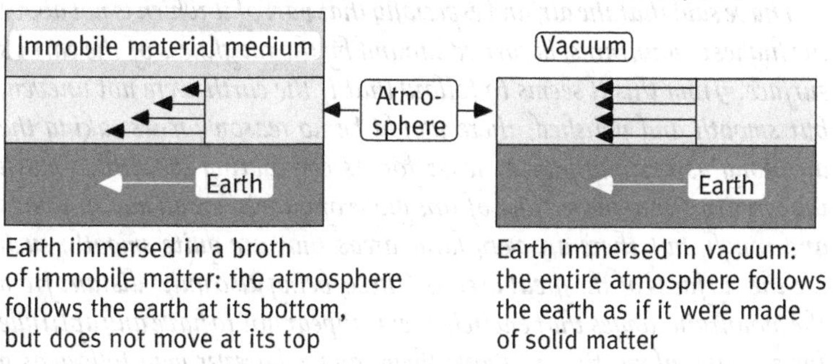

Immobile material medium

Earth immersed in a broth of immobile matter: the atmosphere follows the earth at its bottom, but does not move at its top
Earth immersed in vacuum: the entire atmosphere follows the earth as if it were made of solid matter

The jet stream

Today, experience confirms that the situation, instead, is as shown in the diagram on the right. Ascending to altitudes above the highest mountains—which we can now do thanks to jet planes—we find the velocity of the air remains virtually unchanged, or, if anything, is opposite to that foreseen by Galileo's hypothesis. While at ground level we have the weak effect of the trade-winds, at higher levels the velocity of these winds drops to zero, and then changes sign (*antitrade-winds* at high altitude). The reason is that at high altitudes the difference in atmospheric pressure between the equator and the tropics is necessarily reversed and the air moves in a direction opposite to the trade-winds, that is, in the Northern Hemisphere, from southwest to northeast. This flow of air can, in certain circumstances, become concentrated into a thin, fast stream, called, in fact, the *jet stream*. It is because of this phenomenon that airplanes take less time to go from New York to Rome than vice versa. Galileo, of course, had no knowledge of the behavior of the atmosphere at high altitudes.

The following excerpt is also taken from the *Dialogue*:[210]

SALV: [. . .] What I have said so far seems to me to be an adequate reply to Simplicio's objection. But I want to give him more than satisfaction by means of a new objection and another reply, founded upon a remarkable experiment, and at the same time substantiate for Sagredo the mobility of the earth.

I have said that the air, and especially that part of it which is not above the highest mountains, is carried around by the roughness of the earth's surface. From this it seems to follow that if the earth were not uneven, but smooth and polished, there would be no reason for its taking the air along as company, or at least for its conducting it with so much uniformity. Now the surface of this globe of ours is not all mountainous and rough, but there are very large areas that are quite smooth; such are the surfaces of the great oceans. These, being also quite distant from the mountain ranges that encircle them, appear not to have any aptitude for carrying along the air above them; and whatever may follow as a consequence of not carrying it ought therefore to be felt in such places.

[210] *Dialogue Concerning the Two Chief World Systems*, Fourth Day, translated by S. Drake, The Modern Library, New York, 2001, p. 508 and following. In Italian: *Dialogo sopra i due massimi sistemi del mondo*, Giornata Quarta, National Edition, Vol. VII, p. 464 and following.

SIMPL: I also wanted to raise this same objection, which seems to me very powerful.
SALV: You may well say this, Simplicio, in the sense that from no such thing being felt in the air as would result from this globe of ours going around, you argue its immobility. But what if this thing that you think ought to be felt as a necessary consequence were, as a matter of fact, actually felt? Would you accept this as a sign and a very powerful argument of the mobility of this same globe?

Salviati, we see, has anticipated an objection that Simplicio was about to raise. But as soon as Simplicio realizes that his objection may be employed to confirm the movement of the earth, he becomes concerned not so much with understanding, as with digging deeper defenses for his stubborn Ptolemaic faith. If he is not capable of explaining the causes of the perpetual winds over the seas himself, there will most certainly be someone, he assures his interlocutors, who can explain them, without needing to bring the rotational motion of the earth into the picture.

Simplicio's last two statements at the end of the chapter are particularly interesting. In his penultimate statement he summarizes, with scholastic clarity, the opposing views in two skillful, antithetically structured, sentences. Then in the final statement, already incapable of objecting further, he no longer appeals to generic "others", but to a wider court of philosophers who, in his opinion, would not fail to cast doubt upon Salviati's claims.

Perpetual winds

SIMPL: In that case it would not be a matter of dealing with me alone; for if this should happen and its cause were hidden from me, perhaps it might be known to others.
SALV: So no one can ever win against you, but must always lose; then it would be better not to play. Nevertheless, in order not to cheat our umpire, I shall go on.

We have just said, and will now repeat with some additions, that the air, as a tenuous and fluid body which is not solidly attached to the earth, seems to have no need of obeying the earth's motion, except in so far as the roughness of the terrestrial surface catches and carries along with it that part of the air which is contiguous to it, or does not exceed by any great distance the greatest altitude of the mountains. This portion of the air ought to be least resistant to the earth's rotation, being filled with vapors, fumes, and exhalations, which are materials that participate in the earthy properties and are consequently naturally adapted to these same movements. But where the cause for motion is lacking—that is, where the earth's surface has large flat spaces and where there would be less admixture of earthy vapors—the reason for the surrounding air to obey entirely the seizure of the terrestrial rotation would be partly removed. Hence, while the earth is revolving toward the east, a beating wind blowing from east to west ought to be continually felt in such places, and this blowing should be most perceptible where the earth whirls most rapidly; this would be in the places most distant from the poles and closest to the great circle of the diurnal rotation.

Now the fact is that actual experience strongly confirms this philosophical argument. For within the Torrid Zone (that is, between the tropics), in the open seas, at those parts of them remote from land, just where earthy vapors are absent, a perpetual breeze is felt moving from the east with so constant a tenor that, thanks to this, ships prosper in their voyages to the West Indies. Similarly, departing from the Mexican coast, they plow the waves of the Pacific Ocean with the same ease toward the East Indies, which are east to us but west to them. On the other hand, voyages from the Indies eastward are difficult and uncertain, nor may they in any case be made along the same routes, but must be piloted more toward the land so as to find other occasional and variable winds [. . .].

SAGR: Yet in order to cap all this, I wish also to tell you one particular which seems to me to be unknown to you, yet which confirms this same conclusion. You, Salviati, have mentioned that phenomenon which sailors encounter in the tropics; I mean that constant wind blowing from the east, of which I have heard accounts from those who have made the voyage quite often. Moreover, it is an interesting fact that sailors do not call this a 'wind', but have some other name for it which slips my mind, taken perhaps from its even tenor. When they encounter it, they tie up their shrouds and the other cordage of the sails, and without ever again having any need to touch these, they can continue their voyage in security, or even asleep. Now this perpetual breeze has been known and recognized by reason of its blowing continuously without interruption;

for if other winds had interrupted it, it would not have been recognized as a singular effect different from all the others. From this I may infer that the Mediterranean Sea might also participate in such a phenomenon, but that this escapes unobserved because it is frequently interrupted by other supervening winds. I say this advisedly, and upon very probable theories which occurred to me from what I had occasion to learn during the voyage I made to Syria when I went to Aleppo as consul of our nation. Keeping a special record and account of the days of departure and arrival of ships at the ports of Alexandria, Alexandretta, and here at Venice, I discovered in these again and again that, to my great interest, the returns here (that is, the voyages from east to west over the Mediterranean) were made in proportionately less time than those in the opposite direction, in a ratio of 25 per cent. Thus we see that on the whole the east winds are stronger than those from the west.
SALV: I am glad to know of this detail, which contributes not a little confirmation to the mobility of the earth. [. . .]
SAGR: I, who unlike Simplicio, have not been worrying about convincing anybody besides myself, am satisfied with what has been said regarding this first part. Therefore, Salviati, if you wish to proceed, I am ready to listen.
SALV: I am yours to command; but I should like to hear also how it looks to Simplicio, for from his judgment I can estimate how much I may expect from these arguments of mine in the Peripatetic schools, should they ever reach those ears.
[. . .]
SIMPL: [. . .] Thus, just as you declared that the air surrounding the mountain ranges is carried around by the roughness of the moving earth, we say the converse—that all the element of air is carried around by the motion of the heavens except that part which is lower than the mountain peaks, this being impeded by the roughness of the immovable earth. And where you would say that if such roughness were removed, this would also free the air from being caught, we may say that if this roughness were removed, all the air would proceed in this movement. [. . .]

Simplicio, therefore, seems totally unshaken by Salviati's logic, almost as if he had not been listening to him, and sets off along the usual Aristotelian paths. We will spare the reader this repetition and also ignore Salviati's reply, lucid and exhaustive as ever. The escape route Simplicio chooses when he is left without arguments is the one traditionally adopted by mediocre minds—the appeal to a superior Word.

SIMPL: I have nothing further to say; neither on my own account, because of my lack of inventiveness, nor on that of others, because of the novelty of the opinion. But I do indeed believe that if this were broadcast among the schools, there would be no lack of philosophers who would be able to cast doubt upon it.
SAGR: Then let us wait until that happens. In the meantime, if it is satisfactory with you, Salviati, let us proceed.

Satellite Skylab above earth

11

Stand Still, Ever Moving Sun

The Bible and Science*

nella Scrittura si trovano molte proposizioni le quali,
quanto al nudo senso delle parole, hanno aspetto diverso dal vero,
ma son poste in cotal guisa per accomodarsi all'incapacità del vulgo

in the Scriptures are found many propositions which,
as to the bare senses of the words, have an appearance
different from truth, but were so put to accommodate
the incapacity of the common people

The astronomical and epochal revolution of the heliocentric system. In three letters—the best-known being the one addressed to his pupil Father Benedetto Castelli—Galileo, prompted by objections that had been raised by the ecclesiastical authorities, proposes an honorable compromise which would permit the coexistence of the Copernican theory and the words of the Bible.

Three letters which Galileo wrote between 1613 and 1615 are known collectively as the "Copernican letters". Galileo addressed these to various individuals—Father Benedetto Castelli, *Monsignor* Piero Dini, Grand Duchess Christina of Lorraine—but his intention was that they should receive wide circulation. Common to all three is the very sensitive theme of the relationship between science and faith; in particular, the thesis that there is no disagreement, apart

* From *Copernican letters.*

from superficially, between Biblical text and Copernican theory. Galileo had been convinced of the validity of the heliocentric hypothesis for almost twenty years, as we know from a letter to Kepler dated 1597. But he had become an active supporter of it only after he had made the astronomical discoveries described in the *Sidereus Nuncius*. The Copernican letters represented an important stage in the cultural campaign to which he devoted himself with extraordinary vigor (to the extent that he even diverted time and energy away from his scientific research) following his return to Florence in 1610 (see the Historical Note at the end of this chapter).

Galileo and the Holy Scriptures

In a letter to Galileo, Benedetto Castelli[211] described the arguments he had used in a reply to the Grand Duchess Christina of Lorraine, mother of the Grand Duke of Tuscany. Christina had asked him how it was possible to reconcile the theory that it is the earth that moves about the sun, and not vice versa, with the passage in the Bible in which Joshua commands the sun to stop: "Stand still, you sun, in Gabaon and you, moon, over the valley of Aialon. The sun halted and the moon stood still until the people took vengeance on their enemies" (Joshua, X, 12–13).

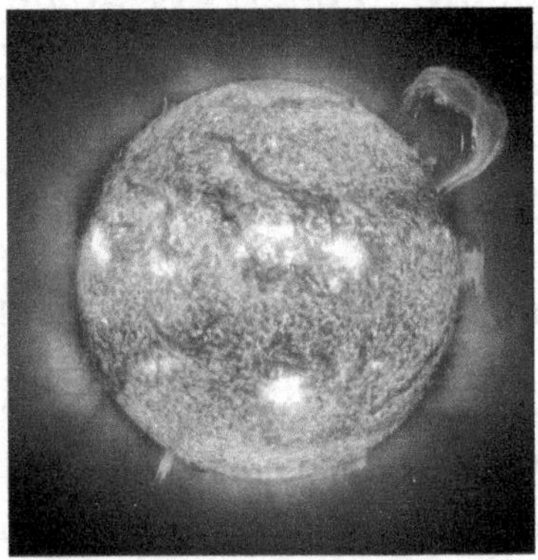

Surface of the sun with large eruptive prominence

[211] Benedictine Father Benedetto Castelli from Brescia (1578–1634) was one of the most important and devoted of Galileo's disciples. From 1626 he held the

In his letter replying to Castelli, Galileo basically says that there is only one absolute Truth and that because the laws of Nature and the Bible both come from God, they cannot be in contradiction. The Holy Scriptures do not lie and do not err, but interpreters of them do err when they keep to the literal meaning of the text, which contains statements that may be far-removed from the truth, owing to their having been dictated by the Holy Ghost for the spiritual salvation of primitive and ignorant men.

Galileo did not go as far as to declare, as Giordano Bruno had done, that the Bible was a book of fairy tales. But he did observe that the Bible had renounced educational aims in order to persuade men of "propositions concerning salvation" of the soul. In so doing it had fallen into (and these were heavy words indeed) not only "various contradictions", but also "grave heresies and blasphemies". For such, said Galileo, is what attributing to God qualities contrary to his essence amounts to (and one cannot avoid doing this if one sticks to the literal meaning of words). This being the case, Galileo suggested, and seeing that the interpreters of the Bible went to the trouble of pointing to hidden meanings in order to free it from evident contradictions and blasphemies when it came to the way God was represented, why should they not do the same regarding the few statements in the Bible concerning the motion of the earth and the sun?

It is abundantly clear that Galileo did not intend to enter into questions *de Fide*,[212] owing, first, to the need for caution, second, to his fundamental lack of interest in problems of a metaphysical nature, and, third (and most importantly), to the imperative he felt, as a scientist working in Catholic, counter-reformist Italy of the early 1600s, to rely on the Church for disseminating his ideas. And his desire that they should be spread was strong, because, finding himself discovering truths that were great and simple, but had nevertheless been long-ignored, he felt that he was the 'keeper of the flame'. Following the approach of humanistic philology he confined himself to locating the Bible in history, so as to be able to justify what he

Chair of Mathematics at the University "La Sapienza" in Rome. Author of the book *Della natura delle acque correnti* (*About the Nature of Flowing Waters*), he is credited with having been the founder of a new science, hydraulics.

212 *de Fide*—Latin for "concerning faith".

wanted to assert. This was that in scientific discussions the Holy Scriptures 'should be reserved to the last place', and that 'physical effects placed before our eyes by sensible experience, or concluded by necessary demonstrations, should not in any circumstances be called into doubt by passages in Scripture that verbally have a different semblance'. However, in asserting this, he effectively stripped the Scriptures of all sacredness, ranking them alongside any other book that could be examined and presented for discussion. He thus set foot upon extremely dangerous ground.

Here, Galileo seems like someone advancing boldly across a minefield they intend to clear, relying on just a transparent shield for protection. He could see the problem lucidly, both in respect of its theological implications and of the perils it presented—he certainly did not underestimate it. He was fully aware of the rigid positions taken by the Council of Trent concerning the interpretation of the Scriptures. He knew what Bruno had suffered as a result of his belief that, so long as he kept within the "terms of Nature",[213] theologians would not have been able to accuse him of anything, since theology deals with matters other than Nature. However, what Galileo did not foresee was the obtuse obstinacy the Church would display when it came to defending the literal meaning of the passages in question. He loved a challenge more than anything, as we well know, and as a consequence he made a mistake. A mistake he kept making stubbornly throughout his life, and which is one of his major legacies: he trusted in the power of reason to persuade. But he was to be defeated.

The text

Let us now read this most famous passage carefully. The reader, by now aware of the relentlessly logical way Galileo built his arguments, cannot fail to be struck here by the way he takes as given a series of statements about the Scriptures (e.g. their dictation by the Holy

[213] Giordano Bruno, *De la causa, principio e uno*, Sansoni, Florence, 1942, p. 92. In 1600, Bruno was condemned by the Inquisition to be burned at the stake.

Ghost). If he had applied his method of scientific analysis to them, they would certainly not have merited such an acceptance.

Note also how the tone of the discourse, which at the outset is colloquial and calm, gradually becomes more and more passionate ("And who wants to set bounds to the human mind? Who wants to assert that everything is known that can be known to the world?"), until it rises to biting sarcasm ("Hence, apart from articles concerning salvation and the establishment of the Faith, against the solidity of which there is no danger that anyone may ever raise a more valid and efficacious doctrine, it would be the best counsel never to add more . . . at the request of persons who, beside the fact that we do not know whether they speak inspired by divine power, are clearly seen to be completely devoid of the information that would be required . . ."). The progression here is typical of Galileo; cautious respect, propriety, and dialectic ability are unable to curb the heat of passion.

Extended excerpts from the letter to Castelli now follow.[214]

Heresies and blasphemies in the Bible

[. . .] As to the first general question of Madame Christina, it seems to me that it was most prudently propounded to you by her and conceded and established by you, that Holy Scripture could never lie or err, but that its decrees are of absolute and inviolable truth. I should only have added that although Scripture can indeed not err, nevertheless some of its interpreters and expositors may sometimes err in various ways, one of which may be very serious and quite frequent, [that is,] when they would base themselves always on the literal meaning of words. For in that way there would appear to be [in the Bible] not only various contradictions, but even grave heresies and blasphemies, since [literally] it would be necessary to give to God feet and hands and eyes, and no less corporeal and human feelings, like wrath, regret, and hatred, or sometimes

[214] Letter of 21 December, 1613 to Benedetto Castelli, in S. Drake, *Letter to Castelli*, in *Galileo at Work: His Scientific Biography*, The University of Chicago Press, Chicago and London 1978, p. 224. In Italian: Edizione Nazionale, Vol. V, p. 281 and following.

even forgetfulness of things gone by and ignorance of the future. Hence, just as in the Scriptures are found many propositions which, as to the bare senses of the words, have an appearance different from truth, but were so put to accommodate the incapacity of the common people, so, for those few who deserve to be separated from the herd, it is necessary that wise expositors should produce the true senses and give particular reasons why they were offered in those words.

Scripture being therefore in many places not only accessible to, but necessarily requiring, expositions differing from the apparent meaning of the words, it seems to me that in physical disputes it should be reserved to the last place, [such questions] proceeding equally from the divine word of the Holy Scripture and from Nature, the former as dictated by the Holy Ghost and the latter as the observant executrix of God's orders. It was moreover necessary in Scripture, in order that it be accommodated to the general understanding, to say things quite diverse, in appearance and by the [literal] meaning of the words, from absolute truth; yet on the other hand, Nature being inexorable and immutable and caring nothing whether her hidden reasons and modes of operating are or are not revealed to the capacities of men,[215] *she never transgresses the bounds of the laws imposed on her. Hence it appears that physical effects placed before our eyes by sensible experience,*[216] *or concluded by necessary demonstrations,*[217] *should not in any circumstances be called in doubt by passages in Scripture that verbally have a different semblance, since not everything in Scripture is linked to such severe obligations as is every physical effect. Rather, if in this respect alone (in order to be accommodated to the capacity of rough and undisciplined people), Scripture has not abstained from adumbrating its principal doctrines by attributing to God himself attributes very far from and contrary to his essence, who is then going to sustain rigidly, ignoring the above consideration, that in speaking incidentally of the earth or the sun*

[215] This concept was much admired by Giacomo Leopardi (1798–1837), who recalled it, for example, in one of his *Operette Morali, Dialogo della natura e di un islandese*. In this treatise, Nature is represented as a huge statue, inexorable, immutable, and uncaring; an image of "sculptured evidence" ("*scolpitezza evidente*"), which was also precisely how the poet described Galileo's writing style.

[216] *sensible experience*—evidence based on the senses, that is, experimental.

[217] *necessary*—that is, rigorous.

or some other created thing, Scripture must be contained rigorously within the limits and restraints of the meanings of words? Especially when it pronounces on such matters things that are far from the primary purpose of Holy Writ, and even such that when said and received as the bare and revealed truth, they would rather injure that primary purpose by rendering the common people contumacious to persuasion by [its] propositions concerning salvation.

Invitation to flexibility

Hence the two exhortations to caution which Galileo addressed to the "wise expositors" of the Scriptures: first, to make hidden meanings clear, as they cannot be in disagreement with the findings of scientific research; second, not to assign an absolute and fixed value to any statement in the Bible regarding natural sciences, because some day, with the advance of knowledge, it might be disproved. Galileo trusted that his advice would be accepted, certain as he was that "we must not fear assaults from anyone, as long as we are given the possibility to speak and to be listened to by intelligent people who are not too biased by personal passions and interests". In Galileo's words:

That being the case, and it being moreover manifest that two truths can never contradict each other, it is the office of wise expositors to work to find the true senses of passages in the Bible that accord with those physical conclusions of which we have first become sure and certain by manifest sense or necessary demonstrations. Indeed, as I have said, granted that Scripture, though dictated by the Holy Ghost, admits for the above reasons expositions that are in many places far from the literal sound [of its words], and moreover as we are unable to assert with certainty that all interpreters speak with divine inspiration, I should think it would be prudent if no one were permitted to oblige Scripture and compel it in a certain way to sustain as true some physical conclusions of which sense and demonstrative and necessary reasons may show the contrary. And who wants to set bounds to the human mind? Who wants to assert that everything is known that can be known to the world? Hence, apart from articles concerning salvation and the establishment of the Faith, against the solidity of which there is no danger that anyone may ever raise a more valid and efficacious doctrine, it would be the best counsel never to add more [articles of faith] without necessity. And if that is so, how much greater disorder [it would be] to add things at the request of persons who, beside the fact that we do not know whether they speak inspired by divine power, are clearly seen to be completely devoid of the information that would be required—I will not say to disprove, but—to understand the demonstrations with which the most acute sciences proceed in confirming some of their conclusions?

I would believe that the authority of Holy Writ had only the aim of persuading men of those articles and propositions which, being necessary for our salvation and overriding all human reason, could not be made credible by any other science, or by other means than the mouth of the Holy Ghost itself. But I do not think it is necessary to believe that the same God who has given us our senses, reason, and intelligence wished us to abandon their use, giving us by some other means the information that we could gain through them—and especially in matters of which only a minimal part, and in partial conclusions, is to be read in Scripture, for such is astronomy, of which there is [in the Bible] so small a part that not even the planets are named. If the original sacred authors had intended to persuade people of the arrangements and movements of the heavenly bodies, they would not have dealt with this so sparingly that it is as nothing in comparison with the infinitely many lofty and admirable conclusions contained within that science.

So you see how disorderly, if I am not mistaken, they would proceed in physical disputes not directly pertaining to faith, by taking at face value passages in Scripture often poorly understood by them. [. . .]

Trust in reason

Galileo, we see, believed that reason, the utilization of which was the basis of scientific research, could also serve to settle eventual conflicts between the new doctrines and the Holy Scriptures. He was encouraged in this belief by the opinion that had been expressed a short time earlier by Cardinal Conti (whom he had consulted) that no irremediable conflict existed between the Bible and his views. But to us, in the light of what was already happening inside the Church and would soon come to maturation, his confidence seems almost naive. All the more so, after all, as it was precisely the increasingly widespread use of reason that was alarming the Church. Had reason remained confined to the narrow world of scientific knowledge, it would not have posed a threat.

But Galileo, as we know, was seeking the broadest possible diffusion for his ideas and had an extensive programme of a cultural kind in mind. So the defenders of the established order could not fail to detect the threat posed to the Church and its dogmas by the rational method of discussing and doubting that Galileo, in full counter-reformist spirit, was having the audacity to apply to the Holy Scriptures. Galileo's approach was indeed a threat, irrespective of his personal religious beliefs. It is completely irrelevant to discuss whether he was, or whether he was not, a good Catholic (even if the evidence provided by his work and his lack of interest in anything that did not fall within the domain of tangible experience, but belonged to a world of paper or to verbose discussions, would lead us to opt for the latter hypothesis). What matters here is certainly not the personal credo of a scientist, albeit a famous one. It was his intellectual independence that represented a threat to a millennia-old edifice founded on dogma.

First offensives

The initial attack on Galileo preceded his letter to Castelli by a few days. It came from a low level and it was launched in such a clumsy way it did not even appear dangerous. Dominican Father Nicolò Lorini—whom history consigns to us solely because of his rage against Galileo—was the first to attack, during the course of a sermon in a church in Florence. He then wrote to Galileo to apologize and to explain that he considered only the opinions of Ippernico (sic!) to be in contradiction with the Holy Scriptures. But on December 20 1614 another Dominican, Tommaso Caccini, accused Galileo more openly in the church of Santa Maria Novella in Florence. Finally, in February 1615, Lorini attacked again, sending to the Holy Office a copy of the letter to Castelli, which in his opinion contained many ideas that were "either suspect or rash". The secret inquiry that was immediately initiated seemed, nonetheless, to reach a conclusion quickly and in Galileo's favour (thanks, among other things, to the high-ranking friends he could count on in the ecclesiastical hierarchy). Yet it was a disquieting signal when—by way of a letter from his friend *Monsignor* Giovanni Ciampoli—Cardinal Maffeo Barberini had an invitation communicated to him not to go beyond the boundaries of physics and mathematics, and not to provoke theologians. In addition, Ciampoli himself advised Galileo to keep in mind that "the salutary clause of submission to the Holy Mother Church etc., can never be repeated often enough . . .".[218]

However this was not the end of it. Father Caccini appeared on the scene again to denounce Galileo personally, together with some of his pupils ("*Galileisti*"), as being the instigators of dangerous opinions—such as one casting doubt upon miracles, in addition to the well-known one concerning the motion of the earth. Caccini also insinuated that Galileo's "orthodoxy" was suspect, because of his friendship with Paolo Sarpi[219] and his correspondence with some Germans[220] (that is to say, heretics). Even so, this new attack, too, appeared to come to nothing, as witnesses gave the lie to Caccini's claims (with the result that he also felt it expedient to visit Galileo and to excuse himself for the plot against him). Nevertheless, the Holy Office, meeting in November 1615, decided that Galileo's text on solar spots should be examined in order to obtain a precise idea of his standpoints.

Meanwhile, Galileo repeated his advice that caution be used when interpreting the Holy Scriptures, in a letter to *Monsignor* Piero Dini, one of his closest friends, who at the time was the Apostolic

[218] Edizione Nazionale, Vol. XII, p. 145.

[219] Paolo Sarpi (1552–1623), friar of the order of the Servites, author of *Istoria del Concilio Tridentino (History of the Council of Trent)*, who was also a brilliant mathematician and a scientific interlocutor and friend of Galileo. He wrote thus about the Jesuits: "I don't believe that there ever existed a genus of men which was more the sworn enemy of goodness and truth". In the jurisdictional dispute between Pope Paul V and Venice, Sarpi sided with the Republic, and for this he suffered excommunication and a murder attempt.

[220] In the first place, Kepler.

Referendary in Rome. In this letter,[221] Galileo warned: "how dangerous a thing it would be to state as an established truth any proposition in the Holy Scriptures that in the future might be demonstrated as false". In this same letter, moreover, Galileo attributed the banning of the Copernican ideas—ideas he supported—to the prejudice and the personal animosity which certain men in the Church harboured in his regard. Finally, with his usual confidence in the persuasive power of his arguments, he asked Dini to read the letter he had sent Castelli to "Jesuit Father Grienberger,[222] a distinguished mathematician and a very good friend and patron of mine", expressing the hope that, through him, it would arrive into the hands of Cardinal Bellarmine.[223]

In fact, Galileo was not underestimating the danger. Even if he was not fully aware of all the machinations that were going on against him, he perceived that they represented a very serious threat, not so much to his person, as to the ideas he supported. However, he was confident as usual in the evidence of reason and in his own dialectical abilities—"I would not despair also to overcome this difficulty, were I in a place where I could avail myself of my tongue instead of my pen"[224]—and early in December 1615 he set out for Rome.[225]

Trusting the Jesuits

The postscript to the letter sent to Dini in February 1615 is of particular interest. In it Galileo expresses his conviction that, in the face of "the enormity of my disgrace, when it is added to the malice and ignorance of my opponents, [. . .] the most immediate remedy

[221] Letter of December 16, 1615 to Piero Dini, Edizione Nazionale, Vol. V, p. 291 and following.

[222] Christopher Grienberger, mathematics teacher at the Roman College.

[223] Cardinal Robert Bellarmine was the top theologist of the Jesuit Order and an interpreter of the Counter-reformist spirit. An Inquisitor in the prosecution of Giordano Bruno, he was made a saint in 1930.

[224] Letter to Piero Dini of May 15, 1615, Edizione Nazionale, Vol. XII, p. 134.

[225] For a description of the complex situation in Rome which Galileo had to negotiate, and also for the minutes of the testimonies against him by Caccini and others before the Congregation of the Holy Office, see: G. de Santillana, *The Crime of Galileo*, University of Chicago Press, Chicago, 1955, in particular Chapters 3 and 6.

would be to approach the Jesuit Fathers, as those whose knowledge is much above the common education of friars". Thus it appears, at this stage, that he believed the clumsy attacks against him by the Dominicans could easily be repelled with the help of the much more cultured Jesuit Fathers.

This optimism, as we know, proved to be short-lived, as the Jesuits were soon to line up alongside the established authority. Many years later, the above-mentioned Father Grienberger would express his opinion on Galileo in these words:[226] "If Galileo had known how to keep the affection of the Fathers of this College, he would now live gloriously in this world and none of his bad times would have come to pass and he would have been able to write as he wished about everything, even, I say, about the motion of the earth, etc".

Here now is an excerpt from the letter to Dini. Once again we find a typical *crescendo*, as the discourse builds up to bitter irony in the final lines.

Letter to Dini

[. . .] these persecutors of mine make efforts to prohibit a book that has long been accepted by the Holy Church,[227] *a book they have not even seen, let alone read or understood. Whereas I keep asking loudly that its doctrine be examined and its reasons be pondered upon by very Catholic and competent people; that its positions be confirmed by sensible experiences; and finally that the book be not condemned before it be found false, if it is the case that a proposition cannot be at the same time both truthful and erroneous. Among Christians there are quite a few experts in this matter, and their opinion about the truth or fallacy of the doctrine should not be placed behind the judgement of people who are not informed, and who even too openly are known to be affected by*

[226] Reported by Galileo in a letter to Elia Diodati dated July 25, 1634, Edizione Nazionale, Vol. XVI, p. 115. Galileo himself makes this comment: "thus Your Lordship can see that it is not this or that opinion which has caused and does cause war for me, but the fact that I am in the disfavor of the Jesuits".

[227] The book referred to is, of course, Copernicus'. *De Revolutionibus Orbium Coelestium*, published in Nuremberg in 1543, the year of the astronomer's death.

partisan feelings. [. . .]. Now these good Fathers, solely out of malicious feelings against me, since they know I think highly of this author,[228] *are proud to reward him for his efforts by having him declared a heretic.*

But what is most worth considering is that their first move against this conception was to allow themselves to be convinced by some of my enemies, who described it as a work of mine, without saying that it had been in print for seventy years. And this same conduct they had with other people, whom they led to form a bad opinion of me. And this to such an extent that, only a few days after I arrived here, Monsignor Gherardini, the Bishop of Fiesole, in one of his first public appearances (at a place where some friends of mine happened to be present) made an outburst of great vehemence against me and, visibly angry, said that he was going to take up the case with Their G.D. Highnesses, because my extravagant and false conception was reason for concern in Rome. And by now he may have already done this, unless he has been halted, having been made aware that the author of this doctrine is not a living Florentine, but a dead German, who printed it already 70 years ago, dedicating the book to the Supreme Pontiff. [. . .]

Cardinal Bellarmine's opinion

In 1615, the same year that he sent the letter to Dini, Galileo also sent a letter, or rather a small treatise of about fifty pages, to Christina of Lorraine.[229] In this, many of the arguments already expounded to Castelli were taken up again, at times word for word. But they were now supplemented by learned quotations from the Fathers of the Church, drawn, in the main, from Augustine. These had been supplied to Galileo by a Barnabite Father who had been consulted for the purpose by his friend Castelli. While he was working on the final draft of the letter to Christina, Galileo learned of a newly published writing by Carmelite Father Foscarini, in which the opinion was

[228] *this author*—that is, Copernicus.

[229] S. Drake, *Letter to the Grand Duchess Christina*, in *Discoveries and Opinions of Galileo*, Doubleday & Co., New York, 1957, p. 175. In Italian: Lettera del 1615 a Cristina di Lorena, Edizione Nazionale, Vol. V, p. 309 and following.

advanced that the motion of the earth was more in accordance with the Scriptures than its immobility. However, the encouragement he drew from this was soon to be dashed by Cardinal Bellarmine, who replied to Foscarini,[230] who had asked him for an opinion on his work, as follows:

First I say that it seems to me that your Paternity and Mr. Galileo are proceeding prudently by limiting yourselves to speaking suppositionally and not absolutely, as I have always believed that Copernicus spoke. For there is no danger in saying that, by assuming the earth moves and the sun stands still, one saves all of the appearances . . . and that is sufficient for the mathematician. However, it is different to want to affirm that in reality the sun is at the center of the world and only turns on itself, without moving from east to west, and the earth is in the third heaven and revolves with great speed around the sun; this is a very dangerous thing, likely not only to irritate all scholastic philosophers and theologians, but also to harm the Holy Faith by rendering Holy Scripture false . . .

Second, I say that, as you know, the Council [of Trent] prohibits interpreting Scripture against the common consensus of the Holy Fathers; and if Your Paternity wants to read not only the Holy Fathers, but also the modern commentaries on Genesis, the Psalms, Ecclesiastes, and Joshua, you will find all agreeing in the literal interpretation that the sun is in heaven and turns around the earth with great speed . . .

Third, I say that if there were a true demonstration that the sun is at the center of the world and the earth in the third heaven, and that the sun does not circle the earth but the earth circles the sun, then one would have to proceed with great care in explaining the Scriptures that appear contrary; and say rather that we do not understand them . . . I add that the one who wrote, "The sun also riseth, and the sun goeth down, and hasteth to his place where he arose", was Solomon, who not only spoke inspired by God, but was a man above all others wise and learned in the human sciences and in the knowledge of created things; he received all this wisdom from God; therefore it is not likely that he was affirming something that was contrary to truth already demonstrated or capable of being demonstrated. . . .

[230] M.A. Finocchiaro, *The Galileo Affair: A Documentary History*, University of California Press, 1989, p. 67 and following. In Italian: Edizione Nazionale, Vol. XII, p. 171.

Thus, according to Bellarmine, Galileo was free to put forward hypotheses that enabled him to make calculations and predictions more easily, but not to claim that these hypotheses could challenge the truths of philosophers and theologians. Bellarmine's words—even though he did concede that he might return to the argument should a decisive proof of the heliocentric thesis be provided—are a classic example of statements that are incontrovertible in that they are not demonstrable, either one way or the other. As an eminent scientist Galileo could not stoop to a level of language and debate such as this.[231] Accordingly, as was his temperament, he did not allow himself to be dissuaded from continuing his advance across territory that was by now clearly treacherous. He finished writing his letter to Christina. In it he persisted in his thesis that there could be no conflict between science and the Scriptures, and that it was in the interests of the Church not to put its reputation at stake by hastily condemning hypotheses and opinions that were scientific in nature and could, in the not-too-distant future, become universally recognized as true.[232]

[231] An alternative strategy to be adopted in such circumstances is illustrated by the following story. During a TV debate, a famous ufologist, renowned for his ability to demolish scientific experts in the eyes of an audience, asked a physicist, Luciano Pietronero: "Can you rule out, professor, that what I have just described, actually occurred?". The physicist replied: "I have a theory explaining how cars move: they are pushed by ten little red devils who have two properties: first, they make cars go; second, they disappear when we look inside the bonnet". The ufologist was suddenly in trouble, because Pietronero had found the appropriate recipe: countering by using the same key, with just the right amount of exaggeration to bring out the absurdity. "Nice joke, professor", to which Pietronero retorted: "It is no joke, I would like you to show me what is wrong with my model".

[232] Centuries later we find Galileo's warning is still topical, for instance when observing the Roman Catholic Church's positions on Darwin's theory of evolution (evolution could not have led to self-consciousness, it had to come from above), as well as on the demographic and environmental sciences. Although, during the recent 'rehabilitation' of Galileo promoted by the Roman Catholic Church, Pope John Paul II said (with praiseworthy intent) 'never again the same mistake made with Galileo', the most frequent mistake the Church has made over the centuries has been to try to impose out-of-date conceptions of life and knowledge, such as to be obliged, sooner or later, to retract them.

Over to Saint Augustine

The excerpts from Augustine which the Barnabite Father picked out seem to offer formidable support for the theses discussed above, so extraordinary is their consonance with Galileo's standpoints. Here are a few of them:

- "It has to be said that our authors knew about the true nature of the sky, but the spirit of God that spoke through them did not want men to be taught about these things that would have been of no help to their salvation".
- "Also concerning the motion of the sky, some brothers ask whether the sky is still or moves: . . . to them I answer that these things have been examined with very subtle and painstaking arguments, in order to understand if it is, or it is not, so; I do not have the time to know and deal with such arguments and those whom we wish to teach about their own salvation and for the necessary benefit of the Holy Church must also not have the time".
- "If the authority of the Scriptures is put in opposition to a sure and clear proof, whoever does this does not understand; they are not opposing to the truth the meaning of the Scriptures, which they are unable to penetrate, but rather their own . . .".

In short, sums up Galileo, quoting a phrase he had heard pronounced by Cardinal Baronio: "the intention of the Holy Ghost is that of teaching us how to go to heaven, not how heaven goes".

New arguments of great interest appear in the letter to Christina. Note, for instance, in the passage which we are now going to read, the conciliatory tone Galileo uses. He acknowledges that it is possible there may be errors on his part, given that the subject is not his own, and he declares that, not wishing "to engage in disputes about this matter with anybody", he is prepared to turn a blind eye to erroneous statements made by others.

Letter to the Grand Duchess Christina of Lorraine[233]

[. . .] I not only agree to freely remove any errors concerning matters of religion which I may incur in this letter, owing to my ignorance, but I also declare that it is not my wish to engage in disputes about these matters with anybody, even if some points may be disputable. In fact, my purpose is this alone: that if, among the errors that might arise in these considerations which are remote from my own profession, there may be something that can help advise the Holy Church in taking a decision about the Copernican system, I wish it to be taken and used as the superiors see fit; otherwise, let my letter be torn and burnt because I do not intend or pretend to acquire any gain from it which is not pious and Catholic. And I add: even though many of the things I quote I heard with my own ears, I will freely admit and concede that those who said them never uttered them, if this is what they desire, confessing that I have misunderstood. Therefore my answers are not meant for them, but for whoever may hold such opinions. [. . .]

What Galileo cannot tolerate is that his opponents, invoking the authority of the Scriptures, should not trouble themselves to provide arguments that demonstrate as invalid those theses they judge to be erroneous; and that they should even go so far as to demand that scientists, who by virtue of experiment and reasoning have reached conclusions of which they are sure, but that are in contradiction to those of theologians, "should take their own proofs apart and discover the fallacies in their own experiments".

[. . .] however, I cannot deny that I am left with some misgiving—which I would like to see removed—when I hear that they expect to be able to compel others, under the authority of the Scriptures, to follow in physical disputes the opinion that appears to them closest to the Bible, while considering themselves not obliged to disprove reasons or experiences which are to the contrary. To justify and support this opinion, they say that theology is the queen of all the sciences, and hence she must,

[233] The translation of these excerpts is original.

under no circumstances stoop to adapt herself to the principles of sciences which are less worthy and subordinate to her. It is the other sciences that must make reference to her, as to a supreme empress, and must change and alter their conclusions according to theological statutes and decrees. Furthermore they add that whenever an inferior science reaches a conclusion which is sure on the basis of demonstrations or experiences, and to which the Scriptures present a contrasting conclusion, then it ought to be the self-same professors of that science who should take their proofs apart and discover the fallacies of their own experiments, without appealing to theologians and exegetes. For, as said above, it is not appropriate for the dignity of theology to abase herself in searching for fallacies in the inferior sciences, it being sufficient for her to decide upon the truth of a given conclusion with absolute authority and in the certainty that it is impossible for her to err. [. . .]

Each to his trade

Galileo's closing words here are an unwitting prediction of the abjuration he was going to be forced to make just a few years later.

In the next passage, Galileo's basic point is "each to his trade". The theologians, if they have no regard for "the lower and humbler speculations of the inferior sciences", should not, acting like absolute despots, arrogate to themselves the authority to make judgements about them. It is quite evident, however, from the impassioned words which follow, that Galileo, when he says that to command men of science not to see what they see and not to understand what they understand is to command "something that is utterly impossible", was well aware that he would be read by people whom—whether out of ignorance, or dishonesty—it would not be possible to persuade, since they held that the intellect could be silenced through imagination or will.

This was certainly an explosive issue and Galileo, as usual, took precautions by stating that he was referring only to scientific, not to religious, themes. Increasingly evident in the final lines of the excerpt is his proud self-awareness that he belongs to that group of people who cultivate the demonstrative sciences, rather than practical or commercial affairs. Moreover, his contempt is disclosed for those who, deeming that truth can be concealed or manipulated, command others to behave accordingly—thus violating even the most elementary of ethical principles, whether professional or otherwise.

[. . .] Now, theology deals with the supreme divine contemplation and occupies the regal throne by dignity, which confers upon her the highest authority. Therefore, as she does not descend to the lower and humbler speculations of the inferior sciences—in fact, as said above, she cares nothing about them, because they are not concerned with beatitude—then theology professors and ministers should not arrogate to themselves the authority to take decisions in professions which they have not practiced or studied. Otherwise, it would be as if an absolute despot, who despite being neither a physician nor an architect but knowing that he can freely give orders and ensure he is obeyed, should wish that building techniques and medical treatments were done just as he pleased, with grave risk for the lives of the poor patients and obvious ruin for the buildings.

And then, to order that it should be the professors of astronomy themselves who take precautions against their own observations and demonstrations, as if it were impossible that these could be anything other than fallacies and sophisms, is to order them to do something that is utterly impossible. Because not only are they being ordered not to see what they see and not to understand what they understand but, when researching, to find the opposite of what comes into their hands. For this to be made possible, they would first need to be shown the way to make mental faculties command each other, the inferior ones the superior, so that the imagination and the will were able to believe and would want to believe the contrary of what the mind understands (I am referring of course to pure natural propositions, which are not de Fide, *and not to supernatural ones and* de Fide*). I would beseech these very prudent Fathers to consider with the utmost care the difference existing between doctrines that require proof and those that are a matter of opinion. And I would like them to realize, by thoroughly assessing the compelling force of logical deductions, that it is not in the power of professors of the demonstrative sciences to change their views at will, adhering now to one and then to the other; and also that there is a great difference between giving orders to a mathematician or to a philosopher, and persuading a* *merchant or lawyer; and that demonstrated conclusions drawn about things in nature and in the heavens cannot be reversed with the same ease as opinions about the legality or not of a contract, a bargain or an exchange. [. . .]*

The next passage is imbued with Galileo's burning indignation at those who arrogate to themselves the right to silence an opinion simply by gagging those who profess it—as if it were possible to stop men asking questions and forbid them from looking at the sky. Nevertheless, alongside the indignation, we also find all the passion of a scientist who is venturing into unexplored territory, and to the pleasure of contemplation (which is accessible to anybody) is adding the pleasure of intellectual understanding, "without ever ceasing to marvel and to feel delighted". Exploring the mysteries of the universe is similar to exploring the "marvelous mechanisms" inside the human body (it should not escape the reader that Galileo locates "the receptacles of imagination, memory and understanding" in well-defined sites in the body, without resorting to spiritualistic hypotheses).

Wonders of research

If, in order to erase this opinion and doctrine from the world, it were sufficient to lock the mouth of a single person—as perhaps those who, measuring the judgments of others in terms of their own, find it impossible that the opinion should be capable of surviving and attracting followers—this would be very easy to achieve. But things stand otherwise. In fact, to attain such an end it would be necessary not only to prohibit the book by Copernicus and the books of other authors who follow the same doctrine, but it would be necessary to ban the whole science of astronomy, and in addition to prevent men from looking at the sky [. . .]. And if the whole science were to be banned, what else would this be than a rejection of a hundred passages in the Holy Scriptures which tell us that the glory and greatness of Almighty God can be seen and marveled at in all his works, and divinely read in the open book of the heavens? Nobody should think that the reading of the lofty concepts written in that book ends simply with seeing the splendor of the sun and other stars, their rising and setting, which is the limit reached by the gaze of brutes and vulgar people. In that book there are mysteries so profound and concepts so sublime that the vigils, toils and studies of hundreds and hundreds of very fine minds have not yet penetrated them entirely, even after continual investigations over thousands and thousands of years. When looking at the external features of a human body what the ordinary person sees is almost nothing in comparison to the marvelous mechanisms that an accomplished and diligent anatomist or philosopher finds in it, when investigating the use of so many muscles, tendons, nerves and bones; or when examining the functions of the heart and the other major organs; or when looking for the seats of the vital faculties; or when observing the marvelous structures of the tools of the senses and—never ceasing to be astonished and delighted—when contemplating the receptacles of imagination, memory and understanding. In like manner, what the sense of sight alone reveals to us is as nothing compared to the wonders that, by lengthy and thorough observations, the ingenious minds of learned men may discover in the sky. [. . .]

What Galileo says in the next section of the letter makes clear the distinction (and it is not clear to everybody, even today) between what belongs to the realm of conjecture, plain and simple, and what can be the subject of true science, demonstrable and verifiable. When dealing with conjecture, says Galileo, one can go along with the literal meaning of the Scriptures, whereas when dealing with true science one must look for the deeper meaning of the Scriptures, which cannot but be in agreement with data resulting from facts. It should be noted that Galileo opposes "science" to "just opinion and faith": there can be no doubt that the word "just" here establishes a hierarchical difference between the two fields, with opinion and faith being classed along with each other in the same rank.

Science and faith

[. . .] among natural propositions there are some about which all human speculation and reasoning can only attain a plausible opinion or likely conjecture, rather than a sure and demonstrated knowledge—this is so, for example, concerning whether the stars are animate. There are other

propositions, instead, about which we have (or we can firmly believe we will be capable of having) doubtless certainty, by means of experiences, lengthy observations and necessary demonstrations—for example, whether the earth and the sun move, or not, whether the earth is spherical, or not. As to the first kind of proposition, I have no doubt that where human reasoning cannot reach—and where, therefore, there can be no science, but just opinion and faith—it behooves us to piously comply in full with the strict meaning of the Scriptures. But as to the other kind, I would think, as I have already said, that first and foremost we should ascertain the fact: this will help us find the real meaning of the Scriptures, which will be in complete agreement with the demonstrated fact, even if at first sight the words might ring differently. For two truths can never be in contradiction with one another [. . .].

From this and other passages the intention of the holy Fathers seems to be, if I am not mistaken, that in questions of nature which are not matters de Fide *one must first consider whether they are demonstrated beyond doubt or known by sensible experience, or whether such knowledge or demonstration is possible. And if it is, since this too is a gift from God, it should then be applied to find out the true meanings of the Holy Scripture in those passages which on the surface might seem to be saying something different. These meanings will undoubtedly be discerned by wise theologians, together with the reasons why the Holy Ghost may sometimes have preferred to veil those passages beneath words of different meaning, perhaps to test us, or for other purposes unknown to me.*[234]

Galileo's expectations concerning the three Copernican letters were, all-too-soon, to be thwarted. Likewise, his generous efforts to persuade the Church to renew itself culturally were to prove to have been in vain. Indeed, the censure of the Copernican propositions by the Holy Office came quickly, and hot on its heels the censure (until corrections were made to it) of Copernicus's book, and of Foscarini's text by the Congregation of the Index. No explicit reference was made, however, to Galileo's writings. Nevertheless, at the end of February, by order of Pope Paul V, the scientist was summoned by Cardinal Bellarmine, who admonished him to abandon the censured

[234] Elsewhere Galileo suggested the reason for this lay in the fact that the Scriptures were aimed at uneducated people.

opinions (see Chapter 22 on this matter). Galileo left Rome in May of that year. Not, however, before he had obtained a written statement from the Cardinal that he had neither abjured, nor been condemned. Of all the high hopes he had had of being able to "carry out so great an undertaking", nothing was left to him in the bitterness of his defeat but to meditate on the overwhelming force of "three very powerful craftsmen, ignorance, envy and impiety".

HISTORICAL NOTE

At the beginning of the seventeenth century, Galileo's astronomical discoveries and the incontrovertibility of their evidence were capable of undermining a millennia-old system, placing it in jeopardy. For us, children of an age in which ideas seem to have become light and inconstant, marketable and replaceable like fashion objects, it is hard to conceive the violence of such an upheaval. The sidereal announcements arrived and shook a world, that of academic culture, which was immobile and sleepy, yet which had no intention of relinquishing its prestige and privileges. Moreover, the ideas also called for confrontation and criticism, they demanded, in Galileo's words, that people finally make use of the tools they were endowed with to think for themselves.

The Church, which had embraced the Aristotelian system, was disoriented. The discoveries, after all, concerned the field of science, meaning they needed be paid only mild attention, no big issue should be made of them. The reading of *De Revolutionibus Orbium Coelestium* by Copernicus—a book dedicated to the Pope—had never been prohibited during the more than seventy years since its publication. However, it was Galileo himself who forced the ecclesiastics into taking a stance. He could, by writing in Latin, have confined himself to circulating his ideas within the restricted world of scientists. But he chose, instead, to use Italian. While there is no doubt that in addressing the largest possible audience he was satisfying his strong need for self-promotion, it is truer still that the urge he felt to stimulate wide interest in his ideas stemmed from his deep confidence in humankind. No matter how dull and backward particular individuals may appear to be, they could, for Galileo, always be taught how to reason. The formidable force of reason must, come what may, emerge victorious in the end.

This fundamental confidence in human beings is attested to by all Galileo's writings—both in their form, which is often that of the dialogue, and in the way they return to particular topics over and again, in order to clarify them and get deeper into them. This was intended not only to make it easier for the reader to assimilate the arguments, but also to shed light on the workings of the logical process that leads to knowledge. This strong pedagogic calling as regards subject matter and, even more, educational aims, was a distinctive feature of Galileo's personality, to the extent that he probably devoted more time and energy to it than he did to scientific research itself.

Viewed in this light, one could claim, paradoxically, that Galileo's defense of Copernicanism was, in some respects, a pretext. He chose to fight under the banner of an argument that everybody found fascinating and was at the same time extreme—in that it turned current theories upside-down. His aim in doing this was to attract attention and promote *his own* reformation. Which was that of the scientific method—"sensible experiences and necessary demonstrations"—and also of critical analysis and doubt. There is no truth that can be acquired once and for all; there exists only the painstaking, yet exciting, cognitive process of an ever evolving truth, of a science that advances while correcting itself all the time.

In the counter-reformist climate of Italy of the first decades of the seventeenth century, it was inconceivable to attempt to spread ideas without the support of the Church. The Church represented the main concentration of political and cultural power and its reach extended throughout the national territory (even if this territory was divided into separate states). Still fresh in the memory were the vicissitudes of Giordano Bruno, burned alive in Rome in 1600, and of Paolo Sarpi, Galileo's friend, excommunicated in 1607, after his writings had been sentenced to be publicly burned (and he himself condemned to the same fate, if he so much as set foot outside Venice).

Galileo had opened his letter to Christina with the words: "It is not my wish to engage in disputes about these matters with anybody".

There were many reasons why he did not want to clash with the Church, over and beyond fear for his personal safety. First and foremost, as pointed out earlier, the fact that he had understood all too well that only by involving the Church could he promote his educational reform, his project for the emancipation of humankind. From this point of view he may be seen as the Italian response to Luther. In Germany, Luther had set the individual free of the wardship of the ecclesiastic hierarchy and, acknowledging the individual's authority to approach the Scriptures directly, he had promoted intellectual autonomy, learning, and thus culture. Galileo said: "in a flat and open region only the blind need a guide"; nevertheless, in Italy, one whole century after Luther, a somewhat similar project of re-evaluation of the individual could not but go through the Church.

Galileo was well aware of this, thanks to his great capacity for attention to the real world (both physical phenomena and society)—and to the mechanisms by which it operates. On the one side, dogma and revealed truth; on the other, research, error, correction, research again, and so on: two contrasting, incompatible realms. Galileo's attempt to reconcile them is a highly unusual one. He never discussed the problem of dogma, he never set foot upon theological ground. Thus we have a situation where, on the one hand, he never failed to take any opportunity to pay his formal homage to the Church (as he had been advised to do time and again by those who, knowing him well, feared his impetuous nature might lead him into harm's way); on the other hand, he developed a persuasive system of arguments in which the words of the Bible and scientific truth could be made to coexist. Caution? Caution too, perhaps. As a matter of fact, he had no interest in metaphysical questions. It is absolutely clear that for him only what came within tangible experience was worthy of analysis and study, and was inexhaustible as a source of intellectual interest. All else was just words. One should not forget, for instance, that in the *Sidereus Nuncius*—that fascinating little diary-like treatise bursting with enthusiasm and wonder for the celestial mechanisms that were being revealed—not once did Galileo turn, with awestruck gratitude, to God. He stayed focused solely on what his telescope had revealed to him.

What Galileo tried to do, in order to defend and promote the new ideas, was to offer the Church an honorable way out, an acceptable compromise, putting himself forward as a "go between", a scientist who was internal to the Church system and, as such, trustworthy and acceptable. It may be this strategy of his that explains why he never made reference to Bruno, though drawing ideas from him, never spoke in favour of Sarpi, never publicly acknowledged the defense that Campanella, from prison, was offering him.[235] Moreover, since his arguments might have appeared unconvincing and captious to theologians, Galileo often supported them by very ably quoting Augustine and other Fathers of the Church—as, in particular, in the long, learned letter to Christina of Lorraine. The idea, in short, was this: the few references to the physical world that are to be found in the Scriptures are only apparently incompatible with what experience and mathematics demonstrate to be correct. In fact, Truth, coming from God, is one, and one alone. It is just the language through which Truth is made manifest that may vary: mathematical (the language in which the great Book of Nature is written), metaphorical and poetical (the language of the Scriptures). The ancient texts were addressed to simple and primitive people, who were not capable of understanding that things may be different from the way they appeared to their senses. If these people had had cause to doubt what was written in the Scriptures about Nature, how could they then have trusted their moral teachings, which were their real purpose?

[235] Giorgio Salvini, a contemporary Italian physicist, writes: "Galileo's papers are always cautious and fearful of political and religious authorities. They tell us all about his scientific thinking, but clearly they remain nobly silent on the most intimate and inaccessible part of his philosophical thinking. If, in spite of everything, he was a believer, then behind his writing must lie immense religious faith. Otherwise, there is a secret loneliness that he could not communicate to anybody" (G. Salvini, "Mitologia e verità nella figurazione di Galileo. Il Galileo di Bertolt Brecht", *Saggi su Galileo Galilei*, Barbèra Editore, Florence 1967, p. 24). And Pietro Nonis writes: "One should say that he exercised constant vigilance, at least in his writings, so as to let nothing leak out beyond what would be reassuring for his friends and could maintain at a proper distance those who had not passed the jealously guarded boundaries of his world". (P.G. Nonis, "Galileo e la religione", in *Nel quarto centenario della nascita di Galileo Galilei*, Società Editrice Vita e Pensiero, Milan, 1966).

PART VI

THE INVASION OF THE INFINITESIMALS

12

Infinite Finite Infinitesimal*

una divisione e subdivisione che si possa proseguir perpetuamente, suppone
che le parti siano infinite,
perché altramente la subdivisione sarebbe terminabile

a division and a subdivision which can be carried on indefinitely
presupposes that the parts are infinite in number,
otherwise the subdivision would reach an end

In the dialogue presented here, the concepts of infinitesimal and finite quantities are discussed. Galileo felt a need for these not in an abstract sense, but rather in relation to the way real physical phenomena occur, and for the solution of concrete geometrical problems. In the discourse, which in this case is not always clear, Galileo advances some hypotheses that are not justified. Among these are the assertions that an infinite number cannot be said to be larger than finite numbers, and that it could have aspects in common with unity.

Quanta and infinitesimals

In Chapter 13 we will see how Galileo, in describing accelerated motion, clearly felt the need for infinitesimal quantities and, along with these, for operations to be iterated an infinite number of times. In the discourse we are going to read in this chapter he makes a diversion to render these concepts more comprehensible. He

* From *Discourse*.

chooses as an example the infinite divisibility of a finite geometrical segment into parts of infinitesimal length (i.e. parts with no extension). It is interesting to note that in Galileo's language a finite and measurable quantity is called "*parte quanta*" or "*quanto*", while an infinitesimal, that is, an indivisible or immeasurably small element, is called "*non quanto*". The Italian term "*quanto*" (*quantum* in Latin and in English) is used in modern physics to refer, instead, to the minimum amount of energy (non-divisible) which is associated to a given oscillating field. So a quantum of light, also called a *photon*, is the smallest packet of energy transported by an electromagnetic wave.

Since Galileo's reason for turning to the problem of infinitesimals was strictly physical—the behavior of accelerated motion and the microscopic structure of matter (see Chapters 13 and 17)—the way he handled mathematical concepts was somewhat superficial. In the pages which follow we shall find some curious assertions that do not appear justifiable and, in any case, have little to do with what is taught within the mathematical sciences. One of these assertions is the correspondence between infinity and unity—a viewpoint that is almost mystical in nature, and in which Galileo sought refuge, dismayed as he was by the immensity of the concepts.

The text

The dialogue, taken from the *Discourses*,[236] is a brilliant exercise in logic and is razor-sharp in the way it unfolds. And yet, in the basic gratuitousness of some of its conclusions, it reminds us of the rhetorical disputes of the seventeenth century. Normally, Salviati keeps within the bounds of what is logically verifiable. Here, instead, he seems to be suffering from a kind of vertigo in the face of what is incomprehensible to "our finite mind", due, on the one hand, to its infinite magnitude, and on the other, to its infinite smallness (and to the seductive discovery that these two imply each other).

[236] *Two New Sciences*, translated by H. Crew and A. de Salvio, Dover Publications, New York, 1954, p. 27 and following. In Italian: *Discorsi e dimostrazioni matematiche intorno a due nuove scienze*, Giornata prima, Edizione Nazionale, Vol. VIII, p. 72 and following.

Nevertheless, Galileo was too lucidly in control of his emotions to identify himself here, as he does elsewhere, entirely with his character. So we see Salviati repeatedly distancing himself from these "somewhat startling" ideas,[237] from these "human fancies", from these "subjects freely chosen and not forced upon us". And we also see a very special role assigned to Simplicio, who has a hard time negotiating this "obstacle difficult to avoid", but plunges into the discussion with zest and draws with satisfaction upon his store of firmly fixed notions. He admires Salviati's reasoning, but in the end, overwhelmed, he bursts out with "I do not quite grasp the meaning of this". And with these words it seems Galileo wishes to bring the discussion back to the modest, but safe, level of common sense—allowing Salviati himself to conclude with the sentence "These are marvels which our imagination cannot grasp".

A delightful *divertissement*, therefore, is what this foray into the domain of the non-finite—this exercise in the cognitive potentialities of stimulating the imagination—can provide. Galileo does not claim he can reach sure truths in this way. Instead he points out the importance of the contribution to science that comes from creative thought, which breaks fixed patterns and proceeds by intuition (provided, it needs to be added, that it is not gratuitous and that it can be verified by reason).

A fanciful idea

SIMPL: [. . .] Besides, this building up of lines out of points, divisibles out of indivisibles, and finites out of infinites, offers me an obstacle difficult to avoid; and the necessity of introducing a vacuum, so conclusively refuted by Aristotle, presents the same difficulty.
SALV: These difficulties are real; and they are not the only ones. But let us remember that we are dealing with infinities and indivisibles, both of which transcend our finite understanding, the former on account of their magnitude, the latter because of their smallness. In spite of this,

[237] The taste for "marvelling" (*maraviglia*), so typical of the poetry of the seventeenth century, applies here to scientific research. This term is encountered in these pages over and again.

men cannot refrain from discussing them, even though it must be done in a roundabout way.

Therefore I also should like to take the liberty to present some of my ideas which, though not necessarily convincing, would, on account of their novelty, at least, prove somewhat startling. But such a diversion might perhaps carry us too far away from the subject under discussion and might therefore appear to you inopportune and not very pleasing.

SAGR: Pray let us enjoy the advantages and privileges which come from conversation between friends, especially upon subjects freely chosen and not forced upon us, a matter vastly different from dealing with dead books which give rise to many doubts but remove none. [. . .]

[. . .]

SALV: But I do have something special to say, and will first of all repeat what I said a little while ago, namely, that infinity and indivisibility are in their very nature incornprehensible to us; imagine then what they are when combined. Yet if we wish to build up a line out of indivisible points, we must take an infinite number of them, and are, therefore, bound to understand both the infinite and the indivisible at the same time. Many ideas have passed through my mind concerning this subject, some of which, possibly the more important, I may not be able to recall on the spur of the moment; but in the course of our discussion it may happen that I shall awaken in you, and especially in Simplicio, objections and difficulties which in turn will bring to memory that which, without such stimulus, would have lain dormant in my mind. Allow me therefore the customary liberty of introducing some of our human fancies, for indeed we may so call them in comparison with supernatural truth which furnishes the one true and safe recourse for decision in our discussions and which is an infallible guide in the dark and dubious paths of thought.

In Salviati's last sentence here it is clear that his intention is to soften potential "heretical" implications of his free thinking. He now goes on to discuss the concept, formulated above, that in order to build a finite geometrical segment using indivisible elements—that is, infinitesimals—one must add together an infinite number of them.

One of the main objections urged against this building up of continuous quantities out of indivisible quantities is that the addition of one indivisible to another cannot produce a divisible, for if this were so it would render the indivisible divisible. Thus if two indivisibles, say two points, can be united to form a quantity, say a divisible line, then an even more divisible line might be formed by the union of three, five, seven, or any other odd number of points. Since however these lines can be cut into two equal parts, it becomes possible to cut the indivisible which lies exactly in the middle of the line. In answer to this and other objections of the same type we reply that a divisible magnitude cannot be constructed out of two or ten or a hundred or a thousand indivisibles, but requires an infinite number of them.[238]

SIMPL: Here a difficulty presents itself which appears to me insoluble. Since it is clear that we may have one line greater than another, each containing an infinite number of points, we are forced to admit that, within one and the same class, we may have something greater than infinity, because the infinity of points in the long line is greater than the infinity of points in the short line. This assigning to an infinite quantity a value greater than infinity is quite beyond my comprehension.

SALV: This is one of the difficulties which arise when we attempt, with our finite minds, to discuss the infinite, assigning to it those properties which we give to the finite and limited; but this I think is wrong, for we cannot speak of infinite quantities as being the one greater or less than or equal to another. To prove this I have in mind an argument which, for the sake of clearness, I shall put in the form of questions to Simplicio who raised this difficulty.

The squares trick

I take it for granted that you know which of the numbers are squares and which are not.

SIMPL: I am quite aware that a squared number is one which results from the multiplication of another number by itself; thus 4, 9, etc., are squared numbers which come from multiplying 2, 3, etc., by themselves.

[238] Elsewhere Galileo wrote (Edizione Nazionale, Vol. VII, p. 749): "divisibility and quantity stem from infinity . . .".

SALV: Very well; and you also know that just as the products are called squares so the factors are called sides or roots; while on the other hand those numbers which do not consist of two equal factors are not squares. Therefore if I assert that all numbers, including both squares and non-squares, are more than the squares alone, I shall speak the truth, shall I not?[239]

SIMPL: Most certainly.

SALV: If I should ask further how many squares there are one might reply truly that there are as many as the corresponding number of roots, since every square has its own root and every root its own square, while no square has more than one root and no root more than one square.

SIMPL: Precisely so.

SALV: But if I inquire how many roots there are, it cannot be denied that there are as many as there are numbers because every number is a root of some square. This being granted we must say that there are as many squares as there are numbers because they are just as numerous as their roots, and all the numbers are roots. Yet at the outset we said there are many more numbers than squares, since the larger portion of them are not squares.[240] *Not only so, but the proportionate number of squares diminishes as we pass to larger numbers. Thus up to 100 we have 10 squares, that is, the squares constitute 1/10 part of all the numbers; up to 10000, we find only 1/100 part to be squares; and up to a million only 1/1000 part; on the other hand in an infinite number, if one could conceive of such a thing, he would be forced to admit that there are as many squares as there are numbers all taken together.*

SAGR: What then must one conclude under these circumstances?

SALV: So far as I see we can only infer that the totality of all numbers is infinite, that the number of squares is infinite, and that the number of their roots is infinite; neither is the number of squares less than the totality of all numbers, nor the latter greater than the former; and finally the attributes "equal", "greater", and "less", are not applicable

[239] For example, of all the numbers between 1 and 16, only four are squares (1, 4, 9, 16).

[240] The two statements that both appear to be true, but in reality are in contradiction, are: (1) the number of numbers is larger than that of squares because some numbers are not squares; (2) the number of numbers equals that of squares because the number of roots equals that of numbers and the number of squares equals that of roots.

to infinite, but only to finite quantities. When therefore Simplicio introduces several lines of different lengths and asks me how it is possible that the longer ones do not contain more points than the shorter, I answer him that one line does not contain more or less or just as many points as another, but that each line contains an infinite number.[241] *Or if I had replied to him that the points in one line were equal in number to the squares; in another, greater than the totality of numbers; and in the little one, as many as the number of cubes, might I not, indeed, have satisfied him by thus placing more points in one line than in another and yet maintaining an infinite number in each? So much for the first difficulty.*

Now come the bizarre ideas, proffered by Sagredo, that not only can no inequality be established between two infinities, but also it cannot be said that an infinite number is greater than a finite one. It is not clear where Galileo drew these convictions from. What is clear, though, is that he did not suspect the possibility that infinity could be raised to a power greater than one, meaning that infinity squared is an infinite number of times larger than infinity, and infinitesimal squared is an infinite number of times smaller than infinitesimal. The premise to the discourse we are now going to read is also very curious: 'the larger the number to which we pass, the more we recede from infinity'.

SAGR: Pray stop a moment and let me add to what has already been said an idea which just occurs to me. If the preceding be true, it seems to me impossible to say either that one infinite number is greater than another or even that it is greater than a finite number, because if the infinite number were greater than, say, a million it would follow that on passing from the million to the higher and higher numbers we would be approaching the infinite; but it is not so; on the contrary, the larger the number to which we pass, the more we recede from infinity, because the greater the numbers the fewer are the squares contained in them; but the squares in infinity cannot be less than the totality of all the

[241] Elsewhere Galileo states (Edizione Nazionale, Vol. VII, p. 749): "every part (if part it can be called) of infinity is infinite . . .".

numbers, as we have just agreed; hence the approach to greater and greater numbers means a departure from infinity.

SALV: And thus from your ingenious argument we are led to conclude that the attributes "larger", "smaller" and "equal" have no place either in comparing infinite quantities with each other or in comparing infinite with finite quantities.

I pass now to another consideration. Since lines and all continuous quantities are divisible into parts which are themselves divisible without end, I do not see how it is possible to avoid the conclusion that these lines are built up of an infinite number of indivisible quantities because a division and a subdivision which can be carried on indefinitely presupposes that the parts are infinite in number, otherwise the subdivision would reach an end; and if the parts are infinite in number, we must conclude that they are not finite in size, because an infinite number of finite quantities would give an infinite magnitude. And thus we have a continuous quantity built up of an infinite number of indivisibles.[242]

SIMPL: But if we can carry on indefinitely the division into finite parts what necessity is there then for the introduction of non finite parts?

SALV: The very fact that one is able to continue, without end, the division into finite parts makes it necessary to regard the quantity as composed of an infinite number of immeasurably small elements. Now in order to settle this matter I shall ask you to tell me whether, in your opinion, a continuum *is made up of a finite or of an infinite number of finite parts.*

SIMPL: My answer is that their number is both infinite and finite; potentially infinite but actually finite; that is to say, potentially infinite before division and actually finite after division; because parts cannot be said to exist in a body which is not yet divided or at least marked out; if this is not done we say that they exist potentially.[243]

SALV: So that a line which is, for instance, twenty spans long is not said to contain actually twenty lines each one span in length except after division into twenty equal parts; before division it is said to contain

[242] Elsewhere Galileo writes (Edizione Nazionale, Vol. VII, p. 745): "to say that the continuum is constituted by parts which can always be subdivided and to say that the continuum is constituted by parts which are indivisible is the same thing".

[243] For Aristotelians, infinity and the infinitesimal (i.e. the indivisible part) were conceivable quantities, but not real.

them only potentially. Suppose the facts are as you say; tell me then whether, when the division is once made, the size of the original quantity is thereby increased, diminished, or unaffected.
SIMPL: It neither increases nor diminishes.
SALV: That is my opinion also. Therefore the finite parts in a continuum, whether actually or potentially present, do not make the quantity either larger or smaller; but it is perfectly clear that, if the number of finite parts actually contained in the whole is infinite in number, they will make the magnitude infinite. Hence the number of finite parts, although existing only potentially, cannot be infinite unless the magnitude containing them be infinite; and conversely if the magnitude is finite it cannot contain an infinite number of finite parts either actually or potentially.
[. . .]
SALV: [. . .] I grant therefore, to the philosophers, that the continuum contains as many finite parts as they please and I concede also that it contains them, either actually or potentially, as they may like; but I must add that just as a line ten fathoms in length contains ten lines each of one fathom and forty lines each of one cubit and eighty lines each of half a cubit, etc., so it contains an infinite number of points; call them actual or potential, as you like, for as to this detail, Simplicio, I defer to your opinion and to your judgment.[244]
SIMPL: I cannot help admiring your discussion; but I fear that this parallelism between the points and the finite parts contained in a line will not prove satisfactory, and that you will not find it so easy to divide a given line into an infinite number of points as the philosophers do to cut it into ten fathoms or forty cubits; not only so, but such a division is quite impossible to realize in practice, so that this will be one of those potentialities which cannot be reduced to actuality.

[244] We cite a remark made by Carlo Felice Manara:

I do not think it excessive if one claims to see here, in these Galilean positions, the proudest and most assured declaration of independence of scientific thought from all other kinds of thought. It is proclaimed loudly and clearly that science refuses to be conditioned by external criteria, to see its research shackled by questions raised from outside its province; that it reserves the right to draw such tools and schemes as seem most appropriate to its goals from the domain of pure sciences. (C.F. Manara, La matematica nel pensiero galileiano, in *Nel quarto centenario della nascita di Galileo Galilei*, Società Editrice Vita e Pensiero, Milan, 1966, p. 106).

SALV: The fact that something can be done only with effort or diligence or with great expenditure of time does not render it impossible; for I think that you yourself could not easily divide a line into a thousand parts, and much less if the number of parts were 937 or any other large prime number.[245] *But if I were to accomplish this division which you deem impossible as readily as another person would divide the line into forty parts would you then be more willing, in our discussion, to concede the possibility of such a division?*
SIMPL: In general I enjoy greatly your method; and replying to your query, I answer that it would be more than sufficient if it prove not more difficult to resolve a line into points than to divide it into a thousand parts.

Infinity of unity

SALV: I will now say something which may perhaps astonish you; it refers to the possibility of dividing a line into its infinitely small elements by following the same order which one employs in dividing the same line into forty, sixty, or a hundred parts, that is, by dividing it into two, four, etc. He who thinks that, by following this method, he can reach an infinite number of points is greatly mistaken; for if this process were followed to eternity there would still remain finite parts which were undivided.

Indeed by such a method one is very far from reaching the goal of indivisibility; on the contrary he recedes from it and while he thinks that, by continuing this division and by multiplying the multitude of parts, he will approach infinity, he is, in my opinion, getting farther and farther away from it. My reason is this. In the preceding discussion we concluded that, in an infinite number, it is necessary that the squares and cubes should be as numerous as the totality of the natural numbers, because both of these are as numerous as their roots which constitute the totality of the natural numbers. Next we saw that the larger the numbers taken the more sparsely distributed were the squares, and still more sparsely the cubes; therefore it is clear that the larger the numbers

[245] If the segment to be divided had a length of 1000, its parts should have a length of 1000/937 = 1.0672358591.

to which we pass the farther we recede from the infinite number; hence it follows that, since this process carries us farther and farther from the end sought, if on turning back we shall find that any number can be said to be infinite, it must be unity. Here indeed are satisfied all those conditions which are requisite for an infinite number; I mean that unity contains in itself as many squares as there are cubes and natural numbers.

SIMPL: I do not quite grasp the meaning of this.

SALV: There is no difficulty in the matter because unity is at once a square, a cube, a square of a square and all the other powers; nor is there any essential peculiarity in squares or cubes which does not belong to unity; as, for example, the property of two square numbers that they have between them a mean proportional; take any square number you please as the first term and unity for the other, then you will always find a number which is a mean proportional.[246] *Consider the two square numbers, 9 and 4; then 3 is the mean proportional between 9 and 1; while 2 is a mean proportional between 4 and 1; between 9 and 4 we have 6 as a mean proportional. A property of cubes is that they must have between them two mean proportional numbers;*[247] *take 8 and 27; between them lie 12 and 18; while between 1 and 8 we have 2 and 4 intervening; and between 1 and 27 there lie 3 and 9. Therefore we conclude that unity is the only infinite number. These are marvels which our imagination cannot grasp and which should warn us against the serious error of those who attempt to discuss the infinite by assigning to it the same properties which we employ for the finite, the natures of the two having nothing in common.*

This last statement "if . . . any number can be said to be infinite, it must be unity" is also, to say the least, odd (and will undoubtedly have struck readers as such). The simple consideration that unity, in contrast to infinity, is divisible into finite parts, is enough to destroy the similarity between the two. Galileo was impressed by the fact that unity always stays the same, no matter what power it is raised to, and he implicitly (and incorrectly) assigned this property to infinity,

[246] If the two squares are a and b the mean proportional is c, such that $c^2 = ab$.

[247] If the two cubes are a and b the mean proportionals are c and d, such as $ab = cd$.

too. In the *Dialogue* he also states that the properties of the circle are infinite and that they "being infinite, are perhaps only one in their essence and in the mind of God", which encompasses them all.

This idea, with its vague theological allusion, is most unusual in Galileo's mode of reasoning. He was evidently bewildered by the difficulty of grasping the real meaning of non-finite quantities, and grasping the real meaning was always for him the fundamental part of a problem. He had never held back when grappling with explanations for the most diverse phenomena. But when it came to non-finite quantities, faced with the abstractions to which his analysis had led him, he gave up on delving deeper into the problem and chose to seek refuge in a sort of mystical fancy. It has to be pointed out, of course, that in his time it simply was not possible to investigate these matters—either by theoretical tools, or by experiment. Moreover, he resorted to such a mystical course only extremely rarely, and the manner in which he does so here is highly hesitant, as is evident from the words "if . . . can be said" and "perhaps" in the sentences above. No heed at all should be paid to the misguided attempts of certain commentators who try to peddle these slight mystic yieldings on Galileo's part as the most genuine expression of his thought and his spiritual message to posterity.

Down to earth

Galileo's lack of interest in abstract (experimentally intangible) problems, and in particular for those relating to infinity, is clearly evidenced by the following words(written from his enforced residence in Arcetri) on the questions of the structure of the universe and of the earth supposedly being the center of it.[248]

> *[. . .] However, it is not worth wasting time believing one can prove that fixed stars are positioned in a space which is bounded by a spherical surface, rather than positioned here and there and separated by huge distances. Likewise, to wish to assign a center to that space whose shape*

[248] Letter to Fortunio Liceti of January 1641, Edizione Nazionale, Vol. XVIII, p. 293.

> *is not known and cannot be known (and it is not even known whether it has any kind of shape), is in my opinion an utterly useless and vain undertaking. Hence to believe that the earth occupies a center, when it is not known if such a center exists in the universe, is, as I said, a frustrating undertaking. [...]*

Elsewhere he stated:[249]

> *Searching for the essence I deem to be no less impossible a task, and no less vain a fatigue, regarding elementary and nearby substances, than it is regarding those that are most remote and heavenly.*

Such statements are emblematic of an intellectually independent and pragmatic approach to science. One in which answers are sought only to phenomena that are accessible, and conjectures about what lies outside the domain of investigable facts are avoided. To the wise person this means taking no interest in what is not a problem, in the sense that it cannot be phrased in terms of a solvable question; giving up dogmatic truths and trusting instead that the frontiers of the investigable expand endlessly, along with progress in science and technology.[250] History shows that people, at the appropriate time, have been able to find answers to questions that previously had been shrouded in mystique, or the subject of cult worship. The process is a perennial evolutionary one, made up of small steps (we are tempted to call them infinitesimals), and will require infinite time to reach completion.

MATHEMATICAL NOTE

In infinitesimal calculus an infinitesimal change in a given quantity—as well as an infinitesimal, or elemental, part of this quantity—is expressed by its differential. In the case of a length x, its infinitesimal variation is indicated by dx. Such variation is defined through the operation of subdividing the given

[249] *Lettere sulle macchie solari*, Edizione Nazionale, Vol. V, p. 187. An excerpted English translation can be found in S. Drake, *Discoveries and Opinions of Galileo*, Anchor Books/Doubleday, Garden City, NY, 1957.

[250] Seneca said: "The time will come when our descendants will be amazed we did not know things that are so obvious".

quantity into smaller and smaller parts, all the way up to the limit of an infinite number of subdivisions. The infinitesimal of order 2, immediately higher, is given by $(dx)^2$, that is, by the square of dx, which obviously represents the area of a square with side dx. The infinitesimal of order 3 is $(dx)^3$, that is, the volume of a cube of side dx, and so forth. Proceeding now in reverse order, if the infinitesimal element of length dx is given, the length x of a finite segment is obtained by adding all the dx elements contained in it, which are infinite in number. Such an operation represents the extension of a sum of a finite number of terms and is called *integral*:

$$x=\sum_{n=1}^{N} x_n \qquad\qquad x=\int_0^x dx$$

(sum of N segments of finite length x_n) (integral of an infinite number of segments of infinitesimal length dx)

The need for the concept of infinitesimals in physical problems immediately becomes evident when one wants to express the instantaneous velocity of a body mathematically. One proceeds as follows. Let us start from the diagram below, representing the behavior in time t of the space x covered by a material

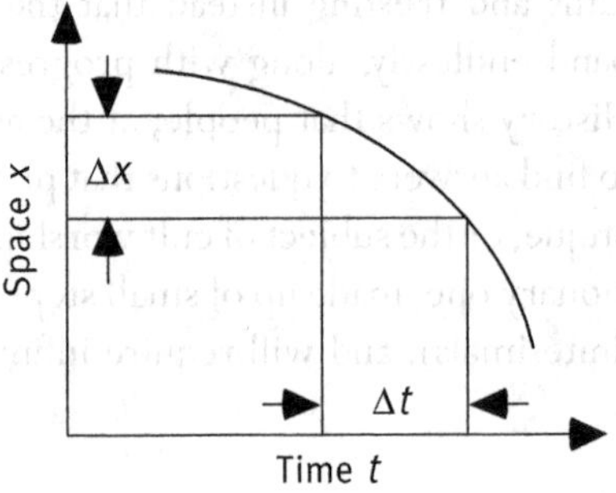

point (space–time curve). The average velocity in a finite interval of time, let us say Δt, is given by the ratio $\Delta x/\Delta t$, where Δx is the space covered in the time interval Δt. The smaller Δt is, the closer the average velocity is to the instantaneous velocity. It is therefore necessary to gradually reduce the interval Δt, and correspondingly the spatial interval Δx, until one gets to the infinitesimal values dt and dx. This operation is called going to the limit for Δt tending to zero. The instantaneous velocity $v(t)$ is given by the limit value of the ratio $\Delta x/\Delta t$ as Δt tends to zero:

$$v(t)=\lim_{\Delta t\to 0}\frac{\Delta x}{\Delta t}=\frac{dx}{dt}$$

The ratio dx/dt between differentials is called the *derivative* of space with respect to time. Using similar reasoning one defines instantaneous acceleration, which is $a(t) = dv/dt$, that is, the derivative of velocity with respect to time.

The integral is the inverse operation of the derivative. Thus, if at any instant in a given interval of time between zero and t the instantaneous velocity is known, the space covered after time t is obtained by integration of its infinitesimal increments $dx = v(t)dt$

$$x = x_0 + \int_0^t v(t)dt$$

where x_0 is the space at the initial instant. Similarly, for the velocity one has

$$v = v_0 + \int_0^t a(t)dt$$

where v_0 is the velocity at the initial instant.

13

Down the Slope

Acceleration and the Inclined Plane Experiment*

l'istessa esperienza che pareva nel primo aspetto mostrare una cosa, meglio considerata ci assicura del contrario

the same experiment which at first glance seemed to show one thing, when more carefully examined, assures us of the contrary

Described here is the famous experiment of a ball which rolls down an inclined plane, devised to demonstrate that in uniformly accelerated motion distance covered is proportional to the square of the time elapsed. The experiment was intended as support for the hypothesis that speed increases in proportion to time, not to distance covered, as many of Galileo's contemporaries held—and as did Galileo too, somewhat inexplicably, during the first phase of his studies. Although he was later to change his mind on this, the conceptual mistake he made when attempting to prove the impossibility of the proportionality of distance to time using logic alone, was substantial. Before coming to the inclined plane experiment (but did Galileo actually perform it?) the three friends Salviati, Sagredo, and Simplicio discuss the matter from various viewpoints, touching upon subtle concepts which are the prelude to differential calculus, such as the relationship between infinitesimal, finite and infinite.

* From *Discourses*.

Uniformly accelerated motion

"Motum aequabiliter, seu uniformiter, acceleratum dico illum, qui, a quiete recedens, temporibus aequalibus aequalia celeritatis momenta sibi superaddit", "I call equally, (i.e. uniformly) accelerated motion, that motion which, starting from rest, in equal times acquires equal increases in speed"—with this clear definition, the discussion on uniformly accelerated motion, entitled *De motu locali* (*About local motion*), gets under way at the start of the third day of the *Discourses*. The discussion sees the three friends Salviati, Sagredo, and Simplicio exploring the modalities of such motion, in particular as regards falling bodies, before going on to study it experimentally by means of a ball rolling down a groove in an inclined plane.

During the course of the dialogue, different fundamental aspects of motion characterized by constant acceleration come under examination. First and foremost that introduced by Sagredo when he expresses the belief, shared by many scientists of the time, that a continuous transition is not possible from an instant in which no motion is occurring to a subsequent instant characterized by a finite value of speed; in other words, that there cannot exist a stage of infinite slowness, or rather of infinitesimal speed, such as is required to allow continuity with zero. A logical demonstration that a body starting from rest passes through all the gradations of velocity without pausing in any of them had already been provided by Galileo in his *Dialogue*.[251] This, in effect, amounted to a precise intuition of the concept of continuous function. What was difficult to grasp at that time was the idea of instantaneous velocity, that is, a velocity defined in an infinitely small interval of time (today expressed by the derivative of space with respect to time). Velocity passes from the initial zero value to a subsequent value different from zero, describing, by infinitesimal increases, all the infinite infinitesimal intervals of time contained in the finite interval of time.

[251] *Dialogue Concerning the Two Chief World Systems*, First Day, translated by S. Drake, The Modern Library, New York, 2001, p. 24 and following. In Italian: *Dialogo sopra i due massimi sistemi del mondo*, Giornata prima, Edizione Nazionale, Vol. VII, p. 46 and following.

The indivisibles

These concepts, which Galileo boldly asserted in a cultural context that was ill-disposed to accepting them, today form the elementary basis of differential calculus (this topic, touched upon in Chapter 12, will be discussed in the Mathematical Note at the end of this chapter). For readers who may not be familiar with the subject, let us say that the fact that a finite interval contains within itself an infinite number of infinitesimal intervals becomes evident when we consider a geometrical analogy. Take a geometrical segment of finite length and divide it into two halves, then divide each of these two halves into two further halves, and so on. There is no limit to this process, because it will always be possible to divide finite segments into two parts. However, if (conceptually speaking) we go on performing this operation an infinite number of times, in the end we will have divided the original segment into an infinite number of segments each of infinitesimal length (i.e. intervals without extension, or points, if we wish, or *indivisibles*, as Galileo calls them when using this same example—see Chapter 12).

Galileo comes up with the brilliant idea of having Salviati use the example of a block falling upon a stake from progressively decreasing heights, vanishing to zero with continuity, in order to get the concept of an infinitely small velocity (a concept rejected by Aristotelians) across to his interlocutors. Another major idea is the statement that the processes of ascent and descent of a body are complementary: falling from a given height, the body will pass through all the continuous values of velocity (though in reverse order) that it would have acquired if it had been launched upwards so as to attain the same height. Of this, thanks to his sharp physical intuition, Galileo was most sure. However, at that time it was not easy to make a confirming experimental measurement capable of convincing the large crowd of skeptics (Aristotelians believed that upward motion differed from spontaneous fall motion because of the residual "impressed force"). Today, the symmetry of the upward and downward motion of a body which has been launched upwards could readily be checked by taking a series of stroboscopic images, that is, by photographing the body at equal time intervals as it describes its parabolic trajectory (see picture below).

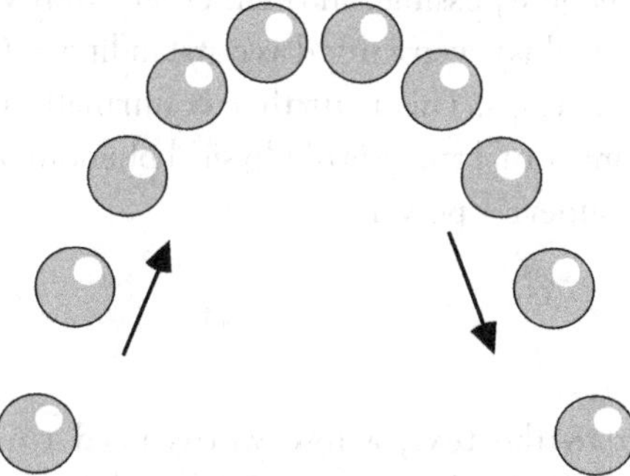

Galileo: "I say that the falling body regains, when passing through all the points of its descent, the same momentum that was sufficient to move it upwards when it passed through those same points while ascending...".

Velocity and time

A second matter of great interest is the question of how the velocity of uniformly accelerated motion increases in time. In his early writings Galileo had shown that he shared the opinion of his contemporaries that velocity increases in direct proportion to the space covered. This was surprising indeed, given that the idea was in no way compatible with the notion, which he already possessed, that the relation between space and time is quadratic. It is rather obvious, in fact, that, since space is dependent on both time and velocity, its quadratic growth with time implies that velocity grows in proportion to time.

That Galileo was able, for a period, to reconcile two contradictory behaviors[252] was due to faults in his reasoning which he was later to

[252] This inconsistency is found, for instance, in the famous letter of 1604 to Paolo Sarpi, in which Galileo defines, for the first time in history, the features of uniformly accelerated motion (Edizione Nazionale, Vol. X, p. 115): "... I demonstrate moreover that the spaces covered by natural motion are in - quadratic relationship with times", in the case that "the velocity of the body

straighten out.[253] What is of particular interest about this matter is that, in the absence of a mathematical treatment of the problem (which was to become possible, and elementary, only with the advent of differential calculus) even a mind as outstanding as Galileo's could fall into insidious traps. This is further confirmation that, when it comes to the conceptual analysis of physical phenomena, mathematical tools are of superior power.

The text

Before coming to the text, a few words need to be said about Galileo's use of Latin during the third and fourth days of the *Discourses*. Latin is used for parts that are in treatise form, for axioms and for geometrical deductions, while the conversational parts and the illustration of experiments continue to be in Italian. Various interpretations have been offered on this choice. A rather trivial hypothesis is that, with a view to rapid publication, Galileo simply copied into his book Latin passages that he had written many years before. More intriguing is Geymonat's idea[254] that Galileo wanted to prove that the two treatments—mathematical and experimental—are mutually irreducible and clearly distinguishable; they operate on completely different levels, yet their roles are complementary and are equally indispensable. More realistic, perhaps, is the opinion expressed in another book by the aforementioned Geymonat, together with Carugo.[255] They argue that in Galileo's overall plans, the work had a twofold purpose: on the one hand, to communicate

increases in proportion to the distance from the origin of its motion", that is the space covered.

[253] For more details, see for instance A. Carugo and L. Geymonat in *Discorsi e dimostrazioni matematiche intorno a due nuove scienze*, *Note*, Boringhieri, Turin, 1958, p. 779 and following. Also texts quoted therein: P. Duhem (*Études sur Léonard de Vinci*, Paris, 1913); P. Tannery (*Mémoires scientifiques*, Paris, 1926); A. Koyré (*Études galiléennes*, Parigi, 1939); R. Marcolongo (*La meccanica di Galileo*, Milan, 1942).

[254] L. Geymonat, *Galileo Galilei*, Einaudi, Turin, 1957. English translation by S. Drake, *Galileo Galilei: A Biography and Inquiry into His Philosophy of Science*, MacGraw-Hill Book Co., New York, 1965.

[255] A. Carugo and L. Geymonat, *ibid.*, p. 756.

to the international academic world the principles and laws he had discovered or perfected—and for this Latin was more appropriate; on the other hand, to spread the basic concepts among Italian engineers and technicians, with a view to their being employed in practical applications—and for this, Italian was better suited. However, as regards the use of Italian, Russo's hypothesis seems still more convincing and plausible:[256] Galileo knew only the Italian language well enough to achieve high literary levels in works where philosophical reflection was being interwoven with scientific arguments, and where irony, allusion, and psychological insight called for the use of a variety of linguistic registers.

But now let us read the original dialogue. Sagredo is discussing the Latin definition of uniformly accelerated motion which opened this chapter, asking whether it applies to free fall motion.[257]

SAGR: Although I can offer no rational objection to this or indeed to any other definition, devised by any author whomsoever, since all definitions are arbitrary, I may nevertheless without offense be allowed to doubt whether such a definition as the above, established in an abstract manner, corresponds to and describes that kind of accelerated motion which we meet in nature in the case of freely falling bodies. And since the Author[258] apparently maintains that the motion described in his definition is that of freely falling bodies, I would like to clear my mind of certain difficulties in order that I may later apply myself more earnestly to the propositions and their demonstrations.
SALV: It is well that you and Simplicio raise these difficulties. They are, I imagine, the same which occurred to me when I first saw this treatise, and which were removed either by discussion with the Author himself, or by turning the matter over in my own mind.

[256] L. Russo, *Segmenti e bastocini*, Feltrinelli, Milan, 1998, p. 121. The reference to Galileo is made in drawing an analogy with the present-day dichotomy, for Italian authors, between using Italian and English. The latter is adopted in scientific and technical communication, while the former allows cultural and expressive richness.

[257] *Two New Sciences*, Third Day, translated by H. Crew and A. de Salvio, Dover Publications, New York, 1954, p. 162 and following. In Italian: *Discorsi e dimostrazioni matematiche intorno a due nuove scienze*, Giornata terza, Edizione Nazionale, Vol. VIII, p. 198 and following.

[258] *The Author*—that is, Galileo.

Infinite slowness

SAGR: When I think of a heavy body falling from rest, that is starting with zero speed and gaining speed in proportion to the time from the beginning of the motion; such a motion as would, for instance, in eight beats of the pulse acquire eight degrees of speed; having at the end of the fourth beat acquired four degrees; at the end of the second, two; at the end of the first, one:[259] *and since time is divisible without limit, it follows from all these considerations that if the earlier speed of a body is less than its present speed in a constant ratio, then there is no degree of speed however small (or, one may say, no degree of slowness however great) with which we may not find this body travelling after starting from infinite slowness, i.e., from rest. So that if that speed which it had at the end of the fourth beat was such that, if kept uniform, the body would traverse two miles in an hour, and if keeping the speed which it had at the end of the second beat, it would traverse one mile an hour, we must infer that, as the instant of starting is more and more nearly approached, the body moves so slowly that, if it kept on moving at this rate, it would not traverse a mile in an hour, or in a day, or in a year or in a thousand years; indeed, it would not traverse a span in an even greater time; a phenomenon which baffles the imagination, while our senses show us that a heavy falling body suddenly acquires great speed.*

Ascent and descent of a body

Salviati's discourse now turns to the question of the perfect symmetry between the slowing down of the ascending motion of a body launched upwards, and the speeding up of its descending motion during its fall. It provides an exemplary lesson in method—one to be used not only in science, but also in everyday life. In defending his opinion, this time Salviati is not confronting a preconceived idea, but an event which takes place before everybody's eyes, and at first sight seems to prove something, but in reality does not ("You say the

[259] The series of values given for the velocity are deduced from the equation $v = at$ (with v = velocity, a = acceleration, t = time), where the simple case $a = 1$ has been considered.

experiment appears to show . . ."). Using the examples of "a heavy body upon a yielding material" and of "a block allowed to fall upon a stake" he comes to the conclusion that a phenomenon, if it is correctly analyzed in its individual details, can prove the exact contrary of what it "at first glance seemed to show". In the case in point, it proves that, contrary to appearances, the initial velocity of a falling body is infinitely small.

While this first step of Salviati's is within everybody's grasp, his next step—to arrive at a truth "by reasoning alone", that is, just using logic—is more demanding. In fact, he had no other choice regarding the thesis that a body thrown upwards touches "the infinite degrees of diminishing velocity" as it passes from a finite velocity to zero velocity. It was not possible for him to carry out a relevant experiment, nor was he in possession of mathematical tools he could draw upon. But his reasoning is so rigorous and penetrating that he not only persuades his interlocutors, but even surprises us with the sharpness of the insight, especially if we bear in mind the cultural context of the time. A time when, in Italy, as Milton wrote in his *Areopagitica*, owing to "the servil condition into which learning . . . was brought. . . nothing had bin there writt'n now these many years but flattery and fustian".

SALV: This is one of the difficulties which I also at the beginning, experienced, but which I shortly afterwards removed; and the removal was effected by the very experiment which creates the difficulty for you. You say the experiment appears to show that immediately after a heavy body starts from rest it acquires a very considerable speed: and I say that the same experiment makes clear the fact that the initial motions of a falling body, no matter how heavy, are very slow and gentle. Place a heavy body upon a yielding material, and leave it there without any pressure except that owing to its own weight; it is clear that if one lifts this body a cubit or two and allows it to fall upon the same material, it will, with this impulse, exert a new and greater pressure than that caused by its mere weight; and this effect is brought about by the [weight of the] falling body together with the velocity acquired during the fall, an effect which will be greater and greater according to the height of the fall, that is according as the velocity of the falling body becomes greater.

From the quality and intensity of the blow we are thus enabled to accurately estimate the speed of a falling body. But tell me, gentlemen, is it not true that if a block be allowed to fall upon a stake from a height of four cubits and drives it into the earth, say, four finger-breadths, that coming from a height of two cubits it will drive the stake a much less distance, and from the height of one cubit a still less distance; and finally if the block be lifted only one finger-breadth how much more will it accomplish than if merely laid on top of the stake without percussion? Certainly very little. If it be lifted only the thickness of a leaf, the effect will be altogether imperceptible. And since the effect of the blow depends upon the velocity of this striking body, can any one doubt the motion is very slow and the speed more than small whenever the effect [of the blow] is imperceptible? See now the power of truth; the same experiment which at first glance seemed to show one thing, when more carefully examined, assures us of the contrary.

But without depending upon the above experiment, which is doubtless very conclusive, it seems to me that it ought not to be difficult to establish such a fact by reasoning alone. Imagine a heavy stone held in the air at rest; the support is removed and the stone set free; then since it is heavier than the air it begins to fall, and not with uniform motion but slowly at the beginning and with a continuously accelerated motion. Now since velocity can be increased and diminished without limit, what reason is there to believe that such a moving body starting with infinite slowness, that is, from rest, immediately acquires a speed of ten degrees rather than one of four, or of two, or of one, or of a half, or of a hundredth; or, indeed, of any of the infinite number of small values [of speed]? Pray listen. I hardly think you will refuse to grant that the gain of speed of the stone falling from rest follows the same sequence as the diminution and loss of this same speed when, by some impelling force, the stone is thrown to its former elevation: but even if you do not grant this, I do not see how you can doubt that the ascending stone, diminishing in speed, must before coming to rest pass through every possible degree of slowness.

SIMPL: But if the number of degrees of greater and greater slowness is limitless, they will never be all exhausted, therefore such an ascending heavy body will never reach rest, but will continue to move without limit always at a slower rate; but this is not the observed fact.

SALV: This would happen, Simplicio, if the moving body were to maintain its speed for any length of time at each degree of velocity; but it merely passes each point without delaying more than an instant: and since each time-interval however small may be divided into an infinite number of instants, these will always be sufficient [in number] to correspond to the infinite degrees of diminished velocity.[260]

That such a heavy rising body does not remain for any length of time at any given degree of velocity is evident from the following: because if, some time-interval having been assigned, the body moves with the same speed in the last as in the first instant of that time-interval, it could from this second degree of elevation be in like manner raised through an equal height, just as it was transferred from the first elevation to the second, and by the same reasoning would pass from the second to the third and would finally continue in uniform motion forever.[261]

[260] The meaning of this is as follows: the infinite number of values assumed by velocity between the initial value and its final zero match the infinite instants contained in the time duration of the ascent. Which is precisely that revolutionary concept of divisibility of a finite interval into an infinite number of infinitesimal parts discussed earlier in this chapter.

[261] Galileo had already expounded these concepts, and in a very similar manner, in his *Dialogue Concerning the Two Chief World Systems (*First Day, translated by S. Drake, The Modern Library, New York, 2001, p. 24. In Italian: *Dialogo sopra i due massimi sistemi del mondo*, Giornata prima, Edizione Nazionale, Vol. VII, p. 46). In the *Dialogue* Salviati had replied as follows to Simplicio's doubts:

I seem to gather from your remarks that a great part of your difficulty consists in accepting this very rapid passage of the movable body through the infinite gradations of slowness antecedent to the velocity acquired during the given time. [. . .] I tell you that the moving body does pass through the gradations, but without pausing in any of them. So that even if the passage requires but a single instant of time, still, since every small time contains infinite instants, we shall not lack a sufficiency of them to assign to each its own part of the infinite degrees of slowness, though the time be as short as you please.

Furthermore, the symmetry of the upward and downward motion of a body had also been clearly and correctly explained in the *Dialogue*. The speaker now is Sagredo (*Dialogue Concerning the Two Chief World Systems*, First Day, translated by S. Drake, The Modern Library, New York, 2001, p. 34. In Italian: *Dialogo sopra i due massimi sistemi del mondo*, Giornata prima, Edizione Nazionale, Vol. VII, p. 55):

Does not this cannon ball, sent perpendicularly upward by the force of the charge, continually decelerate in its motion until finally it reaches its ultimate height, where it comes to rest? And in diminishing its velocity—or I mean in increasing its slowness—is it not reasonable that it makes the change from 10 degrees to 11 sooner that from

Impetus and Galilean "*impeto*"

At this point comes a digression, introduced by Sagredo, regarding the causes of "natural motion" (i.e. the spontaneous fall of bodies) and of "violent motion" of bodies thrown upwards. Simplicio, of course, believes that the fall of a body from a state of complete rest occurs in a manner which is different from the fall of a body that, having reached its maximum height after being launched, begins to descend. And his reason? In the case of the former, the only operating force is weight, in the case of the latter, it is weight plus the remaining impressed force—that force which pre-Galilean philosophers referred to by the Latin word *impetus* and which was invoked to ensure the continuation of the motion of a body after it had been launched. The concept of impressed force associated to "violent motions" (i.e. non-spontaneous) was an entrenched Aristotelian prejudice, already discussed in Chapter 4. Let us recall Aristotle's words: "If a given force or power moves a given body at a given speed, twice as great a force or power will be needed to move the same body at twice as great a speed". The statement implies that a body moves only if a force is applied to it—a position which is extremely demanding, since, in the case of accelerated motion, it calls for a force which is not constant, but keeps changing in time. In the specific case of natural fall, it means assuming, as was done before Galileo, that the weight of a body increases as it approaches the ground.[262]

The Galilean principle of inertia, discussed in detail in Chapters 4 and 5, sweeps away any idea of the kind: natural motion is uniformly

10 to 12? [. . .] And, in short, from any degree to a closer one rather than to one more remote? [. . .] But what degree of slowness is there that is so distant from any degree of motion that the state of rest (which is infinite slowness) is not still farther from it? Whence no doubt can remain that the body, before reaching the point of rest, passes through all the greater and greater gradations of slowness and consequently through that one at which it would not traverse the distance of one inch in a thousand years. Such being the case, as it certainly is, it should not seem improbable to you, Simplicio, that the same ball in returning downward, leaving rest, recovers the velocity of its motion by returning through those same degrees of slowness through which it passed going up; nor should it, on leaving the larger degrees of slowness which are closer to the state of rest, pass by a jump to those farther away.

[262] For an examination of the various pre-Galilean conceptions, see for instance A. Carugo and L. Geymonat, *ibid.*, p. 764 and following.

accelerated, and this goes along with an acting force—that is, weight—which does not change during motion. Weight is also the only acting force in the ascent process of a body thrown upwards. However, the body is subject to two simultaneous and independent motions: fall, which is uniformly accelerated, and ascent, which is due to inertia and is characterized by a constant velocity (see diagram below). This latter prevails initially, but when the velocity related to the fall reaches and exceeds the velocity of ascent, descent begins. The process of ascent is exactly complementary to that of fall from a state of rest, as was illustrated in the diagram at the beginning of this chapter.

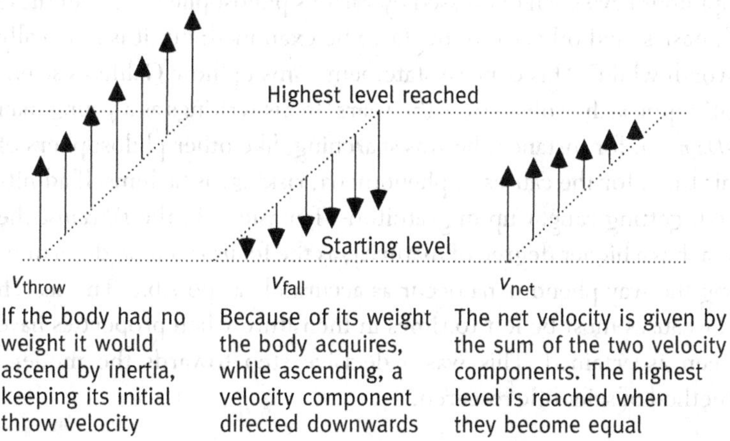
Highest level reached
Starting level
v_{throw}
If the body had no weight it would ascend by inertia, keeping its initial throw velocity
v_{fall}
Because of its weight the body acquires, while ascending, a velocity component directed downwards
v_{net}
The net velocity is given by the sum of the two velocity components. The highest level is reached when they become equal

It is important here to clarify the use Galileo makes (right from the first day of the *Dialogue*) of the Italian term "*impeto*", which we have come across several times in the original text. Although he never pauses to explain it, the meaning he ascribes to the term is quite different from the *impetus* of his Aristotelian predecessors. For him "*impeto*" is usually the velocity of a body in a general sense, with no dynamic implications. In Galileo's system, therefore, "*impeto*" is no longer the cause of motion, but rather the effect. This transformation is emblematic of the Galilean revolution, in which motion is no longer a process linked to a force, but instead is a *status* (state or condition) of the body, and it conserves itself.[263]

[263] In modern mechanics, Galilean *impeto* is best expressed in English by the term *momentum*, which is the product of the mass and the velocity of the body

There are two highlights in the excerpt we are about to read: Sagredo's elegant argument against Simplicio's Aristotelian view, culminating with the phrase: "it makes no difference whatever whether the fall of the stone is preceded by a period of rest which is long, short, or instantaneous provided only the fall does not take place so long as the stone is acted upon by a force opposed to its weight and sufficient to hold it at rest". And then Salviati's words immediately following this are also of great importance: "The present does not seem to be the proper time to investigate the cause of the acceleration of natural motion concerning which various opinions have been expressed by various philosophers . . . all these fantasies, and others too, ought to be examined; but it is not really worth while". This concise statement sums up how Galileo's scientific approach evolved over the years. In his writings as a young man (*De motu*, for instance) he was searching, like other philosophers of his time, for the causes of phenomena, risking, as he himself admitted, getting caught up in gratuitous "fantasies". In the *Discourses* he reaches a higher degree of maturity, as the focus is put on determining the way phenomena occur as accurately as possible. The search for causes must be left to times in the future when properties have been ascertained. This was a decisive step towards the modern method of scientific research.

Weight and violent action

SAGR: From these considerations it appears to me that we may obtain a proper solution of the problem discussed by philosophers, namely, what causes the acceleration in the natural motion of heavy bodies? Since, as it seems to me, the force[264] *impressed by the agent*

(mv). If the body is initially at rest, when it is subjected to a force F for a time duration t, its momentum after time t is equal to the product Ft, called the *impulse*. When the force is removed, the momentum acquired by the body is conserved; therefore, if the mass does not change, the velocity remains constant (unless effects of slowing down due to friction were present).

[264] A comment is essential. The word used by Galileo here and in the following lines is *virtù*—literally "virtue". He uses it as an equivalent to *impeto*, to refer to the

projecting the body upwards diminishes continuously, this force, so long as it was greater than the contrary force[265] *of gravitation, impelled the body upwards; when the two are in equilibrium, the body ceases to rise and passes through the state of rest in which the impressed impetus is not destroyed, but only its excess over that produced by gravitation has been consumed—the excess which caused the body to rise. Then as the diminution of the outside impetus continues, and gravitation gains the upper hand, the fall begins, but slowly at first on account of the opposing impetus, a large portion of which still remains in the body; but as this continues to diminish it also continues to be more and more overcome by gravity, hence the continuous acceleration of motion.*

SIMPL: The idea is clever, yet more subtle than sound; for even if the argument were conclusive, it would explain only the case in which a natural motion is preceded by a violent motion, in which there still remains active a portion of the external force; but where there is no such remaining portion and the body starts from an antecedent state of rest the cogency of the whole argument fails.

SAGR: I believe that you are mistaken and that this distinction between cases which you make is superfluous or rather non-existent. But, tell me, cannot a projectile receive from the projector either a large or a small force such as will throw it to a height of a hundred cubits, and even twenty or four or one?

SIMPL: Undoubtedly, yes.

SAGR: So therefore this impressed force[266] *may exceed the resistance of gravity so slightly as to raise it only a finger-breadth; and finally the force of the projector may be just large enough to exactly balance the resistance of gravity so that the body is not lifted at all but merely sustained. When one holds a stone in his hand does he do anything but give it a force impelling it upwards equal to the power of gravity*

momentum of a body in motion. Crew and de Salvio's translation of *virtù* by "force" is quite misleading, except when it is put in Simplicio's mouth, since it seems to reintroduce the Aristotelian view of upward motion as being due to an "impressed force" which gradually fades out—precisely the point that Galileo is trying to disprove.

[265] *Contrary force*—again, to be understood as the contrary momentum (or velocity) acquired by action of the gravitational force.

[266] See footnote 264.

drawing it downwards? And do you not continuously impress this force upon the stone as long as you hold it in the hand? Does it perhaps diminish with the time during which one holds the stone?

And what does it matter whether this support which prevents the stone from falling is furnished by one's hand or by a table or by a rope from which it hangs? Certainly nothing at all. You must conclude, therefore, Simplicio, that it makes no difference whatever whether the fall of the stone is preceded by a period of rest which is long, short, or instantaneous provided only the fall does not take place so long as the stone is acted upon by a force[267] *opposed to its weight and sufficient to hold it at rest.*

SALV: The present does not seem to be the proper time to investigate the cause of the acceleration of natural motion concerning which various opinions have been expressed by various philosophers, some explaining it by attraction to the center, others to repulsion between the very small parts of the body,[268] *while still others attribute it to a certain stress in the surrounding medium which closes in behind the falling body and drives it from one of its positions to another. Now, all these fantasies, and others too, ought to be examined; but it is not really worth while. At present it is the purpose of our Author*[269] *merely to investigate and to demonstrate some of the properties of accelerated motion (whatever the cause of this acceleration may be)—meaning thereby a motion, such that the momentum of its velocity goes on increasing after departure from rest, in simple proportionality to the time, which is the same as saying that in equal time-intervals the body receives equal increments of velocity [. . .].*

Salviati's last words here are the introduction to the discussion about the law governing, in uniformly accelerated motion, the continuous increase in velocity. As noted at the beginning of this chapter, Galileo changed his mind with respect to his initial stance on this matter, but in doing so, he made a trivial mistake in reasoning.

[267] See footnote 264.

[268] Galileo's original phrase here was "*altri al restar successivamente manco parti del mezo da fendersi*". A more precise translation of this would be: "others to the fact that there remains an ever thinner part of the medium to pass through".

[269] *Our Author*—that is, Galileo.

Error upon error

SAGR: So far as I see at present, the definition might have been put a little more clearly perhaps without changing the fundamental idea, namely, uniformly accelerated motion is such that its speed increases in proportion to the space traversed; so that, for example, the speed acquired by a body in falling four cubits would be double that acquired in falling two cubits and this latter speed would be double that acquired in the first cubit. [. . .]

SALV: It is very comforting to me to have had such a companion in error; and moreover let me tell you that your proposition seems so highly probable that our Author himself admitted, when I advanced this opinion to him, that he had for some time shared the same fallacy. But what most surprised me was to see two propositions so inherently probable that they commanded the assent of everyone to whom they were presented, proven in a few simple words to be not only false, but impossible.

SIMPL: I am one of those who accept the proposition, and believe that a falling body acquires force in its descent, its velocity increasing in proportion to the space, and that the momentum[270] *of the falling body is doubled when it falls from a doubled height; these propositions, it appears to me, ought to be conceded without hesitation or controversy.*

SALV: And yet they are as false and impossible as that motion should be completed instantaneously; and here is a very clear demonstration of it. If the velocities are in proportion to the spaces traversed, or to be traversed, then these spaces are traversed in equal intervals of time; if, therefore, the velocity with which the falling body traverses a space of eight feet were double that with which it covered the first four feet just as the one distance is double the other) then the time intervals required for these passages would be equal. But for one and the same body to fall eight feet and four feet in the same time is possible only in the case of instantaneous [discontinuous] motion; but observation shows us that the motion of a falling body occupies time, and less of it in covering a distance of four feet than of eight feet; therefore it is not true that its velocity increases in proportion to the space.

The falsity of the other proposition may be shown with equal clearness. For if we consider a single striking body the difference of

[270] At last the term *momentum* is being used! Simplicio uses it to mean velocity.

momentum in its blows can depend only upon difference of velocity; for if the striking body falling from a double height were to deliver a blow of double momentum, it would be necessary for this body to strike with a doubled velocity; but with this doubled speed it would traverse a doubled space in the same time-interval; observation however shows that the time required for fall from the greater height is longer.

Whoa there! We must call a halt here and comment on the mistake in Salviati's reply, which surely will not have escaped the reader. The statement "If the velocities are in proportion to the spaces traversed, or to be traversed, then these spaces are traversed in equal intervals of time" is incorrect, because it is based on the calculation of the elapsed times as simple ratios between the space covered and a velocity taken as proportional to this space. If such were the case, the ratio space/velocity would simply be a constant and the time elapsed would always be the same independently of the space covered, which is manifestly impossible. Galileo's reasoning is the equivalent of taking for the velocity its average value with respect to space, rather than with respect to time, as is correct.[271] Galileo was led into this error by his preference—in his desire to arrange reality into a geometric order—for the domain of space rather than that of time; and also by the fact that he did not appreciate sufficiently the deep, inseparable connection that exists between motion and time.

As Drake pointed out,[272] the concept of instantaneous velocity, which is quite clear to us today, was for Galileo, in the initial stage of his studies on motion (i.e. up to 1608), a self-contradictory notion. For him, velocity implied motion and motion implied the elapsing

[271] Let us take a clarifying example. In covering the distance of 100 km between two cities, a car driver travels half the distance at a velocity of 50 km/h, and the other half at 150 km/h. What are the average velocity and the traveling time? According to Galileo's reasoning the average velocity would be 100 km/h and the traveling time 1 h exactly. One can immediately see that this cannot be the case, since the car requires a full hour to cover just the first half of the distance. The average has to be calculated with respect to times, which are 1 h for the first half and 1/3 of an hour for the second half, making a total of 4/3 of an hour. The average velocity is, then, 100/(4/3) = 75 km/h.

[272] S. Drake, *Galileo: Pioneer Scientist*, University of Toronto Press, Toronto, 1994, p. 103.

of a certain amount of time, no matter how small. And, as a consequence, a velocity that increased by steps rather than continuously. Even if later, as this chapter will show, Galileo reached a position which was conceptually valid, he lacked the mathematical tool—differential calculus—needed for calculating the average velocity and the elapsed time. The problem, dealt with in the Mathematical Note at the end of the chapter, will yield a result which is quite different from Galileo's. It will be found, among other things, that the proportionality between velocity and space is not possible in the case of motion that starts from a state of rest, such as is the motion of the natural fall of a body. This is an interesting example of how, in the absence of mathematical support, logical reasoning may fall into surprising traps.

It is worth noting that the explanation put into Salviati's mouth—though wrong—forces Galileo to look for another behavior for the velocity, and this turns out to be the right one. In doing this Galileo shows us how important it is to have a logical understanding of the experimental behavior. In the absence of such understanding, it is indispensable to go back to the problem using a more rigorous approach, until a law is found which is not in contradiction with the experiment. It is also amusing that Sagredo, normally so sagacious in his replies to Salviati, here accepts his reasoning acritically, and even goes so far as to regret that the extreme simplicity of the argumentations may, in the view of many, subtract from the importance of the scientific issue.

Sagredo's words are nonetheless interesting, because they shed light on the way science was conceived and practiced in the academic world of the time (but have such bad habits disappeared today?). What counted was not so much content, as lengthy disquisition which allowed ostentation of learning and verbose dialectic abilities. Salviati's search for brevity and clarity is the polar opposite of practices in that world where knowledge, produced and used in closed circles, was the patrimony and privilege of a select few, a source of individual power rather than a discovery for the common good. Salviati appears to be the ideal prototype of the science popularizer: a difficult profession that nowadays is practiced by many, sometimes, unfortunately, by persons who would be better dubbed science vulgarizers.

A bitter invective

Encouraged by Sagredo, Salviati launches into an invective against those who cannot accept that their mistakes should be revealed in a clear and incontrovertible manner (ironically, precisely on an occasion when he himself is in error!). His words are bitter and sorrowful, and show no trace of the corrosive sarcasm Galileo often used in addressing his adversaries. Written when he was interned, following his conviction and abjuration, and highly autobiographical (the mention of the Academician is intended to leave no doubt about the matter), they reflect the frame of mind of a man who has suffered at length and is trying to understand the reasons why he has been persecuted. More than envy, it is a question, says Galileo, of a kind of destructive irrational impulse—"a stimulus and a strong desire"—which damages those who harbor it just as much as those who are the target of it, leading them to deny what, deep down, they know to be true, solely in order that the value of their adversary should not be recognized. Galileo's psychological analysis is no less accurate and no less valid today, than his many scientific analyses: suffice to think of the acrimony with which certain authors reacted to the Nobel prize for literature being awarded to Dario Fo.

SAGR: You present these recondite matters with too much evidence and ease; this great facility makes them less appreciated than they would be had they been presented in a more abstruse manner. For, in my opinion, people esteem more lightly that knowledge which they acquire with so little labor than that acquired through long and obscure discussion.

SALV: If those who demonstrate with brevity and clearness the fallacy of many popular beliefs were treated with contempt instead of gratitude the injury would be quite bearable; but on the other hand it is very unpleasant and annoying to see men, who claim to be peers of anyone in a certain field of study, take for granted certain conclusions which later are quickly and easily shown by another to be false. I do not describe such a feeling as one of envy, which usually degenerates into hatred and anger against those who discover such fallacies; I would call it a stimulus and a strong desire to maintain old errors, rather than accept newly discovered truths. This desire at times induces them to unite against these truths, although at heart believing in them, merely for the purpose of lowering the esteem in which certain others are held by the unthinking crowd. Indeed, I have heard from our Academician[273] *many such fallacies held as true but easily refutable; some of these I have in mind.*

SAGR: You must not withhold them from us, but, at the proper time, tell us about them even though an extra session be necessary. But now, continuing the thread of our talk, it would seem that up to the present we have established the definition of uniformly accelerated motion which is expressed as follows: Motum aequabiliter, seu uniformiter, acceleratum dicimus eum, qui, a quiete recedens, temporibus aequalibus aequalia celeritatis momenta sibi superaddit.[274]

SALV: This definition established, the Author makes a single assumption, namely, Accipio, gradus velocitatis eiusdem mobilis super diversas planorum inclinationes acquisitos tunc esse aequales, cum eorumdem planorum elevationes aequales sint.[275]

[. . .]

The experiment of the inclined plane

Having established that in uniformly accelerated motion proportionality between velocity and space covered does not exist, Galileo turns to the second alternative, namely, that velocity increases with the time elapsed from the moment the motion began. Note that he

[273] *Our Academician*—that is, Galileo.

[274] *Motum . . . superaddit*—A motion is said to be equally or uniformly accelerated when, starting from rest, its momentum receives equal increments in equal times.

[275] *Accipio . . . sint*—The speeds acquired by one and the same body moving down planes of different inclinations are equal when the heights of these planes are equal.

had no differential calculus at his disposal, with which to demonstrate immediately the validity of this choice, and so he needed an experimental verification.

Unfortunately, given the crudity of the means for measuring time that were available to him, it was not possible for him to determine instantaneous velocity. He was therefore restricted to measuring the space covered as a function of time, and from these data he derived the information he was looking for, using as a basis all the intuitions and the reasoning reported above.[276] This experiment, if performed with a body falling along the perpendicular, presented him with insurmountable difficulties because the whole process of fall takes place in much too short a time. In fact, an observer cannot even judge whether the velocity increases or remains constant. The solution was to have the body—a metallic ball—roll down an inclined groove; gently inclined, though, so as to make the times involved much longer and permit their easy measurement, even by the primitive system of a leaking vessel of water (the *water* chronometer, as described by Galileo). The fact that an acceleration is taking place is fully evident in this case, even to the naked eye. Furthermore, as the ball rolls down the noise it produces gradually increases, and this also indicates that its collisions with the microscopic unevenness of the grooves are getting more frequent in time.

Of course it is essential to demonstrate that the two falls, perpendicular and oblique, are governed by the same law. To this end, Salviati makes a second statement in Latin, which Sagredo immediately supports with geometrical argumentations. We will skip the rather long and detailed discussion and limit ourselves to illustrating the sense of the definition by means of a simple diagram (see below). Suffice to say that the equality of the velocity, at a given height along any inclined guide, is due to the conservation of energy: the decrease in the gravitational energy of the ball as it falls from height A to height B is entirely converted into kinetic energy (provided friction

[276] Elsewhere in the text Galileo explicitly states that "spaces covered are in double proportion with respect to times, and consequently to degrees of velocity"—where "double proportion" means quadratic dependence.

is negligible), and therefore it implies equal velocities, both of translation and of rotation.

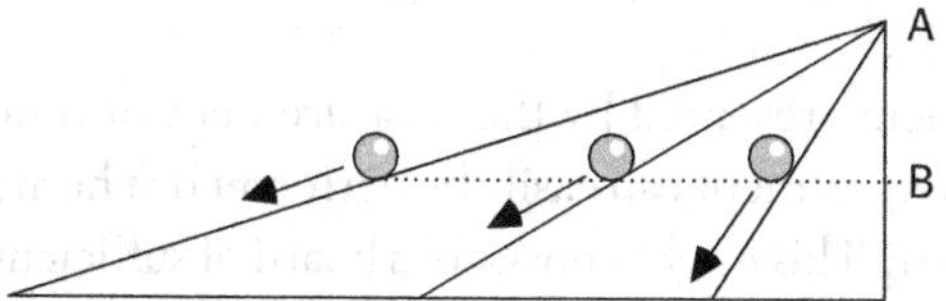

The velocity of the ball at any height is the same for any inclination of the guide

So let us now take a look at this celebrated experiment of the descent of a ball along an inclined plane. It is Simplicio who expresses the need for it, almost as if he had become, no less than Salviati himself, a genuine assertor of the need for experimental verification. Galileo's account of his original experiment has the richness of an illuminated manuscript. Yet some scholars, starting with his contemporaries Mersenne and Descartes, and more recently including Koyré,[277] expressed doubts that he actually carried out the measurements (in which case it would be just one more of the many *gedanken experimenten* Galileo conceived). According to these scholars, the experiment could not have confirmed the assumptions because it would have been affected by serious errors, arising in the main from residual friction and from imprecision in the measurement of time.

However, recently it was demonstrated that the inclined plane experiment can be performed perfectly well, even using Galileo's water chronometer.[278] The experiment was repeated successfully using a water vessel in whose bottom there was a small pipe through

[277] A. Koyré, *ibid*.

[278] T.B. Settle, "An Experiment in the History of Science", *Science*, Vol. 133, 1961, p. 19; "La rete degli esperimenti galileiani", in *Galileo e la scienza sperimentale*, edited by M. Baldo Ceolin, publication of the Physics Department "Galileo Galilei" of Padua Unversity, 1995, p. 13 and following. See also the discussion of the subject by Stillman Drake (S. Drake, *ibid*., Chapter 1). In another of the many repetitions of the inclined plane experiment, Naylor, instead, claimed he had noticed that the parchment Galileo would have used to line the groove was of no help in reducing the effects of friction: in fact, owing to the inevitable joins, it made them worse (R. Naylor, "Galileo and the problem of free fall", *British Journal of History of Science*, No. 7, 1974, p. 105). However, this and other objections by the same author do not seem to us particularly convincing.

which the water flowed during the interval of time to be measured (ending with the noise made by the ball hitting a stop at the finishing line). For a satisfactory result the slope of the guide needed to be between 10° and 40°–45°.

Today, since the obstacle presented by the measurement of time has been overcome, the experiment can easily be performed at home using makeshift equipment. This might consist of a board of sufficient length and smoothness, inclined at about 10°, with finishing lines drawn on it in chalk at varying distances from its upper end. The required times (average of a number of measurements) can be measured by means of an ordinary wrist-stopwatch. Two people need to be involved: one who lets the ball (initially held by a fingertip) go at a given instant; the other who starts the stopwatch as soon as the ball starts to move, and stops it when the ball crosses the chosen finishing line. In order to derive the value of the gravity acceleration—a result which can be achieved with good enough approximation—one must then analyze the data in the manner described in the Mathematical Note at the end of this chapter.

SIMPL: In truth, I find more pleasure in this simple and clear argument of Sagredo than in the Author's demonstration which to me appears rather obscure; so that I am convinced that matters are as described, once having accepted the definition of uniformly accelerated motion. But as to whether this acceleration is that which one meets in nature in the case of falling bodies, I am still doubtful; and it seems to me, not only for my own sake but also for all those who think as I do, that this would be the proper moment to introduce one of those experiments—and there are many of them, I understand—which illustrate in several ways the conclusions reached.

SALV: The request which you, as a man of science, make, is a very reasonable one; for this is the custom—and properly so—in those sciences where mathematical demonstrations are applied to natural phenomena, as is seen in the case of perspective, astronomy, mechanics, music, and others where the principles, once established by well-chosen experiments, become the foundations of the entire superstructure. I hope therefore it will not appear to be a waste of time if we discuss at considerable length this first and most fundamental question upon which hinge numerous consequences of which we have in this book only a small number, placed there by the Author, who has done so much to open a pathway hitherto closed to minds of speculative turn. So far as experiments go they have not been neglected by the Author; and often, in his company, I have attempted in the following manner to assure myself that the acceleration actually experienced by falling bodies is that above described.

A piece of wooden molding or scantling, about 12 cubits long, half a cubit wide, and three finger-breadths thick, was taken; on its edge was cut a channel a little more than one finger in breadth; having made this groove very straight, smooth, and polished, and having lined it with parchment, also as smooth and polished as possible, we rolled along it a hard, smooth, and very round bronze ball. Having placed this board in a sloping position, by lifting one end some one or two cubits above the other, we rolled the ball, as I was just saying, along the channel, noting, in a manner presently to be described, the time required to make the descent. We repeated this experiment more than once in order to measure the time with an accuracy such that the deviation between two observations never exceeded one-tenth of a pulse-beat. Having performed this operation and having assured ourselves of its reliability, we now rolled the ball only one-quarter the length of the channel; and having measured the time of its descent, we found it precisely one-half of the former. Next we tried other distances, comparing the time for the whole length with that for the half, or with that for two-thirds, or three-fourths, or indeed for any fraction; in such experiments, repeated a full hundred times, we always found that the spaces traversed were to each other as the squares of the times, and this was true for all inclinations of the plane, i.e., of the channel, along which we rolled the ball. We also observed that the times of descent, for various inclinations of the plane, bore to one another precisely that ratio which, as we shall see later, the Author had predicted and demonstrated for them.

For the measurement of time, we employed a large vessel of water placed in an elevated position; to the bottom of this vessel was soldered a pipe of small diameter giving a thin jet of water, which we collected in a small glass during the time of each descent, whether for the whole length of the channel or for a part of its length; the water thus collected was weighed, after each descent, on a very accurate balance; the differences and ratios of these weights gave us the differences and ratios of the times, and this with such accuracy that although the operation was repeated many, many times, there was no appreciable discrepancy in the results.

SIMPL: I would like to have been present at these experiments; but feeling confidence in the care with which you performed them, and in the fidelity with which you relate them, I am satisfied and accept them as true and valid.

SALV: Then we can proceed without discussion.

We, instead, shall take a break at this point, so as not to wear the reader out with too many details.

MATHEMATICAL NOTE

First we will restate that in uniformly accelerated motion—for instance "natural motion" of fall with constant gravitational acceleration—velocity increases in proportion to the elapsed time and space in proportion to the square of the elapsed time. We shall then go on to illustrate the consequences of the *a priori* assumption that velocity, instead, increases in proportion to space covered—consequences that are quite different from those arrived at by Galileo's defective self-correcting reasoning.

Uniformly accelerated motion

Let us indicate by t the time elapsed from the initial instant, by x the space covered at time t, by $v = dx/dt$ the velocity, by $a = dv/dt$ the acceleration (constant). The initial conditions are: for $t = 0$, position $x = 0$, and velocity zero. Velocity at time t is obtained by integrating acceleration, and space covered by means of a further integration:

$$v(t) = \int_0^t a\,dt = at \qquad x(t) = \int_0^t v\,dt = \frac{1}{2}at^2 \qquad (13.1)$$

(uniformly accelerated motion)

Average velocity

By reversing the second equation written, one obtains the time elapsed in going from 0 to $x(t)$:

$$t=\sqrt{\frac{2x}{a}} \tag{13.1'}$$

(time needed to reach position x)

The average velocity v_m in the interval from zero to point x is given by

$$v_m=\frac{1}{t}\int_0^{t=\sqrt{2x/a}} v(t)dt=\frac{1}{t}\int_0^t atdt=\frac{at}{2}=\sqrt{\frac{ax}{2}}$$

(average velocity in the interval 0 to x)

Note that this value coincides with the velocity of the body at time $t/2$, that is, at half the time elapsed, not with the value of the velocity at $x/2$, as to all effects is assumed in Galileo's defective reasoning. The time needed for the body to move from 0 to x is then $\sqrt{(2x/a)}$, the time to move from 0 to $2x$ is $\sqrt{(4x/a)}$, that is, greater by a factor $\sqrt{2}$. One can arrive at the same results by calculating the average velocity, as is easily verifiable.

The mistaken conception

Let us see the consequences of the mistaken assumption $v = kx$, where k is a constant. The space covered is obtained by integrating the equation

$$\frac{dx}{dt}=kx$$

namely, for an origin of the motion at x_0,

$$\int_{x_0}^{x}\frac{dx}{x}=k\int_0^t dt$$

which leads immediately to the following expressions for space covered, velocity, and acceleration:

$$x = x_0e^{kt} \qquad v = x_0ke^{kt} \qquad a=x_0k^2e^{kt}$$

(motion with velocity proportional to space covered)

that is to say an exponential increase in all three quantities. The acceleration, in particular, is not constant and therefore the motion is not a uniformly accelerated one. Moreover, if one wanted to describe a free fall motion starting from rest using the written expressions, the initial condition that velocity be zero for $t = 0$ would be possible only if $x_0 = 0$, which implies that space remains always identically zero, that is, that there is no motion. This conclusion is completely different from that which is reached following Galileo's reasoning.[279]

The inclined plane

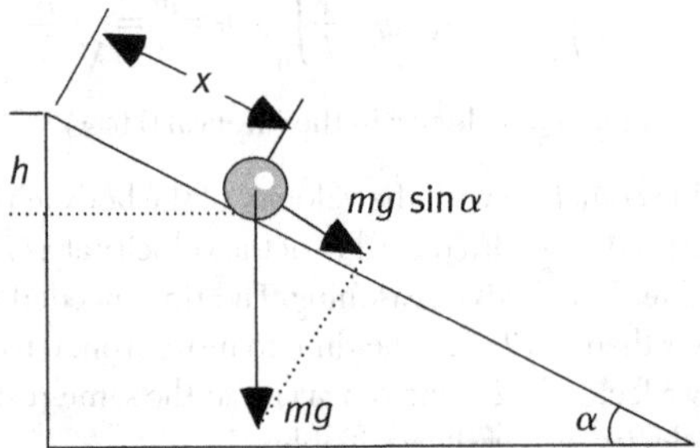

Let us now come to the experiment of the spherical ball that rolls down a groove in an inclined plane, and show first of all that its velocity at a given height is independent of the slope angle. If α is the angle and the top of the groove is taken as the origin of space covered x, in the absence of friction Newton's equation, describing the motion along the guide is the following (let us assume in the first instance that the ball slides without rolling):

$$m\frac{d^2x}{dt^2} = mg \sin \alpha$$

where m is the mass of the ball and g is the acceleration of gravity, then $mg \sin \alpha$ is the projection of the weight force in the direction parallel to the guide. The motion is uniformly accelerated with acceleration $g \sin \alpha$. Taking the origin of time when $x = 0$, by integration one gets velocity and space as a function of time

$$v = \frac{dx}{dt} = gt \sin \alpha \qquad x = \frac{1}{2}gt^2 \sin \alpha$$

Eliminating time one obtains

[279] In his example, if $x_1 = 2$ cubits (velocity $v_1 = kx_1$) and $x_2 = 4$ cubits (velocity $v_2 = kx_2$), the traveling time from x_1 to x_2 is not zero, as Galileo would have it, but rather $t_2 - t_1 = \ln 2/k$.

$$\frac{v^2}{g^2 \sin^2\alpha} = \frac{2x}{g \sin \alpha}$$

and finally, since $h = x \sin \alpha$,

$$v = \sqrt{2gh} \tag{13.2}$$

(velocity depends exclusively on g and on the lowering of height h)

The same result can be achieved more readily by applying the energy conservation principle. For a lowering in height h, the potential energy loss by the ball is mgh; and since it transforms entirely into translational kinetic energy, one can write:

$$mgh = \frac{1}{2}mv^2 \tag{13.3}$$

a relationship that brings us back to (13.2). This latter approach allows the effect of rolling to be introduced. Since, in order to take place, rolling requires energy, instead of (13.3) one must write

$$mgh = \frac{1}{2}mV^2 + \frac{1}{2}I\frac{V^2}{R^2} \tag{13.4}$$

(loss of potential energy = translational kinetic energy of
the center of the ball + rotational kinetic energy
about an axis passing through the center of the ball)

with I = moment of inertia of the spherical ball with respect to an axis passing through its center and R = radius of the ball. Velocity V turns out to be smaller than v in (13.3). By comparison of (13.3) and (13.4), taking into account that $I = (2/5)mR^2$, one has

$$mv^2 = \frac{7}{5}mV^2$$

and from this

$$V = \sqrt{5/7}v = 0.845v = 0.845\sqrt{2gh}$$

The acquired velocity, therefore, still depends solely on height, however it is reduced by about 15% compared to the case when rolling did not occur.

Determination of the gravitational acceleration

In order to determine the gravitational acceleration g, one must first establish a relationship between the velocity V and the acceleration A, after a lowering in the height by h. In uniformly accelerated motion one has

$$V = At \qquad x = \frac{h}{\sin\alpha} = \frac{1}{2}At^2$$

and by combining the two equations

$$A = \frac{V^2}{2x} = \frac{(5/7)2gh}{2(h/\sin\alpha)} = \frac{5}{7}g\sin\alpha$$

Rolling causes a decrease in translational acceleration from the value $g \sin \alpha$ to a value $(5/7)g \sin \alpha$, which corresponds to the following relationship between space and time:

$$x = \frac{1}{2}\left(\frac{5}{7}g\sin\alpha\right)t^2$$

Galileo did not consider this problem of reduction in acceleration, which had no influence on his measurements, however, because these were never absolute, being instead always expressed in terms of ratios and proportions between physical quantities.

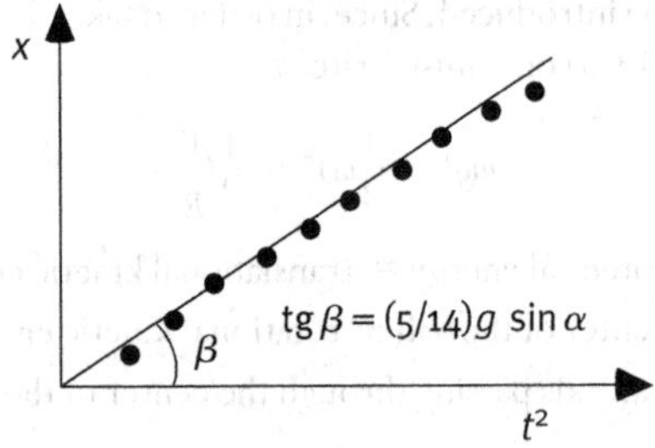

Each point represents the average of various measurements

The gravitational acceleration is determined by looking for the straight line that best fits the experimental data, iterated and represented graphically as x versus t^2. The slope of the line is

$$\tan\beta = \frac{5}{14}g\sin\alpha$$

Accurate measurement of the slope angle α enables g to be readily deduced. An example of data obtainable with the makeshift equipment described earlier is shown in the diagram. The effect of friction is more pronounced for long traveling times, because these correspond to higher velocities. Of course, friction leads to an underestimation of gravitational acceleration.

Galileo's constant

Since the fall motion of a body and that of a pendulum weight of equal mass are both caused by the gravitational force, it is possible to establish an appropriate relationship between them where this force is made to disappear. Let us consider the ratio between the time needed for one quarter of a (very small) oscillation of the pendulum—that is, the simple descent from the highest point of the oscillation to the vertical to ground—and the time for free fall along a distance L, equal to the pendulum length. The former time can be derived from (6.2) in the Mathematical Note of Chapter 6, the latter from (13.1′) above, where we put $a = g$ and $x = L$:

$$t_{\text{pend}} = \frac{\pi}{2}\sqrt{\frac{L}{g}} \qquad t_{\text{fall}} = \sqrt{\frac{2L}{g}}$$

where it is seen that both times correlate to the square root of L/g in the same manner. The ratio of the two times

$$c = \frac{\pi}{2\sqrt{2}} = 1.1110721$$

is therefore a universal constant. Whatever the value of L and of the gravitational acceleration—thus for any pendulum, whether on the moon or on planets other than the earth—the result is exactly the same. Stillman Drake[280] has called this constant c the *constant of Galileo*.

Through thorough analysis of Galileo's work notes, Drake[281] established that the value of the ratio of the two times, as measured for a given L, is in fact $942/850 = 1.11$, which is to all effects identical to the one calculated above using the equations of modern mechanics. Even if such an exact correspondence suggests some chance was involved, it demonstrates that Galileo must have been a first-rate experimenter. It is also evident that, in the face of data obtained for different values of L which always led to the same value for c, the fact that Galileo knew that the pendulum oscillation time depended on the square root of the pendulum length made it rather plain to him that the relationship between space and time in uniformly accelerated motion had to be quadratic.

[280] S. Drake, *ibid.*, p. 237.
[281] S. Drake, *ibid.*, Chapter 1.

PART VII

BODIES AND THE ENVIRONMENT

14

Archimedes and the Weight of Air*

l'otro gonfiato pesa più che sgonfiato

a leather bottle weighs more when inflated than when collapsed

After briefly recalling Archimedes' principle of buoyancy, Galileo discusses in detail a method for determining the weight of air. Since the weight in question is very small, he uses extremely fine grains of sand as a counterweight. It is remarkable that, in spite of the crude instruments available to him, his result is not too far from the correct one.

In this passage from the *Discourses*[282] Galileo describes the ingenious method—in fact, already conceived by Aristotle—by means of which he succeeded in demonstrating that air, contrary to Simplicio's simple-minded suppositions, has weight. He establishes a value for it some 400 times smaller than that of an equal volume of water—a value which is approximately half the correct one. Galileo's description is detailed enough not to require any advance explanation; we shall limit ourselves to illustrating the text with diagrams where the discussion becomes more complicated.

* From *Discourses*.

[282] *Two New Sciences*, translated by H. Crew and A. de Salvio, Dover Publications, New York, 1954, p. 76 and following. In Italian: *Discorsi e dimostrazioni matematiche intorno a due nuove scienze*, Giornata prima, Edizione Nazionale, Vol. VIII, p. 121 and following.

Archimedes' buoyancy

A subtle problem is that the measurement of the weight of air is influenced by buoyancy, which has the effect of diminishing its weight. Since weighings are usually made in air, when it is the air itself that has to be weighed, it is extremely important to clarify the role played by buoyancy. This is the reason why Galileo starts the passage by restating the concept. The principle of buoyancy is well-known, so there is no need for us to dwell on it here: a body immersed in a fluid receives an upward force equal to the weight of the displaced fluid. Therefore, the net force acting on the body is the difference between the weight of the body and this upward force. If the difference is positive, the body sinks; otherwise, it rises to the surface and floats on the fluid, remaining immersed for that fraction of its volume which causes the two opposing forces to balance each other out.

Terminal velocity

Contrary to the Aristotelian view, which asserted that the fall velocity of bodies is proportional to their weight, Galileo demonstrated (as discussed in Chapter 3) that in the definition of velocity, the key role is played by two opposing forces: the weight of the body (whenever buoyancy can be ignored) and the resistance offered by friction in the medium. The weight force is constant, the friction force increases with increasing fall velocity. The motion is initially accelerated, but evolves towards a terminal condition of constant velocity (a saturation velocity, which elsewhere we called the "parachute effect") when the two forces become equal. The more viscous the medium is, the stronger the friction, and the sooner the terminal condition is reached.

And this latter case is precisely what is considered at the beginning of the passage, with the talk focusing on fluids that are viscous enough to cause a very rapid transitory phase and, to all practical effects, a fall with uniform velocity. The successive positions reached at equal intervals of time by a falling body in different media are shown in the diagram below. Now, if the density of the body is not much higher than the density of the medium, buoyancy cannot be ignored. When this is the case, the opposing forces contributing to the establishment of the saturation velocity are, on the one hand, the weight reduced by buoyancy and, on the other hand, the friction force. If the mean density of the fluid and of the body immersed in it should be equal (see the wax-ball game in the next chapter), the total force would be zero and the body would neither sink nor rise to the surface. This is the case with air weighed in air, discussed here—its weight, measured by a balance, is to all appearances zero.

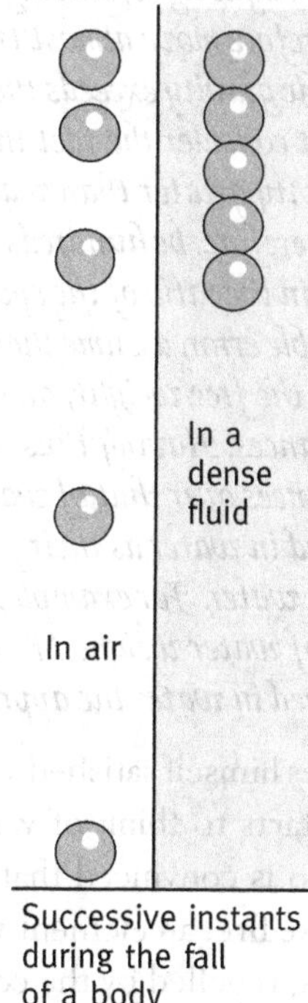

Before coming to the text, one last comment. It is curious that Galileo, despite having guessed and duly verified that air has a weight, did not use this notion to explain the so-called "force of a vacuum", that is, the fact that a force is needed to pull a piston out of a vacuum-sealed cylinder. We now know the reason why this happens is that the atmosphere presses on the external surface of the piston, but not on the internal surface, which is not exposed to it (atmospheric pressure is hydrostatic, and therefore it acts on all surfaces in contact with it, independently of their orientation). This argument will be dealt with in Chapter 17.

SALV: [. . .] Again take a solid a little heavier than water, such as oak, a ball of which will weigh let us say 1000 drachms; suppose an equal volume of water to weigh 950, and an equal volume of air, 2; then it is clear that if the unhindered speed of the ball is 1000, its speed in air will be 998, but in water only 50, seeing that the water removes 950 of the 1000 parts which the body weighs, leaving only 50.

Such a solid would therefore move almost twenty times as fast in air as in water, since its specific gravity exceeds that of water by one part in twenty. And here we must consider the fact that only those substances which have a specific gravity greater than water can fall through it—substances which must, therefore, be hundreds of times heavier than air; hence when we try to obtain the ratio of the speed in air to that in water, we may, without appreciable error, assume that air does not, to any considerable extent, diminish the free weight, and consequently the unhindered speed of such substances. Having thus easily found the excess of the weight of these substances over that of water, we can say that their speed in air is to their speed in water as their free weight is to the excess of this weight over that of water. For example, a ball of ivory weighs 20 ounces; an equal volume of water weighs 17 ounces; hence the speed of ivory in air bears to its speed in water the approximate ratio of 20:3.

Whereas Sagredo declares himself satisfied with his friend's explanations and immediately starts to think of ways they can be verified experimentally, Simplicio is convinced that air has no weight and, indeed, that it behaves like fire, an element which Aristotle deemed "light in absolute", that is, repelled by the center of the earth, or, to put it another way, affected by a negative force of gravity.[283] And this

[283] This belief was still widespread among Galileo's contemporaries, even if it had already been rejected by Giovanni Battista Benedetti in his *Diversarum speculationum mathematicarum liber* of 1585, and by Francesco Buonamici in his *De motu libri* of 1591. Galileo, too, had initially subscribed to the Aristotelian thesis, switching

time, oddly enough, it is left to Salviati to remind him that Aristotle did, in fact, believe that air has weight. But Salviati's invoking of the principle of authority here is just an expedient he adopts, as on other occasions, to help the discussion along in a friendly, playful way.

SAGR: I have made a great step forward in this truly interesting subject upon which I have long labored in vain. In order to put these theories into practice we need only discover a method of determining the specific gravity of air with reference to water and hence with reference to other heavy substances.
SIMPL: But if we find that air has levity instead of gravity what then shall we say of the foregoing discussion which, in other respects, is very clever?
SALV: I should say that it was empty, vain, and trifling. But can you doubt that air has weight when you have the clear testimony of Aristotle affirming that all the elements have weight including air, and excepting only fire? As evidence of this he cites the fact that a leather bottle weighs more when inflated than when collapsed.
SIMPL: I am inclined to believe that the increase of weight observed in the inflated leather bottle or bladder arises, not from the gravity of the air, but from the many thick vapors mingled with it in these lower regions. To this I would attribute the increase of weight in the leather bottle.
SALV: I would not have you say this, and much less attribute it to Aristotle; because, if speaking of the elements, he wished to persuade me by experiment that air has weight and were to say to me: "Take a leather bottle, fill it with heavy vapors and observe how its weight increases". I would reply that the bottle would weigh still more if filled with bran; and would then add that this merely proves that bran and thick vapors are heavy, but in regard to air I should still remain in the same doubt as before. [. . .]

The text

Salviati now goes on to illustrate Aristotle's experiment, which he assures his interlocutors he has repeated with extreme care. This

to the concept of the relativity of weight (e.g. air is light compared to water, but heavy compared to hydrogen) only after reading Benedetti's book.

excerpt provides us with an interesting example of scientific writing, with its specialized lexicon, the meticulous record of the individual operations performed, and the evidence furnished by the descriptions. The experiment "shows precisely the opposite" of what "some have believed", as long as it is performed using an appropriate method and is clearly described, so that others can repeat it, avoiding the mistakes which Aristotle made. Salviati opens by immediately stating his thesis (in the Italian text, this is given particular prominence by the use of a syntactic inversion); next he advances arguments in its favor, and then he outlines the experiment. All this in a single, long sentence, somewhat similar to the abstract of today's scientific papers. Then follows the detailed description of the experiment, by which it is first shown that air has weight, without, however, calculating it; after which the weight of air is determined using two ingenious methods of measurement, the second of which is the simpler and more effective.

Weighing the air

SALV: The experiment with the inflated leather bottle of Aristotle proves conclusively that air possesses positive gravity and not, as some have believed, levity, a property possessed possibly by no substance whatever; for if air did possess this quality of absolute and positive levity, it should on compression exhibit greater levity and, hence, a greater tendency to rise; but experiment shows precisely the opposite.

As to the other question, namely, how to determine the specific gravity of air, I have employed the following method. I took a rather large glass bottle with a narrow neck and attached to it a leather cover, binding it tightly about the neck of the bottle: in the top of this cover I inserted and firmly fastened the valve of a leather bottle, through which I forced into the glass bottle, by means of a syringe, a large quantity of air. And since air is easily condensed one can pump into the bottle two or three times its own volume of air. After this I took an accurate balance and weighed this bottle of compressed air with the utmost precision, adjusting the weight with fine sand. I next opened the valve and allowed the compressed air to escape; then replaced the flask upon the balance and found it perceptibly lighter: from the sand which had been used as a counterweight I now removed and laid aside as much as was necessary to again secure balance. Under these conditions there can be no doubt but that the weight of the sand thus laid aside represents the weight of the air which had been forced into the flask and had afterwards escaped. But after all, this experiment tells me merely that the weight of the compressed air is the same as that of the sand removed from the balance; when, however, it comes to knowing certainly and definitely the weight of air as compared with that of water or any other heavy substance, this I cannot hope to do without first measuring the volume of compressed air; for this measurement, I have devised the two following methods.

According to the first method, one takes a bottle with a narrow neck similar to the previous one; over the mouth of this bottle is slipped a leather tube which is bound tightly about the neck of the flask; the other end of this tube embraces the valve attached to the first flask and is tightly bound about it.[284] *This second flask is provided with a hole in the bottom through which an iron rod can be placed so as to open, at will, the valve above mentioned and thus permit the surplus air of the first to escape after it has once been weighed: but this second bottle must be filled with water. Having prepared everything in the manner above described, open the valve with the rod; the air will rush into the flask containing the water and will drive it through the hole at the bottom, it being clear that*

[284] In reading this excerpt the reader may find it useful to refer to the drawing of the experimental apparatus, below:

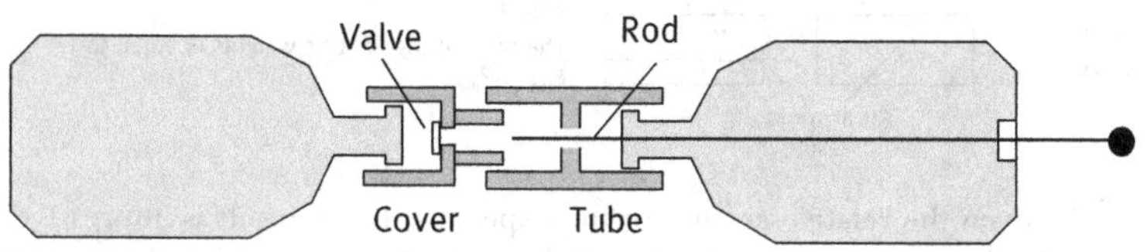

the volume of water thus displaced is equal to the volume of air escaped from the other vessel. Having set aside this displaced water, weigh the vessel from which the air has escaped (which is supposed to have been weighed previously while containing the compressed air), and remove the surplus of sand as described above; it is then manifest that the weight of this sand is precisely the weight of a volume of air equal to the volume of water displaced and set aside; this water we can weigh and find how many times its weight contains the weight of the removed sand, thus determining definitely how many times heavier water is than air;[285] *and we shall find, contrary to the opinion of Aristotle, that this is not 10 times, but, as our experiment shows, more nearly 400 times.*[286]

The second method is more expeditious and can be carried out with a single vessel fitted up as the first was. Here no air is added to that which the vessel naturally contains but water is forced into it without allowing any air to escape; the water thus introduced necessarily compresses the air. Having forced into the vessel as much water as possible, filling it, say, three-fourths full, which does not require any extraordinary effort, place it upon the balance and weigh it accurately; next hold the vessel mouth up, open the valve, and allow the air to escape; the volume of the air thus escaping is precisely equal to the volume of water contained in the flask. Again weigh the vessel which will have diminished in weight on account of the escaped air; this loss in weight represents

[285] To aid understanding, here is a breakdown of the steps of the experiment:

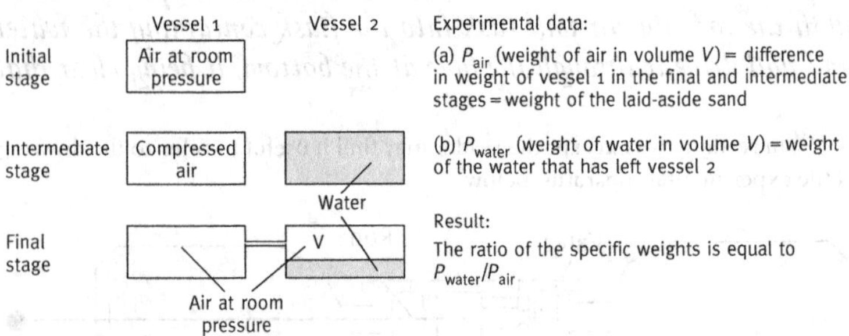

[286] Given the relative crudity of the experiment, the result is impressive, since the density of air is 0.001293, against 1 for water, meaning the correct ratio is 773.

the weight of a volume of air equal to the volume of water contained in the vessel.[287]

SIMPL: No one can deny the cleverness and ingenuity of your devices; but while they appear to give complete intellectual satisfaction they confuse me in another direction. For since it is undoubtedly true that the elements when in their proper places[288] *have neither weight nor levity, I cannot understand how it is possible for that portion of air, which appeared to weigh, say, 4 drachms of sand, should really have such a weight in air as the sand which counterbalances it. It seems to me, therefore, that the experiment should be carried out, not in air, but in a medium in which the air could exhibit its property of weight if such it really has.*

Simplicio's objection is really "to the point", but it is expressed in terms that are not immediately clear. What he means is this. Although, in an air medium, the effect of buoyancy on the sand is negligible, it is not so on the air: thus, one cannot state that the weight measured for the sand laid aside gives the weight of the air that escaped from the vessel. A valid measurement could only be made in a vacuum. This interpretation of Simplicio's meaning becomes immediately evident from Salviati's answer:

SALV: The objection of Simplicio is certainly to the point and must therefore either be unanswerable or demand an equally clear solution. It is perfectly evident that air which, under compression, weighed as much as the sand, loses this weight when once allowed to escape into its own element, while, indeed, the sand retains its weight. Hence for this

[287] To aid understanding, here is a breakdown of the steps of the experiment

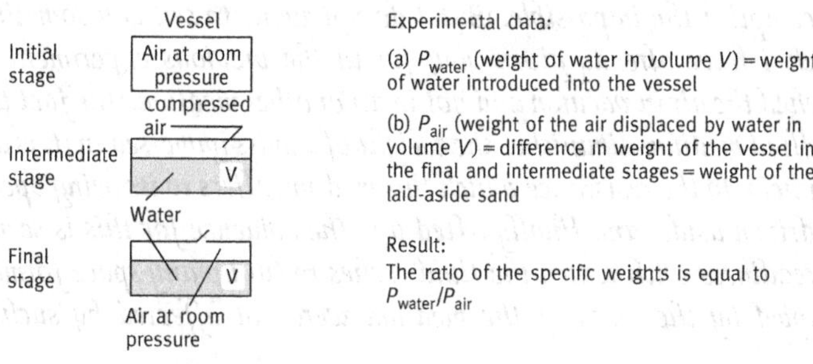

[288] Namely in a medium which is the same as the element itself.

experiment it becomes necessary to select a place where air as well as sand can gravitate; because, as has been often remarked, the medium diminishes the weight of any substance immersed in it by an amount equal to the weight of the displaced medium; so that air in air loses all its weight. If therefore this experiment is to be made with accuracy it should be performed in a vacuum where every heavy body exhibits its momentum[289] *without the slightest diminution. If then, Simplicio, we were to weigh a portion of air in a vacuum would you then be satisfied and assured of the fact?*
SIMPL: Yes truly: but this is to wish or ask the impossible.

Nothing is impossible for Galileo–Salviati, who immediately finds a way to provide an intelligent explanation. This explanation, however, is not a model of clarity and may leave the reader rather dissatisfied, in contrast to Simplicio, who declares himself fully persuaded. Nevertheless, we include it as an example (in truth very rare for Galileo) of excessively complicated reasoning, which even presents difficulties at the level of the language used. In its substance, what Salviati is saying is very simple: buoyancy exercised by the medium is due to its reluctance to admit into its interior an external body, that is, to redistribute itself. But the air (since the amount which is forced into the flask, in addition to that already present, does not modify the volume of the container) is not subject to buoyancy, and therefore it has the same weight as it would have in a vacuum. As for the sand, given its much higher density, it is legitimate in any case to disregard the buoyancy to which it is subject.

SALV: Your obligation will then be very great if, for your sake, I accomplish the impossible. But I do not want to sell you something which I have already given you; for in the previous experiment we weighed the air in vacuum and not in air or other medium. The fact that any fluid medium diminishes the weight of a mass immersed in it, is due, Simplicio, to the resistance which this medium offers to its being opened up, driven aside, and finally lifted up. The evidence for this is seen in the readiness with which the fluid rushes to fill up any space formerly occupied by the mass; if the medium were not affected by such an

[289] *Its momentum*—that is, the action of its weight.

immersion then it would not react against the immersed body. Tell me now, when you have a flask, in air, filled with its natural amount of air and then proceed to pump into the vessel more air, does this extra charge in any way separate or divide or change the circumambient air? Does the vessel perhaps expand so that the surrounding medium is displaced in order to give more room! Certainly not. Therefore one is able to say that this extra charge of air is not immersed in the surrounding medium for it occupies no space in it, but is, as it were, in a vacuum. Indeed, it is really in a vacuum; for it diffuses into the vacuities which are not completely filled by the original and uncondensed air. In fact I do not see any difference between the enclosed and the surrounding media: for the surrounding medium does not press upon the enclosed medium and, vice versa, the enclosed medium exerts no pressure against the surrounding one; this same relationship exists in the case of any matter in a vacuum, as well as in the case of the extra charge of air compressed into the flask. The weight of this condensed air is therefore the same as that which it would have if set free in a vacuum. It is true of course that the weight of the sand used as a counterpoise would be a little greater in vacuo *than in free air. We must, then, say that the air is slightly lighter than the sand required to counterbalance it, that is to say, by an amount equal to the weight in vacuo of a volume of air equal to the volume of the sand.*
SIMPL: The previous experiments, in my opinion, left something to be desired: but now I am fully satisfied.

It seems likely that when he re-read this page some time later, Galileo had doubts himself about the clarity of the explanations he had put into Salviati's mouth. This is suggested by the fact that, to make them clearer (in particular the second method, where a single vessel is used), he added the lines we are now going to read, at the foot of the page in his own copy of the first printing of the *Discourses*. The result does not strike us as noticeably better: on the contrary, in his efforts not to overlook any detail of the demonstration, he merely added further verbosity. Let us treat it as a useful exercise for our mental faculties.

SAGR: A very clever discussion, solving a wonderful problem, because it demonstrates briefly and concisely the manner in which one may find the weight of a body in vacuo by simply weighing it in air. The explanation is as follows: when a heavy body is immersed in air it loses in weight an amount equal to the weight of a volume of air equivalent to the volume of the body itself. Hence if one adds to a body, without expanding it, a quantity of air equal to that which it displaces and weighs it, he will obtain its absolute weight in vacuo, since, without increasing it in size, he has increased its weight by just the amount which it lost through immersion in air.

When therefore we force a quantity of water into a vessel which already contains its normal amount of air, without allowing any of this air to escape it is clear that this normal quantity of air will be compressed and condensed into a smaller space in order to make room for the water which is forced in: it is also clear that the volume of air thus compressed is equal to the volume of water added. If now the vessel be weighed in air in this condition, it is manifest that the weight of the water will be increased by that of an equal volume of air; the total weight of water and air thus obtained is equal to the weight of the water alone in vacuo.

Now record the weight of the entire vessel and then allow the compressed air to escape; weigh the remainder; the difference of these two weights will be the weight of the compressed air which, in volume, is equal to that of the water. Next find the weight of the water alone and add to it that of the compressed air; we shall then have the water alone in vacuo. To find the weight of the water we shall have to remove it from the vessel and weigh the vessel alone; subtract this weight from that of the vessel and water together. It is clear that the remainder will be the weight of the water alone in air.

MATHEMATICAL NOTE

To provide an expression for the descent (or the ascent) velocity, in the saturation regime, of a body immersed in a fluid, let us generalize the formula derived in the Mathematical Note of Chapter 3, so as to include buoyancy

$$v_{\text{terminal}} = \frac{P - A}{b}$$

where $P = \rho V g$ is the weight of the body, $A = \rho' V g$ is Archimedes' buoyancy, b is the friction factor of the medium, ρ is the density of the body, ρ' is the density of the fluid medium, V is the volume of the body, g is the gravitational acceleration.

For $\rho > \rho'$ the body sinks; for $\rho = \rho'$ one has the immobility condition; finally, for $\rho < \rho'$, the body rises to the surface and floats. In this latter case, only a portion of the body is immersed: the immersed volume v settles at a value that satisfies the equality $\rho V = \rho' v$. In the water-ice system, where ρ is only slightly smaller than ρ', the immersed volume v is only a little smaller thant the total volume V of the block of ice (behavior of an iceberg).

If the body immersed in the fluid is now subjected to a further force F of external origin, the equation that gives the velocity of the motion becomes

$$v_{\text{terminal}} = \frac{P - A + F}{b}$$

If we wanted to keep the body immobile inside the fluid, namely $v_{\text{terminal}} = 0$, it would be necessary to apply to it an external force equal to A-P. This allows an alternative formulation of the relationship between the fall velocity and the properties of the body in the medium, which is as follows: all other conditions being equal, the velocities acquired by a body in different media are proportional to the forces that one would need to apply in each case to inhibit its motion. This description is due to Benedetti, who once again preceded Galileo in giving the correct analysis of a physical phenomenon.[290]

[290] Giovanni Battista Benedetti, *ibid*.

15

Wax Balls and Dew Drops*

la giunta di due grani di sale solamente, che si mettino in sei libbre d'acqua, farà risalire dal fondo alla superficie quella palla che vi era pur allora scesa

the addition of two grains of salt to six pounds of water is sufficient to make the ball rise to the surface from the bottom to which it had fallen

Problems concerning bodies in contact with water. How do fish manage to stay suspended in midwater? To study this problem Galileo uses wax balls, which he gradually makes heavier by weighting them with grains of sand. Another question regards dew drops, which do not wet the leaves on which they form. Galileo does not have the right answer, nor could he have it, but amidst the conceptual mistakes, he is still capable of coming up with a brilliant intuition.

The air bladder

And now we have a second passage involving Archimedes' principle of buoyancy. In the introduction to Chapter 14, we saw how buoyancy is used in explaining the particular conditions under which a given body immersed in a liquid stays suspended, neither sinking nor rising

* From *Discourses*.

to the surface. Here let us briefly remind ourselves that this happens when the average densities of the body and the liquid are equal, because if this is the case the body's weight and the buoyancy balance each other out. Now Galileo turns his attention to fish, and attributes their ability to stay suspended in midwater to precisely such a balancing of forces. A colleague of his at the University of Padua, the anatomist Girolamo Fabrici from Acquapendente, had explained in one of his books that freshwater fish are equipped with a small air bladder, attached to their backbone. This enables them to increase their buoyancy, which is weaker in freshwater than in saltwater. Galileo adopts Fabrici's explanation, and in order to study the mechanism better he reproduces this property of fish by means of a simple expedient: he takes a wax ball, which is lighter than water, and weighs it down using grains of sand as ballast, thus permitting its average density to be increased in very small steps.

When the authors of this book were children a similar technique was used in constructing excellent submarines for launching in the goldfish pool. Wood as the base material and pins as ballast—perfection was never achieved, but the result was not bad (at least in terms of the goal, which was not so much to study fish as to confuse them!). Galileo, by his own account, was never completely successful either, which is quite understandable when one considers that equilibrium conditions are highly critical. In view of this, Sagredo's idea of "rigging" the experiment by having two layers of water—a lower one salt and an upper one fresh—was ingenious. The big difference in density makes it easy to find the right ballast needed for the wax ball to become positioned at the interface between the two layers.

Galileo suggests several other little tricks that can be done by altering the density of water—by adding grains of salt to it, for example, or a few drops of warmer or colder water. It seems likely he amused himself for hours playing around with experiments of this type, anticipating the builders of those "toys for grown ups", based on variations in the density of a liquid, which can be purchased today (objects which go up and down in water columns subjected to changes in temperature, pressure, or some other parameter).

It should be remembered that Galileo's pupil, Torricelli, was later to exploit this principle to build a kind of thermometer, in which small balls whose average density was around the same as that of water rose or sank according to changes in the ambient temperature. Considerable improvements to this method of measuring temperature (or "weighing" heat and cold, as it was put at the time) were made by another of Galileo's pupils, Viviani.[291]

Dew on leaves

Galileo was fascinated by the curious behavior some substances display when they come into contact with water. This can be seen in the second half of the passage we are going to read, where he tries to explain the phenomenon of dew drops which form on the leaves of plants, "without spreading out", almost as if held together by some kind of repellence ("they roll themselves up into a ball, and once they have assumed the spherical shape they remain in equilibrium").[292] Galileo noted that the effect does not occur if the air around the drops is substituted with red wine, which causes their immediate collapse. However, he was unable to provide an explanation for the phenomenon, other than talking of a generic antipathy between air and water, a sort of repulsion, which Salviati calls "antagonism" or "incompatibility". Nevertheless, we could still say that, in a sense, Galileo succeeded in arriving at a truth as a result of his mistaken conviction that water molecules do not exert bonding forces upon one another, that is, that the holding together of the drop is due not to volume properties, but instead to surface or, more precisely, to interface properties ("it must follow that the cause of this effect is external"). Indeed, this intuition was to stimulate investigation

[291] Details about these topics can be found in R. Caverni, *Storia del metodo sperimentale in Italia*, Vol. 2, Florence, 1895.

[292] Comment by Galileo upon Aristotle's *De Coelo*.

and research in the field not only by Galileo's pupils, but also by many other scholars.

It was, however, an intuition and nothing more. Certainly it would not have been possible for scientists at that time to come up with the concept of *surface tension*, which is what, in fact, is responsible for drop formation, as well as for capillary action in tubes. A concept such as this could only be formulated later, with the advance of physical and chemical knowledge concerning the structure of condensed matter. Furthermore, systematic experimental studies on capillarity were not to be made until after the middle of the seventeenth century, when the members of the *Accademia del Cimento* discovered, for instance, that the vast majority of liquids form a concave meniscus on the walls of a tube, whereas mercury forms a convex one; and, moreover, that the height of the meniscus depends on the properties of the liquid and on the diameter of the tube.

It should be noted that it did not occur to Galileo to link the mechanism of drop formation to that of capillarity. Worth mentioning (as regards water drops) is the opinion of Leonardo da Vinci, who is believed to have been the first to speak about capillarity. In his *Codice Atlantico* he stated:

> the gravity of water is twofold: one, common to all bodies, attracts it towards the center of the earth, the other, inherent to it, attracts it to the center of its own mass. But I see no possibility for the human mind to give an explanation for this. One can only say what is said of the iron-attracting magnet, namely that such virtue is a secret property, whose nature is infinite.

Leonardo's thinking thus differed from Galileo's in that he accepted the existence of cohesion forces, even if of unknown origin (his internal force seems to anticipate Newton's gravitational interaction). Today we know that the molecules of a liquid—like those of any other state of condensed matter—are held together by cohesion forces of an electromagnetic nature, that is predominantly attractive forces acting between electric charges.[293]

[293] The interested reader will find more details on the problem of surface tension in the Mathematical Note at the end of this chapter.

The text

Let us now read the dialogue,[294] where the turns are taken rapidly and are enlivened by a number of features: first, Salviati's enthusiasm for the subjects being discussed ("These are interesting questions . . ."); second, his admiration for the animal world, capable of teaching so much to whoever observes it; third, his thoroughly Galilean interest in the practical aspects of science (in this case, those concerning medicinal waters); and finally Sagredo's good humor as he laughingly recounts some of his experimental failures and a crafty trick he used to fool his friends. The variety of topics discussed helps make the conversation lively and realistic, but underlying the apparent diversity there is a unifying thread: the phenomena under analysis are never lost from sight, merely considered from different angles.

SAGR: Seeing that Simplicio is silent, I will take the opportunity of saying something. Since you have clearly demonstrated that bodies of different weights do not move in one and the same medium with velocities proportional to their weights, but that they all move with the same speed, understanding of course that they are of the same substance or at least of the same specific gravity; certainly not of different specific gravities, for I hardly think you would have us believe a ball of cork moves with the same speed as one of lead; and again since you have clearly demonstrated that one and the same body moving through differently resisting media does not acquire speeds which are inversely proportional to the resistances, I am curious to learn what are the ratios actually observed in these cases.

Neither up nor down

SALV: These are interesting questions and I have thought much concerning them. I will give you the method of approach and the result

[294] From *Two New Sciences*, First Day, translated by H. Crew and A. de Salvio, Dover Publications, New York, 1954, p. 68 and following. In Italian: *Discorsi e dimostrazioni matematiche intorno a due nuove scienze*, Giornata prima, Edizione Nazionale VIII, p. 112 and following.

which I finally reached. Having once established the falsity of the proposition that one and the same body moving through differently resisting media acquires speeds which are inversely proportional to the resistances of these media, and having also disproved the statement that in the same medium bodies of different weight acquire velocities proportional to their weights (understanding that this applies also to bodies which differ merely in specific gravity), I then began to combine these two facts and to consider what would happen if bodies of different weight were placed in media of different resistances; and I found that the differences in speed were greater in those media which were more resistant, that is, less yielding. This difference was such that two bodies which differed scarcely at all in their speed through air would, in water, fall the one with a speed ten times as great as that of the other. Further, there are bodies which will fall rapidly in air, whereas if placed in water not only will not sink but will remain at rest or will even rise to the top: for it is possible to find some kinds of wood, such as knots and roots, which remain at rest in water but fall rapidly in air.

SAGR: I have often tried with the utmost patience to add grains of sand to a ball of wax until it should acquire the same gravity as water and would therefore remain at rest in this medium. But with all my care I was never able to accomplish this. Indeed, I do not know whether there is any solid substance whose specific gravity is, by nature, so nearly equal to that of water that if placed anywhere in water it will remain at rest.

The skill of fish

SALV: In this, as in a thousand other operations, men are surpassed by animals. In this problem of yours one may learn much from the fish which are very skillful in maintaining their equilibrium not only in one kind of water, but also in waters which are notably different either by their own nature or by some accidental muddiness or through salinity, each of which produces a marked change. So perfectly indeed can fish keep their equilibrium that they are able to remain motionless in any position. This they accomplish, I believe, by means of an apparatus especially provided by nature, namely, a bladder located in the body and communicating with the mouth by means of a narrow tube through which they are able, at will, to expel a portion of the air contained in the bladder: by rising to the surface they can take in more air; thus they make themselves heavier or lighter than water at will and maintain equilibrium.[295]

SAGR: By means of another device I was able to deceive some friends to whom I had boasted that I could make up a ball of wax that would be in equilibrium in water. In the bottom of a vessel I placed some salt water and upon this some fresh water; then I showed them that the ball stopped in the middle of the water, and that, when pushed to the bottom or lifted to the top, would not remain in either of these places but would return to the middle.

Two grains of salt

SALV: This experiment is not without usefulness. For when physicians are testing the various qualities of waters, especially their specific gravities, they employ a ball of this kind so adjusted that, in certain water, it will neither rise nor fall. Then in testing another water, differing ever so slightly in specific gravity, the ball will sink if this water be lighter and rise if it be heavier. And so exact is this experiment that the addition of two grains of salt to six pounds of water is sufficient to make the ball rise to the surface from the bottom to which it had fallen. To illustrate the precision of this experiment and also to clearly demonstrate the non-resistance of water to division, I wish to add that this notable difference in specific gravity can be produced not only by solution of some heavier substance, but also by merely heating or cooling; and so sensitive is water to this process that by simply adding four drops of another water which is slightly warmer or cooler than the six pounds one can cause the ball to sink or rise; it will sink when the warm water is poured in and will rise upon the addition of cold water.[296] *Now you can see how mistaken are those philosophers who ascribe to water*

[295] In reality, a fish does not need to inhale or exhale any air, it merely has to change the volume of its air-bladder using appropriate muscular actions and this will bring about a change in its buoyancy.

[296] The "toys for grown ups" on sale today exploit this and similar mechanisms.

viscosity or some other coherence of parts which offers resistance to separation of parts and to penetration.

Now, what is not clear here is just what the connection is between the lack of viscosity claimed for water, and the preceding observations about its density. In fact, as Sagredo will indicate in the opening to the next excerpt, Galileo is referring to some arguments of his own on the subject, developed against the Peripatetic philosophers. According to these, the particles of a liquid are not held together by forces, but are simply contiguous. From the experimental point of view, Galileo considered strong evidence that water offers no resistance to penetration the fact that, when muddy water is left to clear in a vase, all the particles of soil, no matter how small, end up accumulating at the bottom. His reasoning was, as ever, sharp, but not valid in this case. Only an ideal liquid (i.e. a perfect liquid) has no viscosity: it is certainly not the case with water, or with other real liquids. Galileo's own pupils found themselves having to repudiate his position on this point.[297]

Cabbage leaves

In the following section of dialogue Simplicio stays in the background, silently amused as Salviati flounders trying to explain the behavior of dew drops. Simplicio's suggestion at the end, though made jokingly, is entirely in keeping with his way of thinking: let us simply speak, he says, of an "antipathy" between water and air, and all difficulty will be removed. Salviati, in courteous mood, declares himself ready to agree, but he is perfectly aware that the right answer is not to be found in the choice of a word, and that the day is sure to come when science will be capable of providing an explanation.

SAGR: With regard to this question I have found many convincing arguments in a treatise by our Academician;[298] *but there is one great*

[297] See, for instance, the relevant notes by Carugo and Geymonat in *Discorsi e dimostrazioni matematiche intorno a due nuove scienze*, Boringhieri, Turin, 1958, p. 670 and following.

[298] *Discorso intorno alle cose che stanno in su l'acqua o che in quella si muovono* (*Bodies that Stay Atop Water, or Move in It*), Edizione Nazionale, Vol. VI.

difficulty of which I have not been able to rid myself, namely, if there be no tenacity or coherence between the particles of water how is it possible for those large drops of water to stand out in relief upon cabbage leaves without scattering or spreading out?

SAGR: Although those who are in possession of the truth are able to solve all objections raised, I would not arrogate to myself such power; nevertheless my inability should not be allowed to becloud the truth. To begin with let me confess that I do not understand how these large globules of water stand out and hold themselves up, although I know for a certainty, that it is not owing to any internal tenacity acting between the particles of water; whence it must follow that the cause of this effect is external. Beside the experiments already shown to prove that the cause is not internal, I can offer another which is very convincing. If the particles of water which sustain themselves in a heap, while surrounded by air, did so in virtue of an internal cause then they would sustain themselves much more easily when surrounded by a medium in which they exhibit less tendency to fall than they do in air; such a medium would be any fluid heavier than air, as, for instance, wine: and therefore if some wine be poured about such a drop of water, the wine might rise until the drop was entirely covered, without the particles of water, held together by this internal coherence, ever parting company. But this is not the fact; for as soon as the wine touches the water, the latter without waiting to be covered scatters and spreads out underneath the wine if it be red. The cause of this effect is therefore external and is possibly to be found in the surrounding air.[299] *Indeed there appears to be a considerable antagonism between air and water as I have observed in the following experiment. Having taken a glass globe which had a mouth of about*

[299] A simple diagram illustrates Salviati's explanation:

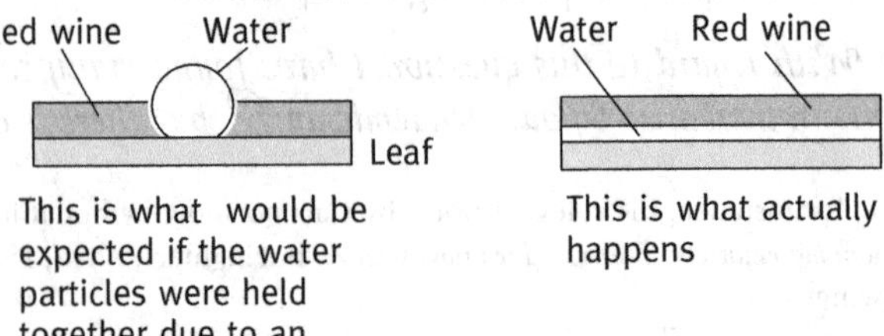
Red wine
Water
Leaf
This is what would be expected if the water particles were held together due to an internal effect
Water
Red wine
This is what actually happens

the same diameter as a straw, I filled it with water and turned it mouth downwards; nevertheless, the water, although quite heavy and prone to descend, and the air, which is very light and disposed to rise through the water, refused, the one to descend and the other to ascend through the opening, but both remained stubborn and defiant. On the other hand, as soon as I apply to this opening a glass of red wine, which is almost inappreciably lighter than water, red streaks are immediately observed to ascend slowly through the water while the water with equal slowness descends through the wine without mixing, until finally the globe is completely filled with wine and the water has all gone down into the vessel below. What then can we say except that there exists, between water and air, a certain incompatibility which I do not understand, but perhaps. . . .
SIMPL: I feel almost like laughing at the great antipathy which Salviati exhibits against the use of the word antipathy; and yet it is excellently adapted to explain the difficulty.
SALV: Alright, if it please Simplicio, let this word antipathy be the solution of our difficulty. Returning from this digression, let us again take up our problem.

MATHEMATICAL NOTE

For details of Archimedes' buoyancy principle and the conditions that make it possible for a body to stay suspended inside a liquid, the reader is referred to the Mathematical Note at the end of Chapter 14. Here, instead, we shall devote a few words to explaining how surface tension comes about. Although it is, as the term itself suggests, a surface effect, contrary to Galileo's opinion it occurs precisely because of the existence of cohesion forces acting among the molecules of a liquid, and because of the fact that such forces become altered in the regions close to the liquid's surface.

Surface tension

Let us consider a molecule A located inside a liquid. Cohesion forces are electromagnetic in nature: they are much stronger than gravitational forces and they act only over very short distances.[300] It can be proved that they

[300] A well-known example is provided by chalk, which consists of extremely fine powder so tightly compressed that its particles are in very close contiguity, and

diminish, as a function of distance r between any pair of adjacent molecules, according to a law of the kind $1/r^n$, where n is a number which is quite a lot larger than the value 2 it takes in Newton's law for gravitational forces.

Let us now draw a small sphere with a radius of few micrometers around molecule A, to represent the sphere of action of cohesive forces. Molecules falling outside this sphere are too far away to produce any effect upon A. For obvious reasons of symmetry, the resultant of all forces acting on molecule A

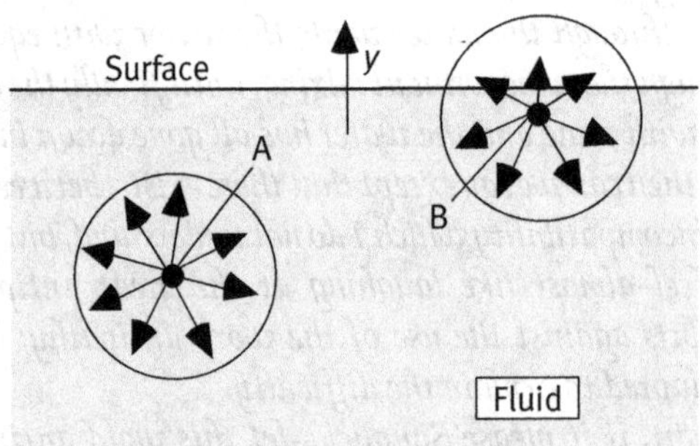

must be zero. We next consider a molecule B, located in the proximity of the surface of the liquid, at a distance less than the radius of the sphere of action. In this case, the forces directed upwards will be weaker, because some of the molecules in the upper part of the sphere are missing (in the case of some medium other than a vacuum, the value of the forces is, anyway, changed). As a result, molecule B finds itself subjected to a net force directed downwards, which is equivalent to a pressure. One therefore has a thin interface layer where the properties of the liquid differ from those in its bulk (for instance, the density is not the same).

The result of this variation in properties is that the potential energy of the system changes and, as a consequence, so does its free energy.[301] If we assume that the total volume of the liquid remains the same, we can calculate the variation. Let χ_0 be the potential energy of a unit mass in the bulk liquid and $\chi(y)$ the same quantity in the thin surface layer, where y is the axis normal to the surface. If S is the area of the layer, for every element dy of the liquid in the

therefore bind to each other. However, if a piece of chalk is broken and the two parts then put together again, they cannot now bind because the distances separating particles across the fracture surfaces have become much larger.

[301] The potential energy of the system corresponds to the work that would be needed to separate all the constituent molecules to infinite distances, so that they no longer exert forces upon one another.

surface layer which replaces a similar element in the bulk, one has a variation of potential energy $dE = (\chi - \chi_0)\rho S dy$, where ρ is the density of the liquid, in first approximation taken to be equal everywhere. For the entire layer, of thickness L, one therefore has

$$\Delta E = S\int_0^L (\chi - \chi_0)\rho dy = S\tau$$

having set the integral equal to a coefficient τ, called *surface tension*. τ can therefore be defined as the variation in the free energy of the system per unit area or, to put it another way, as the work required in order to 'create' a surface of unitary area. Surface tension, as shown by the written equation, has the dimensions of a force per unit length.

Now, a thermodynamic system evolves spontaneously towards the state that takes a minimum value for the free energy, a state which therefore becomes established, the state of equilibrium. All substances, when the external medium is a vacuum, have a positive value for τ. Therefore, surface S tends to reduce to the smallest possible value, meaning that a certain number of molecules passes from the surface region to the bulk region. For a given quantity of liquid which is not subjected to external forces, the smallest surface area would be represented by a sphere. In real drops, this spherical shape is usually modified by the action of gravitational forces or by boundaries (for instance, the plane upon which they sit). Since, for increasing radius, volume increases more rapidly than the surface, there is a limit to the formation of large drops: at a given point, gravity becomes dominant over surface tension and the drop collapses.

The strength of surface tension is different for any pair of substances that are placed in contact with one another. If the two substances are not very different—as was the case of water and wine in the experiment Galileo carried out—the interface tension is very small—less than the 10% of the system air–water. In the case of wine-water it is not sufficient to maintain the drop and the water molecules, as a result of gravity, slip down below the layer of wine.

Capillarity

The presence of a *meniscus* at the surface of a liquid inside a vessel, as well as *capillary action* (owing to which a liquid moves upwards through porous media), are also consequences of surface tension. The meniscus is formed because of the combined actions of two forces: first, the force associated with the surface tension due to the presence of the ambient gas (this force, which is directed towards the interior of the liquid, we shall call the *cohesion force*, because it depends on the intramolecular forces in the liquid); second, the force associated with the material of which the walls of the vessel are made (which we shall call the *adhesion force*). The surface of the liquid must form a 90° angle with the resultant of these two forces, otherwise the liquid would keep moving, without assuming an equilibrium position. If the adhesion force prevails over the cohesion force, the meniscus assumes a concave shape (such is the case with water) and the liquid is said to "wet" the walls. Otherwise, the meniscus is convex (as is the case with mercury).

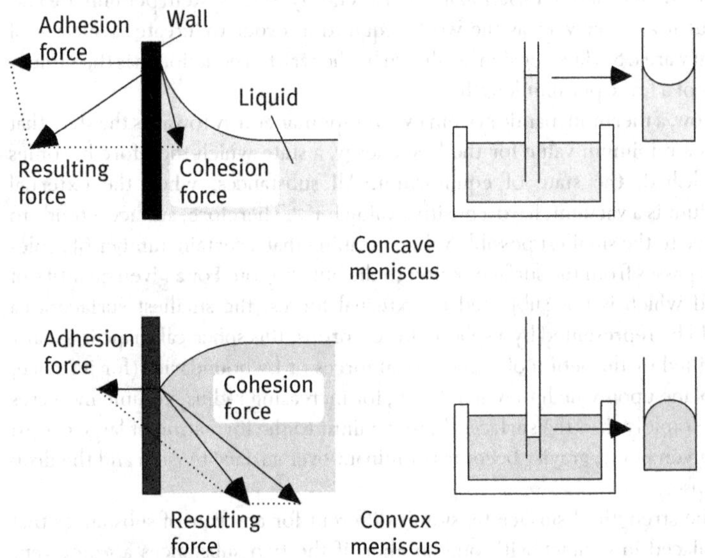

If the vessel is a very thin tube, let us say, with a radius that is not too far from the radius of curvature of the meniscus, the same forces that give rise to the meniscus produce the phenomenon of capillarity: when the thin tube is immersed in a container filled with liquid, if the meniscus is concave the level of the liquid inside the tube will be higher than outside, whereas the opposite will be true if the meniscus is convex (see drawings on the right in the above diagram). It is possible to show that the difference Δh in level between the internal and the external liquid is given by:

$$\Delta h = \frac{2\tau \cos \alpha}{\rho g R}$$

where ρ is the density of the liquid, α is the angle of contact with the wall, R is the radius of the capillary, and τ is the surface tension of the system formed by the liquid and the ambient medium. So it can be seen that the smaller the capillary radius is (or the greater the surface tension), the larger is the difference in levels.

The above equation allows the surface tension of any liquid to be determined experimentally. For example: in a capillary of radius $R = 0.25$ mm, water rises about 62 mm above the external level and the angle α is so small as to be considered negligible ($\cos\alpha \approx 1$). One can therefore deduce that for water in contact with air it is $\tau = 0.076\ \mathrm{N/m}$. Mercury exhibits a convex meniscus because its surface tension is about seven times larger than that of water.

16

Machines Large and Small*

un cavallo cadendo da un'altezza
di tre braccia o quattro si romperà l'ossa,
ma . . . un gatto da una di otto o dieci, non si farà mal nissuno, . . .
né una formica precipitandosi dall'orbe lunare

a horse falling from a height
of three or four cubits will break his bones,
while . . . a cat from a height
of eight or ten cubits will suffer no injury
Equally harmless would be the fall . . . of an ant
from the distance of the moon

The chapter opens with Galileo's famous statement on the importance of applied science. Examples follow regarding mechanics, the resistance of beams, the greater tendency towards breakage in large machines than in small ones, aquatic animals, and land animals.

The importance of technology

SALV: The constant activity which you Venetians display in your arsenal suggests to the studious mind a large field for investigation, especially that part of the work which involves mechanics; for in this department all types of instruments and machines are constantly being

* From *Discourses*.

constructed by many artisans, among whom there must be some who, partly by inherited experience and partly by their own observations, have become highly expert and clever in explanation.
SAGR: You are quite right. Indeed, I myself, being curious by nature, frequently visit this place for the mere pleasure of observing the work of those who, on account of their superiority over other artisans, we call "first rank men" Conference with them has often helped me in the investigation of certain effects including not only those which are striking, but also those which are recondite and almost incredible.

This exchange between Salviati and Sagredo[302] opens Galileo's final book, *Discourses about Two New Sciences*. Galileo began writing the book only a few days after the trial which had ended in his abjuration and his enforced undertaking not to disseminate the principles of Copernicanism. An undertaking which he brazenly did not stick to, seeing that the *Discourses* is nothing less than the natural follow-on to the *Dialogue*—the offending item in the trial. It has to be asked why this further act of insubordination in the face of the religious authorities did not bring down upon him additional and more severe punishment. It was, beyond any shadow of a doubt, a deliberate challenge to the authority of the Church. Among historians of science there are some who maintain no new punishment was meted out simply because the book was not understood. This version, however, is highly unlikely, as the Church could count on highly qualified scientists who would most certainly have been keeping Galileo's work under strict surveillance. The most plausible explanation is that Urban VIII, a friend and admirer of Galileo's prior to being elected to the pontifical throne, did not wish to act cruelly towards a sick, blind old man, against whom he had been induced into taking action in order to appear an upright defender of Catholic orthodoxy, and to avoid being accused of protecting a heretic. Galileo must have been perfectly aware of this, thanks to the friendly complicity of Ascanio Piccolomini, the Archbishop of Siena, and of the Grand Duke of

302 *Two New Sciences*, translated by H. Crew and A. de Salvio, Dover Publications, New York, 1954, p. 1. In Italian: *Discorsi e dimostrazioni matematiche intorno a due nuove scienze*, Edizione Nazionale, Vol. VIII, p. 49.

Tuscany, in whose court he still held the post of "philosopher and mathematician".[303]

Applied science

Activity, instruments, machines, observations—key words in the opening page of the *Discourses*—they amount, in effect, to a recognition of the important role technology plays, with regard both to pure science and to intellectual activity in general. Appearing at the very beginning of the book, which is probably the greatest product of Galileo's mind, these words amount to a declaration in support of a new conception of science. Knowledge comes out of the closed space of libraries, it is democratically enriched by contributions based on the practical experience of people who are not educated. The value of *doing* is acknowledged.[304] Engaging in dispute with the haughty official culture, which pontificates "from the throne of Her Peripatetic Majesty",[305] Galileo champions this new way of conceiving knowledge, extending a line of thinking that went back to Leonardo da Vinci and the Renaissance. On the one hand we have mathematics, logic, and experimentation, the three cornerstones of scientific research; on the other hand we have the resolution of practical engineering problems: two approaches which have different aims, but which are deeply interrelated, and which cross-fertilize. This is the idea that forms the basis of the *Discourses*, and it represents the most innovative feature of all Galileo's work.

On a completely different level are those philosophers who, in Simplicio's words, "occupy themselves principally about universals.

[303] Galileo's great moral strength during this unhappy period, and his unshakable devotion to science, are evident in a letter written in 1634: "And yet, in this distress, my spirit does not feel mortified, nor depressed, and I always nurture thoughts that are free and worthy of a man" (originally in Latin, Edizione Nazionale, Vol. XVI, p. 112).

[304] Shortly later G.B. Vico (Italian philosopher and man of letters, 1668–1744) was to say that we know only what we make.

[305] In his *Life of Galileo*, Bertolt Brecht brings out well the revolutionary character of this approach in the scene depicting the haughtiness of the Florentine university doctors towards Federzoni, Galileo's technician and aid.

They find definitions and criteria, leaving to the mathematicians certain fragments and subtleties, which are then rather curiosities" and "to mechanics or other low artisans the investigation of . . . more detailed features".[306] Which is as much as to say, leaving them to the new physicists as Galileo conceives them. These, as a matter of fact, make precisely such fragments, subtleties and details the object of their investigation. On the one hand, they anchor their research firmly in the world of experience, on the other hand they give due recognition to the absolute cognitive value of "subtleties" and "curiosities". There is no phenomenon that is less worthy of being investigated than others, because all phenomena are subject to the same natural laws.[307]

Galileo, as is well known, was awake to the possibilities of using the discoveries of science in honing the performance of machines for civil engineering, for hydraulics, for military installations, and so forth. This attitude of his, which may be seen as the culmination of a process of convergence between physico-mathematical theory and practical experience that had been taking shape during the course of the Renaissance, clearly emerges in the rich correspondence he had with artisans, engineers, technicians, public, and military supervisors on a wide range of specific items. These include: lifting water by means of pumps and pipe networks, canalization of rivers, problems of the motion of projectiles, fortification works, the construction and refinement of scientific instruments. Among his many achievements in this latter field, worthy of mention (at the very least) are the geometric compass (which he built in a small workshop at his home

[306] *Dialogue Concerning the Two Chief World Systems*, Second Day, translated by S. Drake, The Modern Library, New York, 2001, p. 190. In Italian: *Dialogo sopra i due massimi sistemi del mondo*, Edizione Nazionale, Vol. VII, p. 189.

[307] We report a remark made by physicist Giorgio Salvini:

If one does not get involved with the fragments and the subtleties, one cannot hope to do true science: the physicist knows that in the fragments are to be found not only Foucault's pendulum, but also Michelson's experiment, Fraunhofer's lines, the anomalies of Mercury, X-ray and electron diffraction, the corrections pertaining to the hydrogen atom, i.e. the very seeds of the scientific revolution that began in the seventeenth century. (G. Salvini, 'Mitologia e verità nella figurazione di Galileo. Il Galileo di Bertolt Brecht', *Saggi su Galileo Galilei*, Barbèra Editore, Firenze, 1967, p. 9).

in Padua), the telescope (which certainly owed much to Galileo's contacts with the glass manufacturers in Murano), and the clock based on the isochronism of the pendulum (a task he assigned to his son Vincenzo, which was possibly never brought to completion).

It does not appear that Galileo felt ill at ease with work oriented towards military ends, as, instead, had been the case with Leonardo da Vinci, and indeed is the case today with the majority of scientists. Quite the contrary, in fact. During his teaching in Padua—between 1592 and 1610—Galileo did all he could to improve the war resources of the Republic of Venice and gave lectures on military science and engineering. His willingness to become involved in these matters, rather surprising for a man of broad mind and passionate heart, might be indicative of a desire to please his employers, his foremost and fundamental concern if he was to be able to continue with research characterized by freedom of thought and action, and by adequate means of support.

This process of strong interaction between science and technology, which Galileo set in motion, pervades every aspect of modern life. In the exact sciences, in particular, the two dimensions are nothing other than two faces of a single reality. Each draws considerable benefit from the other, in a continuous process of reciprocal stimulus and mutual amplification, to which no end is in sight. Not everything about this is for the good, but there was no way Galileo, at the dawn of such a revolution, could have had any inkling of this. Today, in addition to the military field, where the most negative spin-offs of science are to be found, technological consumerism threatens to become a real social evil. It induces artificial needs in people, it establishes scales of false values, it promotes petty forms of crime and rewards organized ones, it leads to the destruction of the ecological patrimony and the resources of the planet. Even if science is not directly, and not intentionally, to be blamed for these tendencies, it has to be held jointly responsible.[308]

[308] We cite the view of Pope John Paul II, expressed in a speech at the UNESCO headquarters in 1980:

A most serious threat hangs over the future of humankind and of the world, in spite of the undisputedly noble intentions of scholars and scientists. The reason for this is that

Over and against all this, however, have to be set the enormous benefits which science and technology have produced for the health and the material and spiritual well-being of humankind; they have widened education, ameliorated working conditions, facilitated women's emancipation, and equipped minds fit to tackle problems in more intelligent and more equitable ways. In short, science has made it possible—one should even say, it has promoted—the advent of democracy and greater equality among human beings. What will the outcome of this battle be? Will we, in the long run, be having to thank, or blame, those who, like Galileo, bear the responsibility for this grand and ever faster transformation of human life?

Scale factor

Let us now turn to the content of the excerpt presented in this chapter. The matter under discussion is what today we call "scaling properties": any machine, whether it is a natural machine, such as the body of an animal, or a manufactured machine, such as an architectural structure, has a degree of mechanical resistance which does not scale along with its dimensions. The larger a machine is, the less it is resistant to stress (relatively speaking, of course). The thickness of animal bones, for instance, has to increase in greater proportion compared to increases in the dimensions of other parts of the body: "Who does not know that a horse falling from a height of three or four cubits will break his bones, while a dog falling from the same height or a cat from a height of eight or ten cubits will suffer no injury? Equally harmless would be the fall of a grasshopper from a tower or the fall of an ant from the distance of the moon". And as for the mechanical properties

the astounding results of their research and discoveries, in particular in the field of the natural sciences, have been (and still are being) exploited for ends that have nothing to do with the needs of science [. . .], to the detriment of ethical imperatives. And whereas science ought to be used for the benefit of humankind, all too often we discover that it is subordinated to ends that extinguish the true dignity of the person and of human life.

This is not the usual aversion to science because it is the harbinger of reason and the foe of dogma and superstition, but perfectly justified alarm about the fate of humankind.

of manufactured goods, a reduced-scale model is of little use in establishing the resistance limits of the final product.

In water, instead, things can be very different, because the weight of a body—which is what the frame of the skeleton has to support—is greatly reduced by buoyancy. This permits huge animals, such as whales, to exist. In fact, Galileo says: "In aquatic animals therefore circumstances are just reversed from what they are with land animals inasmuch as, in the latter, the bones sustain not only their own weight but also that of the flesh, while in the former it is the flesh which supports not only its own weight but also that of the bones".

The text

As pointed out earlier when commenting on excerpts from the book, an evolution of Simplicio's character is evident in the *Discourses*. From the passive, sometimes ridiculous, defender of tradition he is portrayed as in the *Dialogue*, he turns into an interlocutor who is often worthy and sharp. It may be that this was a belated act of reparation on Galileo's part, intended to appease Urban VIII, who had been incensed at seeing himself, as he thought, portrayed in the character of Simplicio. The transformation also extends to Sagredo, who appears more hesitant, less frequently aligned with Salviati, and who at times uses formal and metaphorical language, which was never his style in the *Dialogue*.

The following passage is taken from the *Discourses*.[309]

SALV: [. . .] Therefore, Sagredo, you would do well to change the opinion which you, and perhaps also many other students of mechanics, have entertained concerning the ability of machines and structures to resist external disturbances, thinking that when they are built of the same material and maintain the same ratio between parts, they are able equally, or rather proportionally, to resist or yield to such external

[309] *Two New Sciences*, First Day, translated by H. Crew and A. de Salvio, Dover Publication, New York, 1954, p. 3 and following. In Italian: *Discorsi e dimostrazioni matematiche intorno a due nuove scienze*, Giornata prima, Edizione Nazionale, Vol. VIII, p. 51 and following.

disturbances and blows. For we can demonstrate by geometry that the large machine is not proportionately stronger than the small. Finally, we may say that, for every machine and structure, whether artificial or natural, there is set a necessary limit beyond which neither art nor nature can pass; it is here understood, of course, that the material is the same and the proportion preserved.

SAGR: My brain already reels. My mind, like a cloud momentarily illuminated by a lightning flash, is for an instant filled with an unusual light, which now beckons to me and which now suddenly mingles and obscures strange, crude ideas. From what you have said it appears to me impossible to build two similar structures of the same material but of different size and have them proportionately strong [...].

SALV: [...] Who does not know that a horse falling from a height of three or four cubits will break his bone,[310] *while a dog falling from the same height or a cat from a height of eight or ten cubits will suffer no injury? Equally harmless would be the fall of a grasshopper from a tower or the fall of an ant from the distance of the moon. Do not children fall with impunity from heights which would cost their elders a broken leg or perhaps a fractured skull? And just as smaller animals are proportionately stronger and more robust than the larger, so also smaller plants are able to stand up better than larger. I am certain you both know that an oak two hundred cubits high would not be able to sustain its own branches if they were distributed as in a tree of ordinary size; and that nature cannot produce a horse as large as twenty ordinary horses or a giant ten times taller than an ordinary man unless by miracle or by greatly altering the proportions of his limbs and especially of his bones, which would have to be considerably enlarged over the ordinary. Likewise the current belief that, in the case of artificial machines the very large and the small are equally feasible and lasting is a manifest*

[310] In a study on the use of metaphor in Galileo's writings, Italo Calvino

counted at least eleven significant examples where Galileo talks about horses: as an image for motion, and thus as a tool for kinetics experiments, as a symbol of Nature in all its complexity and also beauty, as a figure that excites the imagination, when horses are subjected to the most arduous trials or are made to grow to gigantic sizes; in addition to the analogy between reasoning and running . . .

(I. Calvino, *Lezioni americane*, Garzanti, Milano 1988, p. 43. English translation by Patrick Creagh, *Six Memos for the Next Millennium / the Charles Eliot Norton Lectures 1985–86*, Vintage International).

error. Thus, for example, a small obelisk or column or other solid figure can certainly be laid down or set up without danger of breaking, while the very large ones will go to pieces under the slightest provocation, and that purely on account of their own weight.

The broken column

And here I must relate a circumstance which is worthy of your attention as indeed are all events which happen contrary to expectation, especially when a precautionary measure turns out to be a cause of disaster. A large marble column was laid out so that its two ends rested each upon a piece of beam;[311] *a little later it occurred to a mechanic that, in order to be doubly sure of its not breaking in the middle by its own weight, it would be wise to lay a third support midway; this seemed to all an excellent idea; but the sequel showed that it was quite the opposite, for not many months passed before the column was found cracked and broken exactly above the new middle support.*

SIMPL: A very remarkable and thoroughly unexpected accident, especially if caused by placing that new support in the middle.

SALV: Surely this is the explanation, and the moment the cause is known our surprise vanishes; for when the two pieces of the column were placed on level ground it was observed that one of the end beams had, after a long while, become decayed and sunken, but that the middle one remained hard and strong, thus causing one half of the column to project in the air without any support. Under these circumstances the body therefore behaved differently from what it would have done if supported only upon the first beams; because no matter how much they might have sunken the column would have gone with them. This is an

[311] It may be helpful to illustrate the discussion with a diagram:

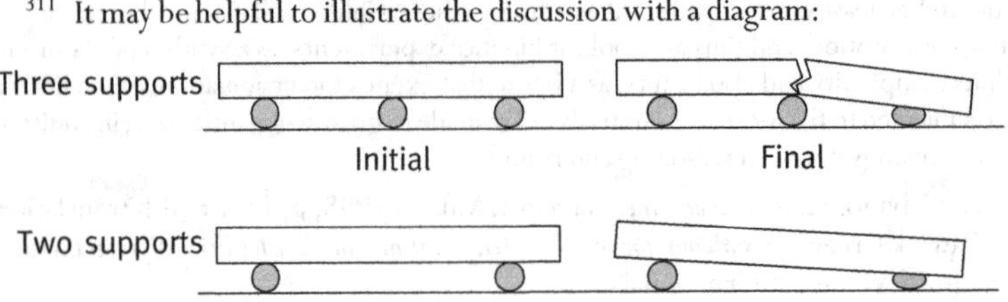

accident which could not possibly have happened to a small column, even though made of the same stone and having a length corresponding to its thickness, i.e., preserving the ratio between thickness and length found in the large pillar.

At this point comes one of Galileo's typical digressions, which his mind, fervent and attentive to the smallest subtleties of surrounding reality, prompts him to make with joy and ever-fresh wonder. In fact, he will not return to the phenomenon of the broken column, to provide an explanation for it, until much later in the dialogue[312] (we are not going to follow him in that argumentation, because it is not particularly enthralling).

However, with regard to Galileo's digressions, it is worth underlining his characteristic striving to embrace a diversity of experimental observations within a single unifying physical mechanism, as if to say the fundamental laws of Nature are just a few, and they are intrinsically connected. A feature which, in the centuries to come, would characterize science to an increasingly marked degree. In science, one must never "be content with an *ad hoc* solution for a given problem: one must try to find the laws that rule not only the particular case but also many others!"—states David Speiser,[313] paraphrasing a teaching applied to music by pianist Alfred Cortot.

A new science

SAGR: I am quite convinced of the facts of the case, but I do not understand why the strength and resistance are not multiplied in the same proportion as the material; and I am the more puzzled because, on the contrary, I have noticed in other cases that the strength and resistance against breaking increase in a larger ratio than the amount of material. Thus, for instance, if two nails be driven into a wall, the one which is

[312] *Two New Sciences*, translated by H. Crew and A. de Salvio, Dover Publishing, New York, 1954, p. 134 and following. In Italian: *Discorsi e dimostrazioni matematiche intorno a due nuove scienze*, Giornata prima, Edizione Nazionale, Vol. VIII, p. 172 and following.

[313] D. Speiser, "Galileo and the beginning of the theory of elasticity", in *Galileo Scientist*, edited by M. Baldo Ceolin, IV Galilean Centenary, Venice, 1992.

twice as big as the other will support not only twice as much weight as the other, but three or four times as much.
SALV: Indeed you will not be far wrong if you say eight times as much; nor does this phenomenon contradict the other even though in appearance they seem so different.
SAGR: Will you not then, Salviati, remove these difficulties and clear away these obscurities if possible: for I imagine that this problem of resistance opens up a field of beautiful and useful ideas and if you are pleased to make this the subject of to-day's discourse you will place Simplicio and me under many obligations.
SALV: I am at your service if only I can call to mind what I learned from our Academician[314] *who had thought much upon this subject and according to his custom had demonstrated everything by geometrical methods*[315] *so that one might fairly call this a new science. For, although some of his conclusions had been reached by others, first of all by Aristotle, these are not the most beautiful and, what is more important, they had not been proven in a rigid manner from fundamental principles. Now, since I wish to convince you by demonstrative reasoning rather than to persuade you by mere probabilities, I shall suppose that you are familiar with present-day mechanics so far as it is needed in our discussion. First of all it is necessary to consider what happens when a piece of wood or any other solid which coheres firmly is broken; for this is the fundamental fact, involving the first and simple principle which we must take for granted as well known. [...]*

Materials science

Salviati now goes on to deal with the details of the problem, starting, as he has just said, with the material structure of bodies. This is the very first step in a new science, materials science, which today plays a role of primary importance in physical disciplines and in technology. Materials for microelectronics, for information science,

[314] *Academician*—Galileo, of course.
[315] Galileo confines himself to making comparisons among bodies which are of different sizes, but are composed of the same material, since at that time a theory of elasticity was not available.

for robotics, for communications, plastic materials, compound materials, construction materials, special alloys. The civilization (if we may call it thus) in which we live is built on the study and the application of electric, chemical, and mechanical properties of matter, many of them still unexplored.

Much further on in the dialogue, Salviati brings the focus of the discussion to natural machines—that is, animals—which are subject to the same limitations as artificial constructions.[316]

Land animals

SALV: [...] From what has already been demonstrated, you can plainly see the impossibility of increasing the size of structures to vast dimensions either in art or in nature; likewise the impossibility of building ships, palaces, or temples of enormous size in such a way that their oars, yards, beams, iron-bolts, and, in short, all their other parts will hold together; nor can nature produce trees of extraordinary size because the branches would break down under their own weight; so also it would be impossible to build up the bony structures of men, horses, or other animals so as to hold together and perform their normal functions if these animals were to be increased enormously in height; for this increase in height can be accomplished only by employing a material which is harder and stronger than usual, or by enlarging the size of the bones, thus changing their shape until the form and appearance of the animals suggest a monstrosity. This is perhaps what our wise Poet[317] had in mind, when he says, in describing a huge giant:

Impossible it is to reckon his height,
So beyond measure is his size.

To illustrate briefly, I have sketched a bone whose natural length has been increased three times and whose thickness has been multiplied until,

[316] *Two New Sciences*, Second Day, translated by H. Crew and A. de Salvio, Dover Publishing, New York, 1954, p. 130 and following. In Italian: *Discorsi e dimostrazioni matematiche intorno a due nuove scienze*, Giornata seconda, Edizione Nazionale, Vol. VIII, p. 169 and following.

[317] Ludovico Ariosto (1474–1533), Italian poet and author of the poem *Orlando Furioso* (the lines are taken from Canto XVII).

for a correspondingly large animal, it would perform the same function which the small bone performs for its small animal. From the figures here shown you can see how out of proportion the enlarged bone appears.[318]

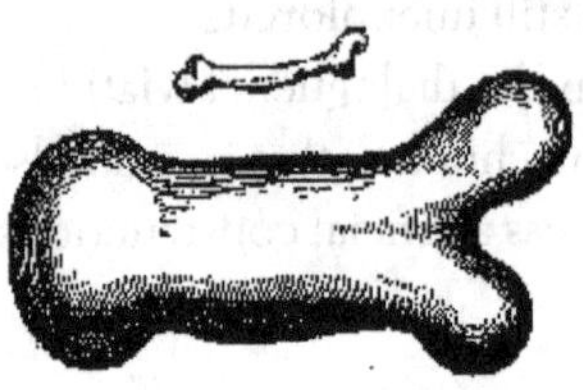

Clearly then if one wishes to maintain in a great giant the same proportion of limb as that found in an ordinary man he must either find a harder and stronger material for making the bones, or he must admit a diminution of strength in comparison with men of medium stature; for if his height be increased inordinately he will fall and be crushed under his own weight. Whereas, if the size of a body be diminished, the strength of that body is not diminished in the same proportion; indeed the smaller the body the greater its relative strength. Thus a small dog could probably carry on his back two or three dogs of his own size; but I believe that a horse could not carry even one of his own size.

Aquatic animals

SIMPL: This may be so; but I am led to doubt it on account of the enormous size reached by certain fish, such as the whale which, I understand, is ten times as large as an elephant; yet they all support themselves.

SALV: Your question, Simplicio, suggests another principle, one which had hitherto escaped my attention and which enables giants and other animals of vast size to support themselves and to move about as well as smaller animals do. This result may be secured either by increasing the strength of the bones and other parts intended to carry not only their

[318] While the ratio between the two lengths is 3, the ratio between the widths is 7. Galileo does not specify which animal he is referring to.

weight but also the superincumbent load; or, keeping the proportions of the bony structure constant, the skeleton will hold together in the same manner or even more easily, provided one diminishes in the proper proportion, the weight of the bony material, of the flesh, and of anything else which the skeleton has to carry. It is this second principle which is employed by nature in the structure of fish, making their bones and muscles not merely light but entirely devoid of weight.

SIMPL: The trend of your argument, Salviati, is evident. Since fish live in water which on account of its density or, as others would say, heaviness diminishes the weight of bodies immersed in it, you mean to say that, for this reason, the bodies of fish will be devoid of weight and will be supported without injury to their bones.[319] *But this is not all; for although the remainder of the body of the fish may be without weight, there can be no question but that their bones have weight. Take the case of a whale's rib, having the dimensions of a beam; who can deny its great weight or its tendency to go to the bottom when placed in water? One would, therefore hardly expect these great masses to sustain themselves.*

SALV: A very shrewd objection! And now, in reply, tell me whether you have ever seen fish stand motionless at will under water, neither descending to the bottom nor rising to the top, without the exertion of force by swimming?

SIMPL: This is a well-known phenomenon.

SALV: The fact then that fish are able to remain motionless under water is a conclusive reason for thinking that the material of their bodies has the same specific gravity as that of water; accordingly, if in their make-up there are certain parts which are heavier than water there must be others which are lighter, for otherwise they would not produce equilibrium.

Hence, if the bones are heavier, it is necessary that the muscles or other constituents of the body should be lighter in order that their buoyancy may counterbalance the weight of the bones. In aquatic animals therefore circumstances are just reversed from what they are with land animals inasmuch as, in the latter, the bones sustain not only their own weight but also that of the flesh, while in the former it is the flesh which

[319] In other words, the buoyancy (see Chapter 14) is virtually equal to the weight of the flesh of the fish.

supports not only its own weight but also that of the bones. We must therefore cease to wonder why these enormously large animals inhabit the water rather than the land, that is to say, the air.

SIMPL: I am convinced and I only wish to add that what we call land animals ought really to be called air animals, seeing that they live in the air, are surrounded by air, and breathe air.

SAGR: I have enjoyed Simplicio's discussion including both the question raised and its answer. Moreover I can easily understand that one of these giant fish, if pulled ashore, would not perhaps sustain itself for any great length of time, but would be crushed under its own mass as soon as the connections between the bones gave way.[320]

SALV: I am inclined to your opinion; and, indeed, I almost think that the same thing would happen in the case of a very big ship which floats on the sea without going to pieces under its load of merchandise and armament, but which on dry land and in air would probably fall apart. [...]

A ship can therefore float on water thanks to buoyancy, which is equally distributed over the whole of the immersed part of the hull. It would not float if the weight of the hull were to rest on a limited number of spots, because very high pressures would act on them.

[320] It is well known, in fact, that a whale does not survive when it becomes stranded on a sand bank owing to crushing of its internal organs.

PART VIII

FROM MATTER TO LIGHT

17

In Horror at a Vacuum

Or, in the Recesses of Matter*

Dirò prima del vacuo, mostrando con chiare esperienze quale e quanta sia la sua virtù

First I shall speak of the vacuum, demonstrating by definite experiment the quality and quantity of its force

A celebrated passage, not without its difficulties, in which Galileo advances the hypothesis that within the interior of a body of condensed matter there are an infinite number of empty microspaces, and that it is these which are responsible for the bonding between its various constituent particles. This idea was inspired by the experimental observation that nature seems to oppose the creation of a vacuum—the famous *horror vacui*—a concept which Galileo here extends to the microscopic level. The key experiment is described, involving the measurement of the force needed to extract an airtight piston from a cylinder. Far from proving that a vacuum exercises forces, the measurement gives—without Galileo realizing it—the weight of the column of air in the atmosphere above us. In support of his model, he puts forward examples using metals, liquids, and powders.

Horror vacui

The idea that a vacuum is present among "the most minute particles" of which condensed matter is constituted is intriguing, and was most

* From *Discourses*.

certainly original. In some senses it anticipated our modern conception of atoms, which may be seen as being formed by individual particles, the electrons and the nuclei, immersed in a vacuum.[321] Proceeding by sheer imagination, Galileo postulated the existence of microscopic vacua to explain the cohesion between the particles of matter. These vacua, he ventured, would have the effect of binding the particles together, owing to the *horror vacui*—the "repugnance" Nature has for a vacuum. This was thought to be seen at work in macroscopic bodies; for example, in the case of two highly polished stone or metal plates set in close contact with one another, where the application of a force is required in order to separate them. Given that no air can penetrate and flow between the two plates in contact, in the initial moment of the opening up of a gap between them a vacuum should be formed (at least temporarily). However Nature has a repugnance for this, and as a consequence she tries to prevent its formation by putting up a resistance to our action.

Such was the misconception prevailing in Galileo's time, and he also clearly subscribed to it. It is still quite common today—in answer to the question "what is the cause that allows one to suck liquid through a straw?" many say (even among university students) that by sucking we produce a vacuum in the straw, and that this attracts the liquid. The correct explanation, of course, is otherwise. Forces can only be exerted among particles. A vacuum, precisely because it is such, is not capable of exerting forces on matter (interestingly, in this same section of dialogue, Galileo contradicts himself, in that he shows he accepts Aristotle's maxim that "the non-existent can produce no effects"). In fact, the reason liquid rises in the straw is that atmospheric pressure acts on the exterior of the straw, but not

[321] The atom is made of a very small dense nucleus, whose radius falls, for the different elements, between 0.1 and 1×10^{-12} cm (10^{-12} cm = one thousandth of a billionth of a centimeter), surrounded by a shell of electrons, which have a much smaller mass and are located at a large distance from the nucleus, that is, at about 10^{-8} cm. An atom, therefore, can be said to contain a large amount of empty space. If all these constituent particles were collapsed together, eliminating the empty spaces—as happens with so-called neutron stars—the volume of a body would decrease by at least 10,000 to a 100,000 times (the Coliseum would be reduced to the size of a glass bead).

on its interior, where the sucking has eliminated the air. The same can be said of the famous experiment of the Magdeburg evacuated hemispheres, which "not even the force of sixteen horses could take apart".[322] These problems were only to be correctly dealt with a few years later, by other scientists, for example, the inventor of the barometer Evangelista Torricelli (who was a pupil of Galileo and of Benedetto Castelli[323]) and Blaise Pascal.

The weight of the atmosphere

The same mechanism can be used to explain the effect of the two highly polished plates. Atmospheric pressure acts upon all surfaces exposed to the air. Since such pressure is hydrostatic, according to Pascal's principle it will act in an equal manner on all surfaces, independently of their orientation—but it will not act on those surfaces placed in contact with each other in between the plates. As a result, the plates are kept pressed together, giving the impression that vacuum is behaving as a glue.

Although he did not provide a valid interpretation of the effect, Galileo has to receive the credit for having demonstrated experimentally that wherever vacua are created, forces become manifest. To measure the presumed "force of a vacuum", he performed an experiment on the extraction of an airtight piston from a hollow cylinder, finding that for this to occur a well-defined threshold force must be applied. As we pointed out in Chapter 14, it is surprising that Galileo, having devised a sophisticated experiment for weighing air (which had enabled him to obtain a reasonable measurement of its density), did not realize that the behavior of the piston was an

[322] The experiment was performed in 1654 by Otto von Guericke, one of the earliest builders of pneumatic pumps, in order to demonstrate that it was possible to produce a vacuum on the earth. The hemispheres, 80 cm in diameter, were held together by a weight of atmospheric air equal to a few tons.

[323] The experiment to determine atmospheric pressure using a column of mercury was conceived by Evangelista Torricelli and performed in 1643 with the help of Vincenzo Viviani, another pupil of Galileo (and also his biographer). See, for instance, M. Gliozzi, *Origine e sviluppi dell'esperienza torricelliana*, Turin, 1931.

alternative method for verifying that air has weight. A simple calculation, even if based on his inaccurate measurement of the specific weight of air (about half the correct one) and on his poor knowledge of the height of the atmosphere, should have suggested to him that the weight of the column of air acting externally on the underside of the piston could not be much less than 1 kg per square centimeter—the weight which would eventually be established as that of atmospheric pressure.

Fusion of a solid

The discussion later touches on the mechanism of the fusion of solids, which Galileo, with remarkable insight, ascribes to the fact that "the extremely fine particles of fire, penetrating the slender pores of the metal . . . would fill the small intervening vacua and would set free these small particles from the attraction which these same vacua exert upon them and which prevents their separation . . .". This description is not far from the reality. Today we say that the atoms in a solid are held together as a result of electromagnetic forces that act between the constituent charges—electrons and protons in the nuclei. These forces give rise to a well-defined binding energy. Thermal agitation acts in the opposite direction, since it provides the atoms with kinetic energy, which helps them escape from their tightly bound state. The more a body is heated, the higher the kinetic energy of its atoms becomes, such that at a given temperature the "freeing energy" prevails over the "binding" energy.[324] The substance then passes into its liquid phase, where forces between atoms are still present, but they are different from, and weaker than, those characterizing the solid state. At a still higher temperature the substance becomes a vapor, and its atoms disperse, no longer experiencing mutual influences.

Worth highlighting is the argument Salviati uses to persuade his interlocutors that interatomic forces—minute as they are—are

[324] Scientific details on the relationship between temperature and kinetic energy are given in the Mathematical Note of Chapter 18.

capable of producing huge amounts of work when they act in unison: just think, he says, that simply by wetting a rope, that is, by filling its internal vacua, one can make it swell, which in turn shortens it and renders it capable of lifting gigantic weights.

The following passage is taken from the *Discourses*.[325]

Freedom of research

SALV: [. . .] And as in the case of the rope whose strength we know to be derived from a multitude of hemp threads which compose it, so in the case of the wood, we observe its fibres and filaments run lengthwise and render it much stronger than a hemp rope of the same thickness. But in the case of a stone or metallic cylinder where the coherence seems to be still greater the cement which holds the parts together must be something other than filaments and fibres; and yet even this can be broken by a strong pull.

SIMPL: If this matter be as you say I can well understand that the fibres of the wood, being as long as the piece of wood itself, render it strong and resistant against large forces tending to break it. But how can one make a rope one hundred cubits long out of hempen fibres which are not more than two or three cubits long, and still give it so much strength? Besides, I should be glad to hear your opinion as to the manner in which the parts of metal, stone, and other materials not showing a filamentous structure are put together; for, if I mistake not, they exhibit even greater tenacity.

SALV: To solve the problems which you raise it will be necessary to make a digression into subjects which have little bearing upon our present purpose.

SAGR: But if, by digressions, we can reach new truth, what harm is there in making one now, so that we may not lose this knowledge, remembering that such an opportunity, once omitted, may not return; remembering also that we are not tied down to a fixed and brief method

[325] *Two New Sciences*, First Day, translated by H. Crew and A. de Salvio, Dover Publications, New York, 1954, p. 7 and following. In Italian: *Discorsi e dimostrazioni matematiche intorno a due nuove scienze*, Giornata prima, Edizione Nazionale, Vol. VIII, p. 55 and following.

but that we meet solely for our own entertainment? Indeed, who knows but that we may thus frequently discover something more interesting and beautiful than the solution originally sought? I beg of you, therefore, to grant the request of Simplicio, which is also mine; for I am no less curious and desirous than he to learn what is the binding material which holds together the parts of solids so that they can scarcely be separated. This information is also needed to understand the coherence of the parts of fibres themselves of which some solids are built up.
[. . .]

The two plates

Regarding Galileo's famous digressions, which often take up more room than the main topic itself, we have already mentioned that a spiteful Descartes was unable to appreciate them (see Chapter 6). In truth, however, the digressions are the clearest proof of Galileo's unitary vision of the natural world: all phenomena are interconnected in various ways and it is the task of research to make them fit within common laws. Digressions are the paths a free mind takes, prompted by the stimuli of the moment as it advances in its quest—but without ever losing sight of the final goal. "Who knows but that we may thus frequently discover something more interesting and beautiful than the solution originally sought?" Sagredo asks himself, showing that he is fully aware of their potential fertility. This, we might say, is an appeal for freedom in research, which must not be directed solely towards objectives that have been precisely defined, else we risk sacrificing originality and innovation.

Salviati gladly accepts the invitation to express his viewpoint:[326]

SALV: [. . .] since you are waiting to hear what I think about the breaking strength of other materials which, unlike ropes and most woods, do not show a filamentous structure. The coherence of these

[326] *Two New Sciences*, translated by H. Crew and A. de Salvio, Dover Publications, New York, 1954, p. 11 and following. In Italian: *Discorsi e dimostrazioni matematiche intorno a due nuove scienze*, Giornata prima, Edizione Nazionale, Vol. VIII, p. 59 and following.

bodies is, in my estimation, produced by other causes which may be grouped under two heads. One is that much-talked-of repugnance which nature exhibits towards a vacuum; but this horror of a vacuum not being sufficient, it is necessary to introduce another cause in the form of a gluey or viscous substance which binds firmly together the component parts of the body.

First I shall speak of the vacuum, demonstrating by definite experiment the quality and quantity of its force. If you take two highly polished and smooth plates of marble, metal, or glass and place them face to face, one will slide over the other with the greatest ease, showing conclusively that there is nothing of a viscous nature between them. But when you attempt to separate them and keep them at a constant distance apart, you find the plates exhibit such a repugnance to separation that the upper one will carry the lower one with it and keep it lifted indefinitely, even when the latter is big and heavy.

This experiment shows the aversion of nature for empty space, even during the brief moment required for the outside air to rush in and fill up the region between the two plates. It is also observed that if two plates are not thoroughly polished, their contact is imperfect so that when you attempt to separate them slowly the only resistance offered is that of weight; if, however, the pull be sudden, then the lower plate rises, but quickly falls back, having followed the upper plate only for that very short interval of time required for the expansion of the small amount of air remaining between the plates, in consequence of their not fitting, and for the entrance of the surrounding air. This resistance which is exhibited between the two is doubtless likewise present between the parts of a solid, and enters, at least in part, as a concomitant cause of their coherence.

SAGR: Allow me to interrupt you for a moment, please: for I want to speak of something which just occurs to me [. . .] it appears to me that, in the order of nature, the cause must precede the effect, even though it appears to follow in point of time, and since every positive effect must have a positive cause, I do not see how the adhesion of two plates and their resistance to separation—actual facts—can be referred to a vacuum as cause when this vacuum is yet to follow. According to the infallible maxim of the Philosopher, the non-existent can produce no effects.

SIMPL: Seeing that you accept this axiom of Aristotle, I hardly think you will reject another excellent and reliable maxim of his, namely, Nature undertakes only that which happens without resistance; and in this saying, it appears to me, you will find the solution of your difficulty. Since nature abhors a vacuum, she prevents that from which a vacuum would follow as a necessary consequence. Thus it happens that nature prevents the separation of the two plates.

A vacuum is not sufficient

Having accepted the example of the two plates proffered by Salviati, and reinforced by Simplicio's comment, Sagredo wonders why the same mechanism should not be sufficient to explain the resistance to rupture of all solid bodies. But Salviati states that an additional cohesive force is needed, and he promises to prove its existence with an experiment.

SAGR: Now admitting that what Simplicio says is an adequate solution of my difficulty, it seems to me, if I may be allowed to resume my former argument, that this very resistance to a vacuum ought to be sufficient to hold together the parts either of stone or of metal or the parts of any other solid which is knit together more strongly and which is more resistant to separation. If for one effect there be only one cause, or if, more being assigned, they can be reduced to one, then why is not this vacuum which really exists a sufficient cause for all kinds of resistance?

SALV: I do not wish just now to enter this discussion as to whether the vacuum alone is sufficient to hold together the separate parts of a solid body; but I assure you that the vacuum which acts as a sufficient cause in the case of the two plates is not alone sufficient to bind together the parts of a solid cylinder of marble or metal which, when pulled violently, separates and divides. And now if I find a method of distinguishing this well known resistance, depending upon the vacuum, from every other kind which might increase the coherence, and if I show you that the aforesaid resistance alone is not nearly sufficient for such an effect, will you not grant that we are bound to introduce another cause? Help him, Simplicio, since he does not know what reply to make.

Simplicio's transformation

The next excerpt sees Salviati and Sagredo once again acting in complicity at the expense of Simplicio, who does not perceive the sarcasm that has drawn him into the argument. It is Sagredo's turn now to pretend Simplicio is right about his hesitation, saying that he was distracted by a thought ("I was wondering whether . . . to pay the army . . .") which is seemingly unrelated, but is, in fact, a metaphor in support of Salviati's hypothesis. Despite this, however, it is worth underlining the transformation Simplicio has undergone in the *Discourses*, compared to the *Dialogue*. In the latter he was an obstinate defender of *ipse dixit*, incapable of autonomous thought, whereas in the *Discourses* he has become much more receptive to experimental evidence and to the fascination of the new way of thinking. And the certainties he is in search of, no longer founded on bookish knowledge, have to be provided through reasoning. His sharp comments at times enable Salviati to express his theses in more rigorous terms.[327]

SIMPL: Surely, Sagredo's hesitation must be owing to another reason, for there can be no doubt concerning a conclusion which is at once so clear and logical.

SAGR: You have guessed rightly, Simplicio. I was wondering whether, if a million of gold each year from Spain were not sufficient to pay the army, it might not be necessary to make provision other than small coin for the pay of the soldiers.

But go ahead, Salviati; assume that I admit your conclusion and show us your method of separating the action of the vacuum from other causes; and by measuring it show us how it is not sufficient to produce the effect in question.

SALV: Your good angel assist you. I will tell you how to separate the force of the vacuum from the others and afterwards how to measure it. [. . .]

[327] According to Carugo and Geymonat, Simplicio's transformation is to be attributed to "a gradual change on Galileo's part towards Aristotle; a change which was to bring him, little by little, to realize that a profound continuity existed between his work and that of the great Greek philosopher (betrayed, rather than carried on, by Peripatetic scholars of the 16th and 17th centuries)". See A. Carugo e L. Geymonat, *Discorsi intorno a due nuove scienze*, *Introduzione*, Boringhieri, Turin, 1958, p. XIV.

Measurement of the "force of a vacuum"

Let us omit the description of the experiment itself,[328] enjoyable though it is, because it contains too much detail. We will limit ourselves to a brief summary, accompanied by a simplified diagram of the experimental apparatus. Basically, a kind of syringe has to be constructed, composed of a cylinder, either metal or glass, and of an airtight wooden piston ("fitting perfectly inside the cavity" and, if necessary, "carefully bordered with tow or other yielding material"). Some water is introduced into the cylinder and any residual air is expelled by means of a procedure that Galileo duly describes (this is a critical point because, as the piston starts to be pulled back, the air could expand, complicating the measurement). At this point, by means of an iron hook, a bucket is hung from the piston. Sand is then gradually poured into the bucket, up to the moment when the piston begins to descend, allowing an empty space to

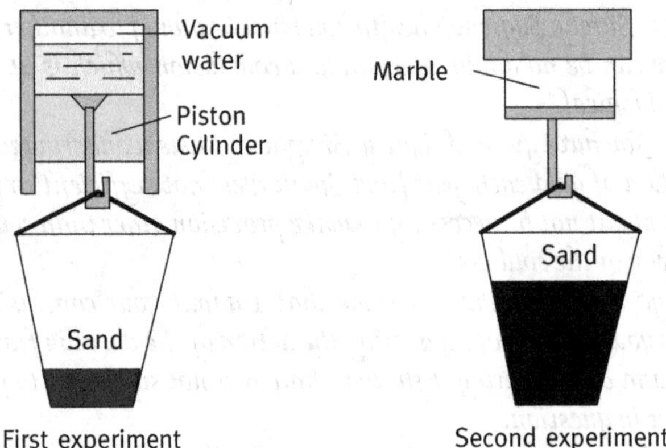

In order to break a marble cylinder equal in shape to the cylinder of water one needs, according to Galileo, a weight five times larger than the "force of a vacuum" alone

[328] This description is to be found on pp. 14–15 of *Two New Sciences*, translated by H. Crew and A. de Salvio, Dover Publications, New York, 1954, p. 14–15. In Italian: *Discorsi e dimostrazioni matematiche intorno a due nuove scienze*, Giornata prima, Edizione Nazionale, Vol. VIII, pp. 62–63.

form above the water. Since, according to Galileo, water is a liquid where "no resistance is opposed to separation other than that originating from the vacuum", in order to determine the magnitude of the "force of a vacuum" it will be sufficient to weigh the system (piston, plus hook, plus bucket with sand) at the moment when the descent begins. The experiment must then be repeated with a cylinder made of solid material, cemented to a supporting structure at its top, replacing the water. This is Galileo's conclusion:

SALV: [...] If one attaches to a cylinder of marble or glass a weight which, together with the weight of the marble or glass itself, is just equal to the sum of the weights before mentioned, and if breaking occurs we shall then be justified in saying that the vacuum alone holds the parts of the marble and glass together; but if this weight does not suffice and if breaking occurs only after adding, say, four times this weight, we shall then be compelled to say that the vacuum furnishes only one fifth of the total resistance. [...]

The conclusion, therefore, is that in the bonds which hold together the atoms of a solid, such as marble or crystal glass, external vacuum (i.e. the vacuum formed above the water in the first experiment, called by Galileo the "great vacuum") is responsible for only 20% of the force, the rest being due to some additional gluey substance. Quite aside from the fact that, as we said earlier, the experiment using water would give, if correctly interpreted, a measure of the weight of the atmospheric column pressing on the underside surface of the piston—that is 1 kg per square centimetre—it is patently clear that Galileo's conclusion has little validity, based as it is on an experiment that was not simple to perform reliably. For instance, at the time it was not easy to turn metal cylinders capable of guaranteeing the required airtightness with wooden pistons. And these pistons would, anyway, have produced strong friction, a factor that Galileo does not seem to have paid much attention to.[329] In spite of

[329] This was immediately pointed out by a skilled engineer of the time, Antonio de Ville, who stated dramatically: "the bucket would have to be filled more than the strength of the apparatus could allow, so everything would fall to pieces before the bucket started descending". Today the experiment can be performed quite easily using a pharmaceutical syringe.

the great care he takes over the description, he does not mention, for example, the use of grease to facilitate sliding and improve air-tightness. For these reasons, according to Koyré[330] the experiment, like other *gedanken experimenten* conceived by Galileo, was never carried out. However, in this particular case Galileo probably knew what he was talking about—that is to say, he must have at least attempted to perform the experiment in his workshop in Padua, seeing that on many occasions in his career he was faced with nearly identical problems in connection with hydraulic pumps, which, of course, were of utmost concern for the Republic of Venice. But let us move on to discover what Salviati thinks of the additional "gluey substance".[331]

The mysterious gluey substance

SAGR: It still remains for you to tell us upon what depends the resistance to breaking, other than that of the vacuum; what is the gluey or viscous substance which cements together the parts of the solid? For I cannot imagine a glue that will not burn up in a highly heated furnace in two or three months, or certainly within ten or a hundred. For if gold, silver and glass are kept for a long while in the molten state and are removed from the furnace, their parts, on cooling, immediately reunite and bind themselves together as before. Not only so, but whatever difficulty arises with respect to the cementation of the parts of the glass arises also with regard to the parts of the glue; in other words, what is that which holds these parts together so firmly?

SALV: A little while ago, I expressed the hope that your good angel might assist you. I now find myself in the same straits. Experiment leaves no doubt that the reason why two plates cannot be separated, except with violent effort, is that they are held together by the resistance of the vacuum; and the same can be said of two large pieces of a

[330] A. Koyré, *Études galiléennes*, Paris, 1966.

[331] *Two New Sciences*, translated by H. Crew and A. de Salvio, Dover Publications, New York, 1954, p. 18 and following. In Italian: *Discorsi e dimostrazioni matematiche intorno a due nuove scienze*, Giornata prima, Edizione Nazionale, Vol. VIII, p. 65 and following.

marble or bronze column. This being so, I do not see why this same cause may not explain the coherence of smaller parts and indeed of the very smallest particles of these materials. Now, since each effect must have one true and sufficient cause and since I find no other cement, am I not justified in trying to discover whether the vacuum is not a sufficient cause?

But just look what Salviati is driving at now! After all this talk about some special "gluey substance", he has brought us back to *horror vacui*, a force he is now claiming acts at all levels, right down to the most microscopic, that is, the cohesion 'of smaller parts and indeed of the very smallest particles of these materials'. We will let readers themselves come to terms with the doubts that will surely have been raised by that 20% contribution to the resistance of the marble cylinder due to the macroscopic external vacuum, which is not relevant in this case. Indeed, we have to appreciate the fact that, a little further on, Galileo–Salviati, too, will acknowledge, in replying to Simplicio's objections, that his idea was to be treated not as "an absolute fact, but rather as a passing thought".

In the section of dialogue which follows the intriguing hypothesis mentioned in the opening to this chapter is put forward, namely, that the fusion of a body is caused by the insertion into the empty spaces in the solid of "extremely fine particles of fire"—*ignicoli*, from Latin *ignis*, in Galileo's actual words—which, it is suggested, have the effect of breaking the bonds that hold the constituent particles together.

The particles of fire

SIMPL: But seeing that you have already proved that the resistance which the large vacuum offers to the separation of two large parts of a solid is really very small in comparison with that cohesive force which binds together the most minute parts, why do you hesitate to regard this latter as something very different from the former?
SALV: Sagredo has already answered this question when he remarked that each individual soldier was being paid from coin collected by a

general tax of pennies and far-things, while even a million of gold would not suffice to pay the entire army. And who knows but that there may be other extremely minute vacua which affect the smallest particles so that which binds together the contiguous parts is throughout of the same mintage? Let me tell you something which has just occurred to me and which I do not offer as an absolute fact, but rather as a passing thought, still immature and calling for more careful consideration. You may take of it what you like; and judge the rest as you see fit. Sometimes when I have observed how fire winds its way in between the most minute particles of this or that metal and, even though these are solidly cemented together, tears them apart and separates them, and when I have observed that, on removing the fire, these particles reunite with the same tenacity as at first, without any loss of quantity in the case of gold and with little loss in the case of other metals, even though these parts have been separated for a long while, I have thought that the explanation might lie in the fact that the extremely fine particles of fire, penetrating the slender pores of the metal (too small to admit even the finest particles of air or of many other fluids), would fill the small intervening vacua and would set free these small particles from the attraction which these same vacua exert upon them and which prevents their separation. Thus the particles are able to move freely so that the mass becomes fluid and remains so as long as the particles of fire remain inside; but if they depart and leave the former vacua then the original attraction returns and the parts are again cemented together.

In reply to the question raised by Simplicio, one may say that although each particular vacuum is exceedingly minute and therefore easily overcome, yet their number is so extraordinarily great that their combined resistance is, so to speak, multiplied almost without limit. The nature and the amount of force which results from adding together an immense number of small forces is clearly illustrated by the fact that a weight of millions of pounds, suspended by great cables, is overcome and lifted, when the south wind carries innumerable atoms of water, suspended, in thin mist, which moving through the air penetrate between the fibres of the tense ropes in spite of the tremendous force of the hanging weight. When these particles enter the narrow pores they swell the ropes, thereby shorten them, and perforce lift the heavy mass.

Strength in unity

SAGR: There can be no doubt that any resistance, so long as it is not infinite, may be overcome by a multitude of minute forces. Thus a vast number of ants might carry ashore a ship laden with grain. And since experience shows us daily that one ant can easily carry one grain, it is clear that the number of grains in the ship is not infinite, but falls below a certain limit. If you take another number four or six times as great, and if you set to work a corresponding number of ants they will carry the grain ashore and the boat also. It is true that this will call for a prodigious number of ants, but in my opinion this is precisely the case with the vacua which bind together the least particles of a metal.
SALV: But even if this demanded an infinite number would you still think it impossible?
SAGR: Not if the mass of metal were infinite; otherwise....
SALV: Otherwise what? Now since we have arrived at paradoxes let us see if we cannot prove that within a finite extent it is possible to discover an infinite number of vacua. [...]

Regarding these last few exchanges of the dialogue, it has to be said that not all Galileo's ideas concerning infinity and infinitesimals are crystal-clear, as we saw in Chapter 12. Making an analogy with a known geometric problem (which we omit here for the sake of brevity) Galileo advances the hypothesis that the "small vacua" inside a body can be infinite in number, which implies that they must be of infinitesimal size, just like the points contained in a geometrical segment of finite length. And this leads him to assert, in truth using rather unconvincing arguments, that the constituent particles of matter are themselves infinite in number and infinitesimal in size, that is, dimensionless.

We pointed out in earlier chapters that Galileo did not treat the "indivisible parts", that is the infinitesimals, as abstract geometrical entities only, but also as entities that are directly associated with physical problems, such as the mechanism of acceleration. Here, instead, in an attempt at unification, he is trying to draw a close parallel between mathematical reality and the structure of matter. In his *Assayer* (as we shall see in the next chapter) he had assigned to the expression "indivisible parts" a meaning which was less daring—simply that of a quantity which cannot be subdivided further, adopting the atomistic view of Democritus and Anassagoras. This meaning has reappeared in modern physics to indicate a particle that is not separable into subparticles (like a *quark*), or an energy packet which can only be observed in its entirety (like a *quantum of light*). The concept of light or matter quantization is much too modern for Galileo to have had any feeling for it. It should not surprise us, therefore, if, as he became more mature, Galileo dropped his original meaning of "indivisible", reminiscent of Democritus, in favour of a conception that seemed to him more revolutionary, more appealing and more consistent with the idea of a mathematized Nature—an idea which treated the indivisible part of matter as equivalent to the indivisible part in geometry, that is, the very last portion of an object, which is arrived at after one has performed an infinite number of subdivisions.

Some science historians have perceived more complex meanings in this silent passing from an "atomistic" conception of matter in the *Assayer*, to the purely "mathematical" conception adopted here—see, for example, the book by Redondi[332] and the authors quoted therein (we shall come back to this point in Chapter 18). It seems to us, however, that the elegance of the parallel between the structure of matter and geometry is more than enough to justify Galileo's final position. This accords with Koyré's view:[333] "Right from the time of his early work in Pisa, the efforts of young Galileo, a follower of Archimedes and Plato, were directed towards a well-defined goal: the mathematization of physics. Nobody before him—not even Benedetti—had pursued this objective so consciously, patiently and stubbornly".

Although the arguments we are about to read are questionable, Galileo has to receive the credit—as Geymonat points out—for having understood that "the investigation of the structure of matter implies the study of very complicated questions concerning *continuity, vacuum* and *atoms*, and, as a consequence, the identification of the

[332] P. Redondi, *Galileo eretico*, Einaudi Editore, Turin, 1983, p. 30 (English translation *Galileo Heretic* by R. Rosenthal, Princeton University Press, Princeton, 1987).

[333] A. Koyré, *ibid.*, p. 96.

similarities and differences between the subdividing done by the mathematician, and that done by the physicist".[334]

A ventured hypothesis

SALV: [. . .] And here I wish you to observe that after dividing and resolving a line into a finite number of parts, that is, into a number which can be counted, it is not possible to arrange them again into a greater length than that which they occupied when they formed a continuum and were connected without the interposition of as many empty spaces. But if we consider the line resolved into an infinite number of infinitely small and indivisible parts, we shall be able to conceive the line extended indefinitely by the interposition, not of a finite, but of an infinite number of infinitely small indivisible empty spaces.

Now this which has been said concerning simple lines must be understood to hold also in the case of surfaces and solid bodies, it being assumed that they are made up of an infinite, not a finite, number of atoms. Such a body once divided into a finite number of parts it is impossible to reassemble them so as to occupy more space than before unless we interpose a finite number of empty spaces, that is to say, spaces free from the substance of which the solid is made. But if we imagine the body, by some extreme and final analysis, resolved into its primary elements, infinite in number, then we shall be able to think of them as indefinitely extended in space, not by the interposition of a finite, but of an infinite number of empty spaces. Thus one can easily imagine a small ball of gold expanded into a very large space without the introduction of a finite number of empty spaces, always provided the gold is made up of an infinite number of indivisible parts.

SIMPL: It seems to me that you are travelling along toward those vacua advocated by a certain ancient philosopher.[335]

[334] L. Geymonat, *Galileo Galilei*, Einaudi, Turin, 1969, p. 316. English translation by S. Drake, MacGraw-Hill Book Co., New York, 1965.

[335] Simplicio is here making a scornful allusion to Democritus, whose atomistic conception was opposed by Aristotle, but was evidently to Galileo's liking and has also found recognition in modern science.

SALV: But you have failed to add, "who denied Divine Providence", an inapt remark made on a similar occasion by a certain antagonist of our Academician.
SIMPL: I noticed, and not without indignation, the rancor of this ill-natured opponent; further references to these affairs I omit, not only as a matter of good form, but also because I know how unpleasant they are to the good tempered and well-ordered mind of one so religious and pious, so orthodox and God-fearing as you. [. . .]

The opponent of Galileo whom Salviati mentions may be Jesuit Father Orazio Grassi, who engaged in bitter disputes with him and went so far as to accuse him of heresy in a matter of faith (see Chapters 18 and 22). Therefore, the testimonial of respectability and fidelity to the Church which Simplicio here confers upon Salviati assumes a very special meaning in comparison to the many other exercises in caution to be found scattered throughout Galileo's writings. Galileo never expressed his intimate beliefs in an explicit way, although his basic mentality, as it emerges in all his writings, exhibits the essential characteristics of rationalistic agnosticism. His displays of orthodoxy—strongly urged upon him by his prelate friends—have to be seen as utterly indispensable when we consider the cultural milieu in which Galileo had to operate.

Following this come a few light diversions, until Salviati, examining the differences between a finely powdered solid and a liquid, proposes, with a remarkable dose of arbitrariness, that liquids are as they are because they happen to be divided into their infinitesimal elements.[336]

Liquids

SALV: [. . .] Having broken up a solid into many parts, having reduced it to the finest of powder and having resolved it into its infinitely small indivisible atoms why may we not say that this solid has been reduced to

[336] *Two New Sciences*, translated by H. Crew and A. de Salvio, Dover Publications, New York, 1954, p. 39 and following. In Italian: *Discorsi e dimostrazioni matematiche intorno a due nuove scienze*, Giornata prima, Edizione Nazionale, Vol. VIII, p. 85 and following.

a single continuum *perhaps a fluid like water or mercury or even a liquefied metal? And do we not see stones melt into glass and the glass itself under strong heat become more fluid than water?*

SAGR: Are we then to believe that substances become fluid in virtue of being resolved into their infinitely small indivisible components?

SALV: I am not able to find any better means of accounting for certain phenomena of which the following is one. When I take a hard substance such as stone or metal and when I reduce it by means of a hammer or fine file to the most minute and impalpable powder, it is clear that its finest particles, although when taken one by one are, on account of their smallness, imperceptible to our sight and touch, are nevertheless finite in size, possess shape, and capability of being counted. It is also true that when once heaped up they remain in a heap; and if an excavation be made within limits the cavity will remain and the surrounding particles will not rush in to fill it; if shaken the particles come to rest immediately after the external disturbing agent is removed; the same effects are observed in all piles of larger and larger particles, of any shape, even if spherical, as is the case with piles of millet, wheat, lead shot, and every other material. But if we attempt to discover such properties in water we do not find them; for when once heaped up it immediately flattens out unless held up by some vessel or other external retaining body; when hollowed out it quickly rushes in to fill the cavity; and when disturbed it fluctuates for a long time and sends out its waves through great distances.

Seeing that water has less firmness than the finest of powder, in fact has no consistence whatever, we may, it seems to me, very reasonably conclude that the smallest particles into which it can be resolved are quite different from finite and divisible particles; indeed the only difference I am able to discover is that the former are indivisible. The exquisite transparency of water also favors this view; for the most transparent crystal when broken and ground and reduced to powder loses its transparency; the finer the grinding the greater the loss; but in the case of water where the attrition is of the highest degree we have extreme transparency. Gold and silver when pulverized with acids more finely than is possible with any file still remain powders, and do not become fluids until the finest particles of fire or of the rays of small components.

[. . .]

Contraction and rarefaction

The dialogue now touches on the problem of the contraction and rarefaction of bodies—their capacity, that is, for changing in density—which according to Salviati finds adequate explanation in his model of a solid constituted of an infinite number of infinitesimal parts separated by an infinite number of vacua, also of infinitesimal size. For him, this model accounts for the fact that gold can be drawn to such an extent that it becomes an extremely thin leaf, "so thin that it almost floats in air". And, once in this form, rolled around a silver wire and drawn again to be reduced "to the fineness of a lady's hair". This is, as a matter of fact, true—but it is hard to see how it can be related to the structure proposed by Galileo, since it is merely a matter of geometry, as Simplicio is careful to point out.

Actually, it would seem doubts enter Galileo's mind right from the outset of the discussion, seeing that Salviati opens with the words: if my idea does not convince you, "regard it, together with my remarks, as idle talk". As for Sagredo, he tries to hide his embarrassment through unusually flowery speech, full of metaphor and alliteration[337] (e.g. "to explain the explanation"). The two contributions Simplicio makes to the following debate are very lucid and confirm our earlier remarks concerning the intellectual growth of the character. Here it is he who gives voice to the doubts Galileo had, and his criticism is unimpeachable when he says: "The arguments and demonstrations which you have advanced are mathematical, abstract, and far removed from concrete matter; and I do not believe that when applied to the physical and natural world these laws will hold". Consistently with this, his style changes, too: previously pompous and affected, it now becomes concise and logical, a new medium that is appropriate for a new way of thinking.

The dialogue that follows is taken, like the preceding ones, from the *Discourses*.[338]

[337] Not all alliterations have been maintained in the English translation.

[338] *Two New Sciences*, translated by H. Crew and A. de Salvio, Dover Publications, New York, 1954, p. 51 and following. In Italian: *Discorsi e dimostrazioni matematiche intorno a due nuove scienze*, Giornata prima, Edizione Nazionale, Vol. VIII, p. 96 and following.

SALV: [. . .] This contraction of an infinite number of infinitely small parts without the interpenetration or overlapping of finite parts and the previously mentioned expansion of an infinite number of indivisible parts by the interposition of indivisible vacua is, in my opinion, the most that can be said concerning the contraction and rarefaction of bodies, unless we give up the impenetrability of matter and introduce empty spaces of finite size. If you find anything here that you consider worth while, pray use it; if not regard it, together with my remarks, as idle talk; but this remember, we are dealing with the infinite and the indivisible.

SAGR: I frankly confess that your idea is subtle and that it impresses me as new and strange; but whether, as a matter of fact, nature actually behaves according to such a law I am unable to determine; however, until I find a more satisfactory explanation I shall hold fast to this one. Perhaps Simplicio can tell us something which I have not yet heard, namely, how to explain the explanation which the philosophers have given of this abstruse matter; for, indeed, all that I have hitherto read concerning contraction is so dense and that concerning expansion so thin that my poor brain can neither penetrate the former nor grasp the latter.

SIMPL: I am all at sea and find difficulties in following either path, especially this new one; because according to this theory an ounce of gold might be rarefied and expanded until its size would exceed that of the earth, while the earth, in turn, might be condensed and reduced until it would become smaller than a walnut, something which I do not believe; nor do I believe that you believe it. The arguments and demonstrations which you have advanced are mathematical, abstract, and far removed from concrete matter; and I do not believe that when applied to the physical and natural world these laws will hold.

SALV: I am not able to render the invisible visible, nor do I think that you will ask this. But now that you mention gold, do not our senses tell us that metal can be immensely expanded? I do not know whether you have observed the method employed by those who are skilled in drawing gold wire, of which really only the surface is gold, the inside material being silver. The way they draw it is as follows: they take a cylinder or, if you please, a rod of silver, about half a cubit long and three or four times as wide as one's thumb; this rod they cover with gold-leaf which is so thin that it almost floats in air, putting on not more than eight or ten thicknesses. Once gilded they begin to pull it, with great force, through the holes of a draw-plate; again and again it is made to pass through

smaller and smaller holes, until, after very many passages, it is reduced to the fineness of a lady's hair, or perhaps even finer; yet the surface remains gilded. Imagine now how the substance of this gold has been expanded and to what fineness it has been reduced.[339]
SIMPL: I do not see that this process would produce, as a consequence, that marvellous thinning of the substance of the gold which you suggest: first, because the original gilding consisting of ten layers of gold-leaf has a sensible thickness; secondly, because in drawing out the silver it grows in length but at the same time diminishes proportionally in thickness; and, since one dimension thus compensates the other, the area will not be so increased as to make it necessary during the process of gilding to reduce the thinness of the gold beyond that of the original leaves.
SALV: You are greatly mistaken, Simplicio, because the surface increases directly as the square root of the length, a fact which I can demonstrate geometrically.
[. . .]
SALV: [. . .] Consider now what degree of fineness it must have and whether one could conceive it to happen in any other way than by enormous expansion of parts; consider also whether this experiment does not suggest that physical bodies are composed of infinitely small indivisible particles, a view which is supported by other more striking and conclusive examples.

Salviati does not specify which examples he has in mind, and indeed it is not at all easy to imagine what they might be, since the hypothesis he has made is patently an arbitrary one. What lingers in the mind after the discussion, however, is the graceful image of artisans drawing gold, testimony to Galileo's keen attention to the world of technology.

MATHEMATICAL NOTE

In order to demonstrate the origin of the cohesive forces that bind atoms together, and explain why they lead to a well-defined value as regards the

[339] Referring to this example, Descartes had objected that gold does not undergo rarefaction, it simply assumes a different geometrical shape.

interatomic separation, let us consider the simple case of a crystal of the ionic group—rock salt. This salt has a lattice made up of ionized atoms, positive sodium ions (Na^+), alternating with negative ions of chlorine (Cl^-), as shown in the diagram below. These two types of ion exert an attractive electrostatic force on each other, which causes each pair of them to take up a potential energy that is inversely proportional to the distance r between them (Coulomb's law):

$$E_{coul} = -kQ^2/r$$

where Q is their charge. The minus sign in front of the energy expresses the mutual attraction between the two ions: in fact, the energy is reduced if they get closer to each other. The system would, therefore, spontaneously seek to evolve towards a condition where the two ions end up overlapping. But the Coulomb force, to the world's good fortune, is not the only one that is acting.

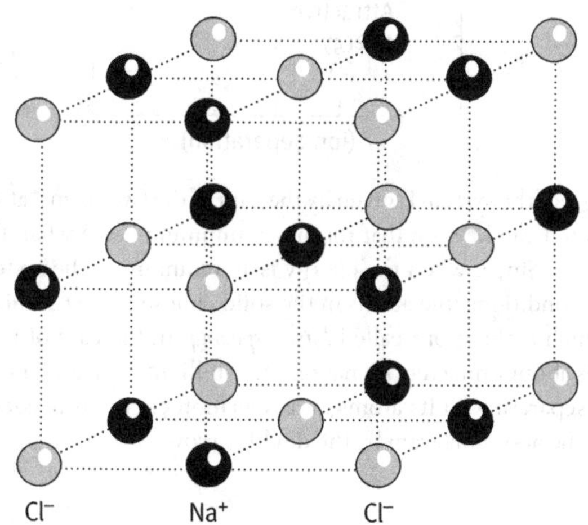

Rock-salt crystal lattice, made of alternating sodium and chlorine ions

Quantum mechanics considerations (which we leave aside here, suffice to say they are related to the *Pauli exclusion principle* which prohibits two electrons from occupying the same state), predict that, when the distance between the two ions becomes very small, a repulsive force develops which prevents the electrons belonging to the two ions from overlapping. In other words, work would be needed to bring them closer together. This repulsion comes into the potential energy as a positive term: it is of little importance if the two ions are far apart, but it increases more rapidly than the decrease in the negative

Coulomb term when the distance between the ions approaches zero, as the diagram shows.

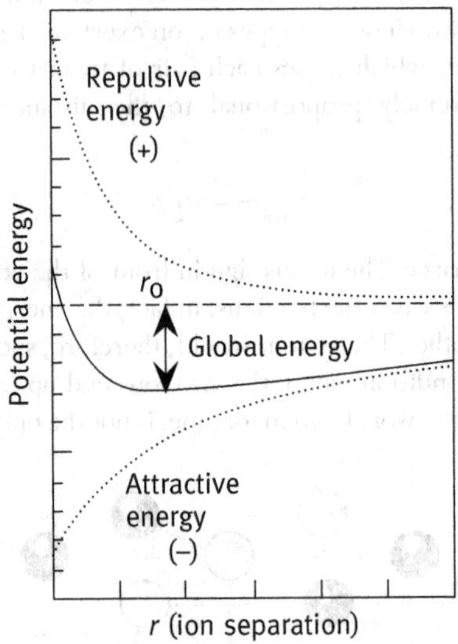

The energy of the system is given by the sum of the two terms, attractive and repulsive, and it can be seen that this has a minimum value when the ions are separated by r_0. Since, when the energy is at a minimum, the system reaches equilibrium conditions, the atoms in the solid choose to stay at this particular distance, which is therefore called *lattice spacing*. In the case of rock salt the value for r_0 is about a quarter of a nanometer (billionth of a meter). To dissolve the crystal, separating all its atoms, a dissociation energy is needed, which is indicated in the above diagram by the double arrow.

18

The Dark Labyrinth of the Senses

Mathematics in Nature and Sensory Perception*

vo io pensando che questi sapori, odori, colori . . .
tengano solamente lor residenza nel corpo sensitivo,
sì che rimosso l'animale, sieno levate
ed annichilate tutte queste qualità

I think that tastes, odors, colors . . . have
their habitation only in the sensorium.
Thus, if the living creature were removed,
all these qualities would be removed and annihilated

Here Galileo's famous statement that mathematics is the cardinal point of science, shedding light on Nature, is examined. So are the controversial statements about secondary properties of bodies—odors, tastes, etc.—which, according to Galileo, are not inherent to bodies themselves, but exist solely insomuch as they are the objects of our perception. The discussion closes with two insights: first, the correspondence between the movement of particles and the sensation of heat, and second, the relationship between light and matter.

The passages we are going to look at in this chapter are taken from *The Assayer*. The brilliant dialectics and typically Tuscan taste for satire to be found in the book (about whose content more will be said in

* From *The Assayer*.

Chapter 20) were particularly to Pope Urban VIII's liking, and he had pages from it read to him at mealtimes. However, over and beyond its incisive, pleasant style, and its witty polemic, *The Assayer* is also valuable as a work "of cultural propaganda, of disruption of the old methods" and a "reminder . . . to the scientist of his duty to reason in mathematical terms, in order not to be led astray by fancies and sophistic equivocations".[340] And it is in this perspective that Galileo's celebrated pronouncement, generally looked upon as the manifesto of his thinking, should be viewed.[341]

The grand Book of Nature

[. . .] Philosophy is written in this grand book—I mean the universe—which stands continually open to our gaze, but it cannot be understood unless one first learns to comprehend the language and interpret the characters in which it is written. It is written in the language of mathematics, and its characters are triangles, circles, and other geometrical figures, without which it is humanly impossible to understand a single word of it; without these, one is wandering about in a dark labyrinth.

This fundamental concept lies at the foundations of the new science, which was set in motion by Galileo and other great physicists of his time, but it had already been expressed by Leonardo da Vinci, who had written:[342] "No human investigation can be called true science unless it is based on mathematical demonstrations", and also: "Whoever doubts the supreme certainty of mathematics feeds himself with confusion and will never silence the contradictions of the sophistic sciences, from which one can only learn an endless uproar".[343]

[340] L. Geymonat, *Galileo Galilei*, Einaudi, Turin, 1969, p. 131, 137 (English translation by S. Drake, *Galileo Galilei: A Biography and Inquiry into His Philosophy of Science*, MacGraw-Hill Book Co., New York, 1965).

[341] S. Drake and O'Malley, *The Assayer*, in *Controversy on the Comets of 1618*, Philadelphia, PA, 1960 (out of print, available in photostatic form), p. 183. In Italian: *Il Saggiatore*, Edizione Nazionale, Vol. VI, p. 232.

[342] See F. Flora, *Antologia leonardesca*, Istituto Editoriale Cisalpino, Milan, 1947, p. 52.

[343] Regarding sophisms, it is worth recalling a funny story told by Galileo himself, that "of the Cretan who said that all Cretans were liars. Therefore, being a

The role of mathematics

The systematic mathematization of Nature represents the most important innovation in Galilean thinking. "Mathematization", of course, is to be understood as "geometrization", since Galileo never made use of formulae, algebra, or even fractions. If, as some would have it, his approach was Platonically inspired, it was so only as regards the rationalistic aspect of Plato's philosophy, certainly not its mystic-magic aspect.[344] In fact, the progenitor of Galileo's conception was Archimedes, a constant point of reference for him. Nothing could be further from Galilean thinking than the setting up of a contraposition between the ever-changing world of phenomena and the incorruptible world of Ideas. Mathematics, for him, was merely a reasoning tool which helped one avoid being led astray by "vain fancies".

Galileo's statement is an open declaration of trust in the rationality of Nature and humankind—and in the capacities of humans for deciphering and understanding reality, provided they know its language. The two metaphors, "open book" and "dark labyrinth" are symbols, respectively, for the new way (Renaissance and humanistic), and for the old way (medieval and dogmatic), of looking at Nature—in which humanity risks losing its way if it tries to do without "Arianne's thread", that is, reason.[345]

Cretan, he had told a lie in saying that Cretans were liars. It follows therefore that Cretans were not liars, and consequently that he, being a Cretan, had spoken truth. And since in saying that Cretans were liars he had spoken truly, including himself as a Cretan, he must consequently be a liar. And thus, in such sophisms, a man may go round and round forever and never come to any conclusion" (*Dialogue Concerning the Two Chief World Systems*, translated by S. Drake, The Modern Library, New York, 2001, p. 47. In Italian: *Dialogo sopra i due massimi sistemi del mondo*, National Edition, Vol. VII, p. 66).

[344] This is pointed out in detail by G. Castellani in *Galileo ritrovato, Prometheus 14*, Franco Angeli, Milan, 1993, p. 62.

[345] The image of a book to represent Nature can be found frequently in pre-Galilean literature. Yet, before him, "book reading was an esoteric act, the privilege of a select few to whom the Creator, at His discretion, permitted access. With the new science the book is no longer '*involutus*' . . . , it is democratically 'open', readable without initiatory preclusions . . .", as pointed out by Battistini (quoted by C. Salinari and C. Ricci, *Storia della letteratura italiana*, Vol. 2, Laterza, Bari, 1995, p. 1467).

With the decisive role it assigns to mathematics, the statement constitutes, today even more than in Galileo's time, the essential prerequisite to doing science. Indeed, one can say that interdependence and cross-fertilization among experimental science, theoretical models and mathematical methods are the most distinctive features of contemporary science—and this becomes the more pronounced the more advanced and mature science becomes.

As regards the layperson (i.e. the non-scientist), are we in a position to say, all these centuries later, that Galileo's statement has become part of the cultural patrimony of those who live in the so-called civilized countries of the West? Unfortunately not. Too many people still rely acritically on the opinions of others and "wander about in the dark labyrinth of senses", yet to be enlightened by reason.

The qualities perceived

Concerning the mathematization of Nature, there is a brief digression in *The Assayer*—almost a *divertissement* that Galileo permits himself in the middle of the main scientific discourse—where the aspects of Nature that the mathematical physicist should abstain from investigating are specified. These are the qualitative aspects: that is, those secondary properties of objects, such as colors, sounds, odors, tastes, that are inherent not to the objects themselves, but, instead, to the subject who perceives them. In the passage we are going to read below, he says:

> I do not believe that, for exciting in us tastes, odors, and sounds there are required in external bodies anything but sizes, shapes, numbers, and slow or fast movements; I think that if ears, tongues, and noses were taken away, shapes and numbers and motions would remain but not odors or tastes or sounds. These, I believe, are nothing but names. . . .

The task of the physicist is therefore to search for the primary qualities—that is, size, shape, number, and movement—which can be expressed quantitatively and defined objectively, thus providing the basis for scientific study.

Despite its marginality in Galileo's writings, this argument deserves looking at, because it has attracted the attention of those who would have it that Galileo upheld that mechanisms of the soul exist, and are distinct from those of a physical and material nature. One commentator, for instance, says "*The Assayer* represents the most important philosophical revolution of modern times, because of the distinction it draws between the primary or physical properties of living bodies (all mathematizable) and the psychological properties (secondary) . . . which exist only in the conscious mind, even though they are determined by the external physical qualities . . .".[346] In fact, however, Galileo talks about an "animate and sensitive body", of qualities that "have their habitation only in the sensorium. Thus, if the living creature were removed, all these qualities would be removed and annihilated". A position that, if anything, appears to be materialistic (and, in fact, he uses the terms sensorium and conscious mind interchangeably[347]).

Moreover, it needs to be stressed that Galileo, in drawing a distinction between physical and psychological qualities, was doing so not from the viewpoint of a philosopher, but from that of a scientist; that is, he was adopting "a mental approach which was more operative than speculative. To put it another way, he dared make this distinction because it was of utility in his investigations, but only within those limits where this utility was clearcut"—he did not care about "analyzing its general cognitive premises, because such analysis lay outside what interested him".[348]

Not only outside what interested him, but also outside what was possible for him. He knew too well that it is unscientific to conjecture about what cannot be submitted to logical investigation and experimental verification, and thus he refrained from enquiring into it, leaving the problem for future generations. Otherwise the risk—worse, even, than the "endless uproar" of the sophists—would have been quackery, mystification, and false science. Four centuries later

[346] D. Galati, *Galileo*, Pagoda, Rome, 1991, p. X.

[347] In the original Italian Galileo's terms for "sensorium" and "conscious mind" were respectively *corpo sensitivo* and *anima sensitiva*.

[348] L. Geymonat, *ibid.*, p. 133.

humankind has still not learned this lesson, and seems content to stew in the groundless hypotheses of astrology, mystery, ufology, and the paranormal.

In contrast, professional science has, made great strides forward. Subjective properties that once might have appeared unsuitable for mathematical treatment and experimental testing have become a field of quantitative studies. The advent of electronics and computers has opened the way to objective analysis of the phenomena of perception, to the artificial simulation of memory and intelligence—in short, to the extraordinary unexplored universe of the cognitive sciences. It may well be that, sooner or later, the mechanisms of the "soul" will be explained through mathematical formulae similar to those used to describe natural phenomena.[349] Phenomena that Galileo succeeded in putting into such precise focus but that, prior to him, had challenged and defeated successive generations of thinkers. Throughout history, even if the objectives have continually changed, becoming higher and more ambitious, they have always been achieved. When, then, can we expect an equation for consciousness? When will a machine be capable of emulating a human being?

Among the arguments dealt with in the following passages are two insights of Galileo's which are of particular importance; first, that there must be some kind of connection between the sensation of heat and the movement of the particles of the thermal source; second, that light and matter are, in some sense, different faces of the same reality. We shall return to these matters in more detail during the course of the reading.

Let us now move on to the celebrated passage on sensory perception.[350]

[349] At the beginning of the twentieth century, a giant of physics, Ludwig Boltzmann, wrote: "Psychical processes coincide with well-defined material processes in our brain . . . spirit and will are not something above the body, but rather complex actions of material parts . . . intuition, will and self-consciousness are merely the stages of highest development of the physical–chemical forces acting in matter . . ." (L. Boltzmann, *Populäre Schriften*, Barth, Leipzig, 1905).

[350] S. Drake and O'Malley, *The Assayer*, in *Controversy on the Comets of 1618*, Philadelphia, PA, 1960 (out of print, available in photostatic form), p. 308. In Italian: *Il Saggiatore*, Edizione Nazionale, Vol. VI, p. 347 and following.

It now remains, in accordance with the promise made above to your Excellency, for me to tell you some of my thoughts about the proposition, 'motion is the cause of heat', and to show in what sense this may be true in my opinion. But first I must give some consideration to what we call heat, for I much suspect that in general people have a conception of this which is very remote from the truth, believing heat to be a real attribute, property and quality which actually resides in the material by which we feel ourselves warmed.

Therefore I say that upon conceiving of a material or corporeal substance, I immediately feel the need to conceive simultaneously that it is bounded and has this or that shape; that it is in this place or that at any given time; that it moves or stays still; that it does or does not touch another body; and that it is one, few, or many. I cannot separate it from these conditions by any stretch of my imagination. But that it must be white or red, bitter or sweet, noisy or silent, of sweet or foul odor, my mind feel no compulsion to understand as necessary accompaniment. Indeed, without the senses to guide us, reason or imagination alone would perhaps never arrive at such qualities. For that reason, I think that tastes, odors, colors, and so forth are no more than mere names so far as pertains to the subject wherein they reside, and that they have their habitation only in the sensorium. Thus, if the living creature were removed, all these qualities would be removed and annihilated. [. . .]

Tickling

I believe I can explain my idea better by means of some examples. I move my hand first over a marble statue and then over a living man. Now as to the action derived from my hand, this is the same with respect to both subjects so far as the hand is concerned; it consists of the primary phenomena of motion and touch which we have not designated by any other names. But the animate body which receives these operations feels diverse sensations according to the various parts which are touched. Being touched on the soles of the feet, for example, or upon the knee or under the armpit, it feels in addition to the general sense of touch another sensation upon which we have conferred a special name, calling it tickling; this sensation belongs entirely to us and not to the hand in any way. It seems to me that anyone would seriously err who might wish to say that the hand had within itself, in addition to the properties of moving and touching, another faculty different from these; that of tickling—as if the tickling were an attribute which resided in the hand. A piece of paper or a feather drawn lightly over any part of our bodies performs what are inherently quite the same operations of moving and touching; by touching the eye, the nose, or the upper lip it excites in us an almost intolerable titillation while in other regions it is scarcely felt. Now, this titillation belongs entirely to us and not to the feather; if the animate and sensitive body were removed, it would remain no more than a mere name. And I believe that many qualities which we come to attribute to natural bodies, such as tastes, odors, colors, and other things may be of similar and no more solid existence.

[. . .] I do not believe that, for exciting in us tastes, odors, and sounds there are required in external bodies anything but sizes, shapes, numbers, and slow or fast movements; and I think that if ears, tongues, and noses were taken away, shapes and numbers and motions would remain but not odors or tastes or sounds. These, I believe, are nothing but names, apart from the living animal—just as tickling and titillation are nothing but names when armpits and the skin around the nose are absent [. . .]

How, then, are these words of Galileo to be taken, if one rejects the animistic interpretation mentioned earlier? We cite a comment made by Enrico Bellone, which appears to us to put the problem into the correct light:

Knowledge had to be . . . independent of the operations of the "sensorium", and therefore of the sensory system of observers, in order to apply only to what really characterized the outside world. This way of conceiving knowledge already had, of course, a long tradition—apart from anything else because it also had roots in religious beliefs. According to these, when creating the world God had ordered its parts "*in mensura et numero et pondere*".[351] Galileo's view, however, differed sharply from this tradition on a fundamental point. What he was underscoring in *The Assayer* was not so much the common view that the whole universe is founded on "sizes,

[351] *in mensura et numero et pondere*—Latin for "according to size, number and weight".

shapes, numbers, and slow or fast movements", as the fact that, since this was the objective architecture of the universe, only science was capable of making discoveries along the path towards truth.[352]

The nature of heat

Returning now to my original purpose, [. . .] I say that I am inclined to believe that heat is of this character. Those materials which produce heat in us and make us feel warmth, which we call by the general name fire, would be a multitude of minute particles having certain shapes and moving with certain velocities. Meeting with our bodies, they penetrate by means of their consummate subtlety, and their touch which we feel, made in their passage through our substance, is the sensation which we call heat. This is pleasant or obnoxious according to the number and the greater or lesser velocity of these particles which thus go pricking and penetrating; that penetration is pleasant which assists our necessary insensible transpiration, and that is obnoxious which makes too great a division and dissolution of our substance. To sum up, the operation of fire by means of its particles is merely that in moving it penetrates all bodies by reason of its great subtlety, dissolving them more quickly or more slowly in proportion to the number and velocity of the fire-corpuscles and the density or rarity of the material of these bodies, of which many are such that in their decomposition the major part of them passes over into further tiny corpuscles, and the dissolution goes on so long as it meets with matter capable of being so resolved. But I do not believe at all that in addition to shape, number, motion, penetration, and touch there is any other quality in fire which is 'heat'; I believe that this belongs to us, and so intimately that when the animate and sensitive body is removed, 'heat' remains nothing but a simple vocable. And since this sensation is produced in us by the passage and touch of the tiny corpuscles through our substance, it is obvious that if they were to remain at rest their operation would remain null. [. . .]

352 E. Bellone, "Galileo", *I grandi della scienza*, no. 1, *Le Scienze* (Italian edition of *Scientific American*), Milan, 1998, pp. 78–9.

Since, then, the presence of the fire-corpuscles does not suffice to excite heat, but we need also their movement, it seems to me that one may very reasonably say that motion is the cause of heat.

And so our sensory experience ends up being explained by the objective properties of external bodies. Or, rather, by their interaction, via their emanations, with other material entities, that is, our sense organs. An interaction which is possible, please note, only if there is action and not mere, motionless, existence.

This insight of Galileo's, establishing as it does a correspondence between the motion of the constituent particles of a substance and the sensation of heat we experience when we touch it, is of great interest, even if it was expressed in a vague way. As Redondi points out,[353] the whole "world of the senses was seen as a dense movement of particles of matter". Galileo, however, put the hypothesis forward with particular caution ("this may be true in my opinion"). In the Mathematical Note at the end of this chapter it will be shown that the temperature of a body, when expressed in the Kelvin absolute scale, is actually in direct proportion to the average kinetic energy of its molecules, that is, with their mean square velocity.

Let us take a closer look at the phenomenon in the case of heat sensation upon contact. When we touch an object we feel it to be hot if its molecules have a high kinetic energy or, as is usually said, if they are in a state of high thermal agitation. When these molecules collide with the molecules of our skin, they transfer mechanical energy to them, thus accelerating their motion. This increase in agitation is detected by our nerve cells, which send signals to the brain: it is these signals which, at the perceptive level, give rise to the *sensation of heat*. If the object is at a lower temperature than that of our skin, the mechanical energy transfer occurs in the opposite direction, producing the *sensation of cold*.

In his intuitive and rather serendipitous description, Galileo calls into play "fire-corpuscles", material particles which he suggests

[353] P. Redondi, *Galileo eretico*, Einaudi Editore, Turin, 1983, p. 17 (English translation *Galileo Heretic* by R. Rosenthal, Princeton University Press, Princeton, NJ, 1987).

detach themselves from the heat source and penetrate other objects ("minute particles . . . Meeting with our bodies, they penetrate . . ."). It needs to be pointed out that the sensation of heat produced by contact heat is in fact a purely surface effect. Penetration is only possible in the case of irradiation heat, where the energy which our molecules acquire, and which increases their degree of motion, is carried by electromagnetic waves. The depth to which these penetrate may differ greatly depending on their wavelength, ranging from the skin alone, in the case of visible and infrared waves (as are most of those coming from the sun, or from fire), to several centimeters in the case of microwaves, which are capable of cooking food immediately in depth.[354]

Fire-corpuscles and quanta

"Hang on a moment!" Galileo might say at this point if he were to make an appearance in the present. Why make a distinction between the collisions of particles and the impinging of waves? Don't we now know that electromagnetic radiation has a twofold character, being at once wave-like and particle-like? Planck and Einstein have explained that light is made of quanta, namely of indivisible energy packets, which bodies exchange with each other in their entirety, almost as if they were material corpuscles.[355] After all, are mass and energy different things, given that they can transform into each other?[356] And the vibrational

[354] Cooking in the traditional oven is based instead on the toasting of the surface and the subsequent slow diffusion of heat to the interior by transfer of the agitation from molecule to molecule.

[355] Max Karl Ernst Ludwig Planck introduced the concept of "energy quanta" in 1900 in order to describe the emission spectrum of a black body (irradiation by incandescence). Albert Einstein used this concept in 1905 to explain the photoelectric effect. If the frequency of the wave field is f, the energy transported is "lumped" in quanta equal to hf, where $h = 6.626 \cdot 10^{-27}$ erg · s is called the Planck constant. It should be noted that the word quanta in modern physics has a somewhat different meaning from that adopted by Galileo, to stand for a finite quantity that cannot be further subdivided.

[356] This transformation takes place according to Einstein's well-known equation $E = mc^2$, that is, energy = mass times velocity of light squared.

energy exchanged between colliding particles, is not this too packeted up in quanta? Just think of my fire-corpuscles as entities similar to quanta as they are conceived today, and everything falls into place.

Obviously, in Galileo's time an idea such as this was not even remotely imaginable. One more century would be needed before Newton, unable to explain certain optical phenomena by his corpuscular theory of light[357] (something very similar, please note, to Galileo's stream of "fire-corpuscles") was to come up with a generic hypothesis on the formation of waves caused by the passage of material corpuscles. The wave-particle duality became established definitively only in the twentieth century, when Louis de Broglie stated the principle, bearing his name, that an electron is indeed a particle, but one having a wavelength associated with it—the larger the momentum of the electron, the shorter its wavelength.[358]

Let us now return to Galileo's discussion, to read his curious, but nonetheless logically consistent, explanation for the fact that, when two bodies are rubbed together, they become hot enough to catch fire and emit not only heat, but also light. Afterwards we will give the explanation accepted today.

Light and matter

The rubbing together and friction of two hard bodies, either by resolving their parts into very subtle flying particles or by opening an exit for the tiny fire-corpuscles within, finally set these in motion and, upon their

[357] Light consists of beams of individual corpuscles which, by striking the retina in our eye like bullets, induce visual perception. It is obvious that this model cannot account, for instance, for the phenomenon of destructive interference (light + light = dark). This effect can, instead, be explained by the wave model—advocated primarily by Christiaan Huygens—where light is described as an undulatory perturbation similar to sound.

[358] For the interested reader we specify that momentum is mv, with m = mass and v = velocity of the electron. De Broglie's principle states that an electron with momentum mv has an associated wavelength $\lambda = h/mv$, where h is Planck's constant given in footnote 355. This twofold character of the electron has made it possible to build the electronic microscope, in which light rays are replaced by electron beams.

encountering our bodies and penetrating and coursing through them, our conscious mind feels that pleasant or obnoxious sensation which we have named heat, burning, or scalding. And perhaps when the thinning and attrition stop at or are confined within the tiniest particles, their motion is temporal and their action is calorific only, but, when their ultimate and highest resolution into truly indivisible atoms[359] *is reached, light is created which has instantaneous motion*[360]*—or let us say instantaneous expansion and diffusion—and is capable of occupying immense spaces by its—I do not know whether to say by its subtlety, its rarity, its immateriality, or yet some other property different from all these, and nameless.*

So rubbing, according to Galileo, has the effect of freeing fire-corpuscles, setting them in motion and sending them to our body, where they activate the perception of the "conscious mind". These fire-corpuscles, however, are of varying sizes, ranging from real corpuscles (in the usual sense of the word) to minute particles that are divisible no further. The corpuscles of larger size produce only heat sensations, from plain warming to burning; the smallest and indivisible fire-corpuscles, instead, cause a luminous sensation, that is, light. They mark the point where the boundary between matter and light disappears.

Light, for Galileo, was a fascinating entity: it was the source of life, the very essence of life itself. In a letter to *Monsignor* Dini he wrote:[361]

[...] It would seem to me that in nature there exists an extremely immaterial substance, very rarefied and very fast, which, spreading about the universe, penetrates everywhere without meeting resistance,

[359] *Indivisible*—that is, which cannot be further split up. In this context, the term does not imply an infinitesimal size, even if it does not rule it out, in contrast to the use that later Galileo was to make of it in his *Discourses* (see discussion on this subject in Chapter 17). Here the meaning appears instead to be the atomistic one, inspired by Democritus or Anassagoras, already accepted through the centuries by poets and philosophers like Epicures, Lucretius, Occam, Telesius, Campanella.

[360] Later Galileo was to radically change his opinion on the fact that the propagation of light happens instantaneously. In Chapter 19, we shall report his description (taken from *Discourses*) of an experiment for measuring the (finite) speed of light.

[361] Letter to Piero Dini of 23 March 1615, Edizione Nazionale, Vol. V, p. 301.

warms up and invigorates all living creatures, making them fecund. And our senses tell us that the receptacle of this immaterial substance is the sun: from it, an immense illumination expands throughout the entire universe, accompanied by a calorific spirit which, penetrating all bodies, gives them life and fertility. [. . .] I said that the solar body is a receptacle of this spirit and of this light, which is to say, he is a depository of what he receives from elsewhere, rather than a principle and a primary source in which they originate. It seems to me that this is made quite clear by the Scriptures, where [. . .] we have the creation of light on the first day, whereas the solar body is created on the fourth day. Hence we can in very plausibly state that this fertilizing spirit and this light spread throughout the whole world come together to combine and strengthen in the solar body (for this reason placed in the center of the universe), from whence they diffuse again, having grown more splendid and vigorous.

There is more. For Galileo, light, too, was to be conceived of as matter. Matter, however, in its more subtle and indivisible form—regular matter being condensed light, no less. Direct confirmation of this comes from Orazio Ricasoli Rucellai,[362] a contemporary of Galileo, who attributes to him the opinion that "light might be the universal starting point of Nature, inasmuch as he believed light to be the extreme expansion, that is to say, the ultimate attainable rarefaction. And from this first principle, upon greater or smaller condensation, all things would be formed up to, in the densest case, the hardest and most impenetrable stones".[363]

What can be said about this idea? In the light of Einstein's equivalence between mass and energy, Galileo's words ring prophetically. Just think that, according to special relativity theory, a body can travel at the speed of light if, and only if, it has zero mass—and such,

[362] Cited by A. Favaro in *Pensieri, sentenze e motti di Galileo Galilei*, F.lli Fusi, Pavia, 1907, p. 105.

[363] In his last years Galileo wrote (letter to Fortunio Liceti, Edizione Nazionale, Vol. XVIII, p. 233): "I had always felt so unable to understand what light is, that I would have gladly spent all my life in jail, fed with bread and water, if only I was assured that I would eventually attain that longed-for understanding".

indeed, is the characteristic of light. However, this insight of Galileo's was obviously more the product of a mental flight of fancy than a scientific approach. And he must have been well aware of this limitation: it is no accident that, in his more mature works he did not elaborate further upon these insights. This was because they were aspects of research that, at his time, could not be investigated with the appropriate logical and experimental methods, which he considered to be the basis of the new science. It was only in the nineteenth century that scientists began penetrating the interior of matter, while knowledge of microscopic phenomena at the atomic level was a conquest of the twentieth century.

The blackbody

Before concluding, readers may like to hear the currently accepted explanation for the emission of heat and light on the part of an incandescent body. It comes under the name *theory of the black body*, and was developed by Max Planck at the end of the nineteenth century. We outline here some of its essential features, referring the interested reader to the Mathematical Note for more details.

The molecules in a body are always in a state of oscillation and when the body is warmer than the surrounding environment they irradiate electromagnetic energy. As noted above, the higher the temperature of the body, the higher the average kinetic energy of the molecules. We also said that the energy is irradiated in the form of quanta, or *photons*, that is, energy packets equal to a multiple of the oscillation frequency. It is therefore obvious that, as temperature increases, besides an increase in total intensity of radiation emitted by the body, the number of more energetic photons (those of high frequency—that is, of short wavelength) becomes larger than the number of less energetic ones (those of low frequency—that is, of long wavelength). As a consequence, the emission peak, which in lukewarm bodies falls within the far infrared (perceived by us as heat), in hot bodies shifts to the near infrared (again perceived as heat), then to the visible light range in incandescent bodies; next to the ultraviolet range in bodies that are at tens of thousands degrees Celsius and finally to X- and gamma-rays. The sun, whose surface is around 6000°C, belongs to the class of incandescent bodies, those that emit white light (the root meaning of the word "incandescent"). So, to conclude: there is no physical difference between the energy emitted in the form of heat and that in the form of light. What differs is just the frequency, in other words the magnitude of the energy packet. Galileo could not have known anything about all this, and yet he treated it as a question of the size of fire-corpuscles: large fire-corpuscles for heat (today: photons of long wavelength, that is, weak), minute fire-corpuscles for light (today: photons of short wavelength, that is, more energetic). The reader may find this an intriguing correspondence, if nothing else.

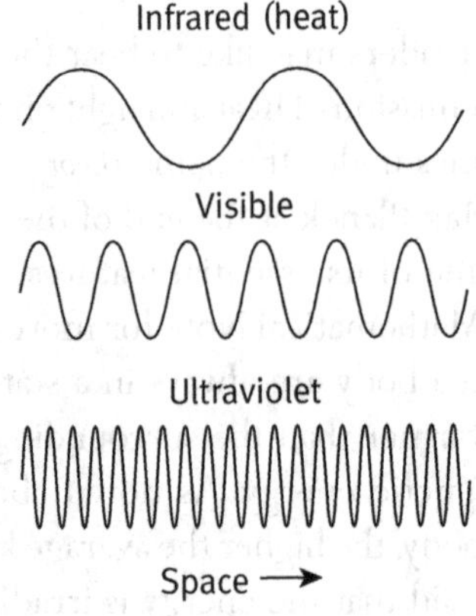

Heat and light differ only in respect of the wavelength of the radiation

An alleged heresy of Galileo

Regarding Galileo's switch from an *atomistic* to a *mathematical* conception of the structure of matter (i.e. from finite particles, following Democritus, to infinitesimal particles), Pietro Redondi advances an original hypothesis.[364] The substantial and materialistic atomism

[364] P. Redondi, *ibid.*, Chapters 4–6.

of the approach expounded by Galileo in *The Assayer* led to him being anonymously denounced to the Holy Office for Eucharistic heresy. To give just a flavor of the arguments used in the denouncement, we quote a few lines from Redondi:

> If Galileo's atoms are substantial, like Anassagoras' homeomeries, Galileo's doctrine is not compatible with the existence of the Eucharistic "accidents"... A great "experimental" principle, of philosophical and theological value, was the miraculous permanence in bread and wine of heat, color, taste, odor—the so-called sensible accidents—after the consecration that transmuted the whole substance into Christ's body and blood. If we interpret those accidents according to the approach of *The Assayer*, namely as "minute particles" of substance, then, even after the consecration it will be particles of the substance of the Eucharistic bread that produce those sensations. Thus there would remain... particles of the substance of bread in the consecrated Host, but this is an error anathematized by the Council of Trent.... the "philosophy" of accidents in *The Assayer* is accused of defying the official theological philosophy and of contradicting the dogma of transubstantiation, by claiming the real permanence of the quantity as figured substance.

Redondi hypothesizes that the denouncement is to be attributed to Father Orazio Grassi, whom Galileo treated very harshly in *The Assayer* (see Chapter 20 of the present book). As a matter of fact, Father Grassi, in his reply to Galileo,[365] devoted space to the question, stating that the qualities of bread and wine remain, even though their substance disappears, "only as a result of divine will... and in a miraculous way". And appalled he adds: "One must then deduce, from what Galileo says, that heat and taste do not survive in the Host. Our soul is horrified at the mere thought".

Of course, as with every other question of doctrine, Galileo never expressed himself on the matter of the Eucharist. It was terrain remote from his interests and, anyhow, full of treacherous pitfalls. It must be said, though, that atomistic conceptions, in postulating that all different substances were made up of just one single type of particle, differing only in its configuration and combination, were at the time held to be, almost by definition, incompatible with Catholic

[365] O. Grassi, *Ratio ponderum Librae et Simbellae*, 1626.

doctrine. It is not easy to understand today how the Eucharistic miracle—being a miracle—could be thought to be influenced by physics models explaining the microscopic structure of matter. In any case, both the argument in the anonymous denouncement and that in Father Grassi's reply, were so twisted and captious that one doubts they could have been taken seriously by the Church. This means it is unlikely they would have played a significant role in Galileo's misfortunes to come. We shall return to this topic in Chapter 22, devoted to the trial of the scientist.

MATHEMATICAL NOTE

Temperature and molecular velocity

We show here that there exists a direct relationship between the temperature of a given substance and the average kinetic energy of its molecules. As a simple example, we consider a perfect gas, for which the Boyle–Mariotte law holds (written for the case of a single gram molecule of substance)[366]

$$pV = RT$$

(at a given fixed temperature, the product pressure times volume is constant) where p is the gas pressure, V the volume of the recipient, R the universal gas constant, and T the temperature in degrees Kelvin (the zero of the Celsius scale corresponds to 273.16 K).

The pressure exercised by a gas on the walls of the recipient is due to the collisions of its molecules against them; now, the more frequent these collisions are, the higher is the pressure, and this means there is a dependence of the pressure on the speed with which the molecules move from one wall to another. However, the velocity of the molecules plays a second role, too: in their collisions, the molecules get stopped, yielding their momentum to the walls. This must therefore lead to a quadratic relationship between pressure and velocity. The square of the velocity, on the other hand, is proportional to the kinetic energy of the particle

$$E_{\text{kin}} = \frac{1}{2} m v^2$$

where m is the mass of the molecule and v is its velocity. We can therefore expect a direct relationship between the pressure exercized by the gas and the

366 For further details on the arguments dealt with here the reader should consult any university-level basic physics text.

average kinetic energy of its molecules. Without entering into details, we shall say that the kinetic theory of gases leads to the following relationship between pressure and mean kinetic energy $<E_{kin}>$:

$$p=\frac{2N<E_{kin}>}{3V}$$

where N is Avogadro's number.[367]

Putting the pressure, as written in the last equation, equal to the pressure appearing in the Boyle–Mariotte law, one finally obtains

$$<E_{kin}>=\frac{3}{2}kT$$

where $k = R/N$ is the Boltzmann constant. Therefore, the higher the average kinetic energy of the particles is, the higher is the temperature in degrees Kelvin.

The blackbody

Let us now take a brief look at *blackbody theory*, which describes the emission of electromagnetic energy by hot bodies (even if bodies, in general, are not black, their behavior, regarding the essential aspects, is not too dissimilar from that discussed here). This theory, due to Max Planck, is of paramount importance because it marks the advent of the concept of quantization of electromagnetic energy.

From the experimental point of view, it was already known prior to the end of the nineteenth century that a body brought to incandescence emits electromagnetic radiation in a wide range of frequencies, as illustrated by the bell-shaped curve in the diagram below. The spectrum, that is, presents an emission peak which shifts to higher and higher frequencies as temperature increases. The empirical law describing this shift is

$$\lambda_{max}=\frac{2.898\cdot 10^6}{T} \tag{18.1}$$

(Wien's law)

where λ_{max} is the wavelength which corresponds to the maximum emission, expressed in nanometers, and T is the absolute temperature in degrees Kelvin. Let us give an example: taking the sun's surface temperature to be about 6000 K, one calculates a λ_{max} around 500 nm, that is, in the green (this is why our eye, as a result of the evolution of the species, has its highest sensitivity in that zone). A tungsten filament, which operates around 2500 K, has its peak of

367 Avogadro's number is the number of molecules contained in one gram molecule of a substance ($6.022\cdot 10^{23}$).

emission a little above 1100 nm, that is, in the near infrared (and in fact, in this case, the bulk of the emission is heat rather than light).

As regards the spectral behavior of the blackbody *emissive power*, the available classical theory in the nineteenth century—which assumed for all frequencies an equal emission probability and an equal average energy kT—resulted in the equation

$$\epsilon(f,T)=\frac{2\pi kTf^2}{c^2} \tag{18.2}$$

(Rayleigh–Jeans law)

where c is the velocity of light and k is the Boltzmann constant again. This spectral behavior is illustrated by the dashed curve in the diagram below. The dependence on the frequency squared implies that, in going from the infrared to the ultraviolet and up to the X- and gamma-rays, the intensity of the emitted radiation keeps increasing until it reaches an infinite value, giving rise to the so-called *ultraviolet catastrophe*. To the great good fortune of living beings, at least the way they are designed, black bodies, and in particular the sun, do not behave at all in this fashion. What was wrong, then, in the classical model?

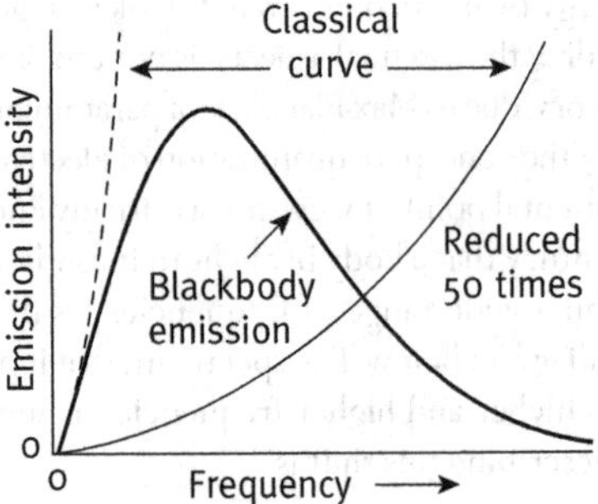

Continuous curve: experimental behavior of blackbody emission. The dashed curve represents the spectrum according to classical theory

It is at this point that Max Planck's ingenious idea comes into the picture: we have to assume that the higher is the frequency involved, the lower the emission probability is. The Rayleigh–Jeans curve, in this case, becomes truncated on the high-frequency side and it is possible, playing around with mathematics, to make it match the experimental curve perfectly. In simple words, Planck's reasoning, which he formulated after having reached the correct result, will now be outlined.

Let us hypothesize that electromagnetic energy does not flow with continuity, but rather that it gathers into packets, which we shall call *energy quanta*. The higher the frequency, the higher the energy associated with the quantum:

Planck postulated the direct proportionality $E = hf$ via the constant h, which took his name. In order for the incandescent body to emit high frequency radiation it is necessary that a large amount of energy gathers into a single packet. At low temperatures, thermal agitation of the body molecules is weak and consequently the energy available in the body is small: it is therefore more likely that quanta of low frequency will be formed than quanta of high frequency, given that the former are built with less energy expenditure. If the temperature increases, the number of more energetic quanta gets larger and the emission shifts towards higher frequencies. In statistical mechanics one says that the 'population' of the different energy levels hf is a function of temperature. Such a population is described to a good approximation by a statistical distribution function of an exponential nature (*Boltzmann function*)

$$F(f,T) = \exp\left(-\frac{hf}{kT}\right)$$

whose behavior is described in the diagram below.

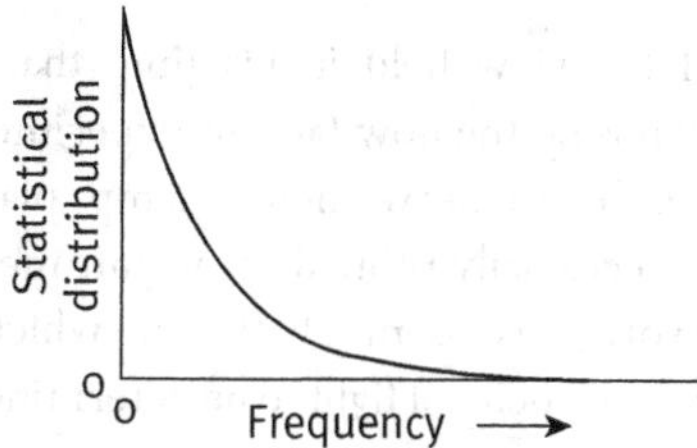

Boltzmann's statistical distribution

Consequently, the average energy corresponding to each frequency f is no longer kT but rather

$$\frac{hf}{\exp^{hf/kT} - 1}$$

This leads to an emissive power different from the classical one—given in (18.2)—and precisely equal to

$$\epsilon(f,T) = \frac{2\pi f^2}{c^2} \frac{hf}{\exp^{hf/kT} - 1} \qquad (18.2')$$

which exhibits a behavior in perfect agreement with the experimental findings, that is, with the bell-shaped curve given in the first diagram shown in this Mathematical Note. It is worth stressing that for $hf << kT$ curve (18.2′) falls back into (18.2) of the classical model, as one would expect.

19

As Fast as Light*

io non saprei intendere che l'azzione della luce, benché purissima, potesse esser senza moto

I do not understand how the action of light, although very pure, can be devoid of motion

Galileo repudiated the view held in his time that light propagates instantaneously, proposing the now famous experiment for measuring its velocity, based on the use of two lighted lamps placed atop separate hills. This method, necessarily refined to permit measurement of the very short times involved, is essentially the one which, a few centuries later, was to enable the velocity of light to be determined over terrestrial distances.

In the excerpt from the *Discourses* examined in this chapter Galileo tackles the problem of whether light propagates instantaneously, or whether it requires a certain interval of time to travel from its source to the point where it is observed. As regards terrestrial distances, it did not become possible to resolve this problem until sophisticated optical laboratory instruments became available, in particular those in use today which, based on lasers, are extremely precise.

As regards interplanetary distances, on the other hand, in 1676 the Danish astronomer Ole Römer, after years of careful observations, succeeded in deducing an approximate value for the velocity of

* From *Discourses*.

light. This was based on his monitoring of the eclipses of Jupiter's moon, Io, as its orbit takes it behind the giant planet. He noted that in the six months during which the earth, travelling along its own orbit around the sun, was moving away from Jupiter, each disappearance of Io behind the planet (i.e. each "extinguishing" of the satellite's light in his telescope) occurred a little later than would have been predicted on the basis of the preceding one; whereas the opposite took place in the six months during which the distance between the earth and Jupiter was reducing. This variation—which amounted, in the two extreme cases of minimum and maximum proximity of the two planets, to as much as 22 min (16.6 min according to later, more precise, measurements)—Römer attributed to the different distances that the light coming from Io had to cover in order to reach the eyes of observers on earth. On the basis of the known distances between the two planets over the course of the year, Römer arrived at a value for the velocity of light of 190,000 km/s, that is two thirds of the velocity accepted today (for more details, see the Mathematical Note at the end of this chapter).

Descartes' opinion

The virtue of Römer's experiment was that it made a comparison between measured data, thus avoiding the need to refer to theoretical models other than knowledge about the orbits of the planets in the solar system, which by then had already become quite well established. Some forty years earlier, Descartes, observing that no discrepancy existed between the times theoretically predicted and those actually measured for the eclipses of the moon and the sun, had, instead, concluded that the propagation of light was instantaneous. Descartes' study was more or less contemporary with the publication, in 1638, of the book containing the proposal for measuring the velocity of light reported here. As a matter of fact, in a letter to Marin Mersenne towards the end of that same year, Descartes showed he knew of the experiment proposed by Galileo, but he considered it useless, as he maintained that a study based on the distance earth–moon was far superior to one based on terrestrial distances.

Given the lively interest Galileo had for any novelty that was circulating within the European scientific community, it is likely that he, in turn, was already aware of Descartes' conclusion. Nevertheless, his solidly "dow to earth" attitude—which made him absolutely sure that no physical phenomenon could take place with instantaneous propagation—meant that he had dismissed that result as of no importance (behavior which was not entirely unusual for him). In reality, Descartes' data—supporting, as they did, the widespread belief held by scientists past and present, Kepler included, that the velocity of light was infinite—presented a strong challenge to him.

Galileo's precursor experiment

Galileo proposes an experiment in which two friends, each holding a lighted but covered lamp, position themselves on hilltops at a distance of a few kilometers from each other. The first, at a given moment, uncovers his lamp; the second, as soon as he sees the light, uncovers his own. The first then has the task of estimating the time-delay between the instant when he sent the light and the instant when the light came back to him. This experiment is, of course, totally inadequate, as (apart from anything else) it ignores the reaction times of the experimenters, which in themselves are already longer by far than the traveling times of the light signals. This indicates that Galileo had not the slightest idea of just how high the speed of light is. He probably thought it would be rather greater than the speed of sound, but still comparable to it, somehow.

The experiment, if he had ever got round to performing it in person, could not have helped but fail. All the same, it was the precursor of modern methods for determining the speed of light, in which far more sophisticated measuring equipment was to be used. Towards the middle of the 1800s, in fact, Fizeau set up his celebrated experiment based on a cog wheel, and this was a clear derivation of the original idea by Galileo (see the Mathematical Note at the end of this chapter). Fizeau obtained a value of approximately 313,000 km/s for the propagation velocity of light in air. After him, Foucault, using a method based on rotating mirrors (still conceptually similar, but rather more refined) made a more precise measurement. This was very close to the best value available today, which is known up to the ninth figure (299,792,457 m/s).

The text

As to the literary value of the passage which now follows, one can only agree in substance with a comment made by Bozzi:[368]

The page is one of Galileo's most exquisite, and a very skilled reading would be required to identify, in a critically competent way, the sources in the text of the unique fragrance of his prose, better, of his scientific poetry. The structuralist and the textual critic would be delighted both, and even some of Croce's or De Sanctis' followers[369] would relish it. The fact is that the literary game is one with a complex game of scientific logic. Masterly, it has to be said.

Here, then, is the experiment of the two friends equipped with lamps, as Galileo describes it:[370]

SALV: [. . .] I do not understand how the action of light, although very pure, can be devoid of motion and that of the swiftest type.
SAGR: But of what kind and how great must we consider this speed of light to be? Is it instantaneous or momentary[371] *or does it like other motions require time? Can we not decide this by experiment?*
SIMPL: Everyday experience shows that the propagation of light is instantaneous; for when we see a piece of artillery fired, at great

[368] P. Bozzi, in *Galileo e la scienza sperimentale*, edited by M. Baldo Ceolin, published by the Department of Physics "Galileo Galilei" of the University of Padua, Padua 1995, p. 115.

[369] Benedetto Croce (1866–1952) was the most important Italian philosopher of the last century; Francesco De Sanctis (1817–83) was one of the greatest critics of Italian literature.

[370] *Two New Sciences*, First Day, translated by H. Crew and A. de Salvio, Dover Publications, New York, 1954, p. 42. In Italian: *Discorsi e dimostrazioni matematiche intorno a due nuove scienze*, Giornata prima, Edizione Nazionale, Vol. VIII, p. 87.

[371] *Momentary*—that is, involving times which are not perceptible.

distance, the flash reaches our eyes without lapse of time; but the sound reaches the ear only after a noticeable interval.
SAGR: Well, Simplicio, the only thing I am able to infer from this familiar bit of experience is that sound, in reaching our ear, travels more slowly than light; it does not inform me whether the coming of the light is instantaneous or whether, although extremely rapid, it still occupies time. An observation of this kind tells us nothing more than one in which it is claimed that "As soon as the sun reaches the horizon its light reaches our eyes"; but who will assure me that these rays had not reached this limit[372] *earlier than they reached our vision?*

Galileo's experiment

SALV: The small conclusiveness of these and other similar observations once led me to devise a method by which one might accurately ascertain whether illumination, i.e., the propagation of light, is really instantaneous. The fact that the speed of sound is as high as it is, assures us that the motion of light cannot fail to be extraordinarily swift. The experiment which I devised was as follows:

Let each of two persons take a light contained in a lantern, or other receptacle, such that by the interposition of the hand, the one can shut off or admit the light to the vision of the other. Next let them stand opposite each other at a distance of a few cubits and practice until they acquire such skill in uncovering and occulting their lights that the instant one sees the light of his companion he will uncover his own. After a few trials the response will be so prompt that without sensible error the uncovering of one light is immediately followed by the uncovering of the other, so that as soon as one exposes his light he will instantly see that of the other. Having acquired skill at this short distance let the two experimenters, equipped as before, take up positions separated by a distance of two or three miles and let them perform the same experiment at night noting carefully whether the exposures and occultations occur in the same manner as at short distances; if they do, we may safely conclude that the propagation of

[372] *This limit*—that is, the horizon.

light is instantaneous; but if time is required at a distance of three miles which, considering the going of one light and the coming of the other, really amounts to six, then the delay ought to be easily observable. If the experiment is to be made at still greater distances, say eight or ten miles, telescopes may be employed, each observer adjusting one for himself at the place where he is to make the experiment at night; then although the lights are not large and are therefore invisible to the naked eye at so great a distance, they can readily be covered and uncovered since by aid of the telescopes, once adjusted and fixed, they will become easily visible.

SAGR: This experiment strikes me as a clever and reliable invention. But tell us what you conclude from the results.

The "head" and "spread" of a lightning flash

SALV: In fact I have tried the experiment only at a short distance, less than a mile, from which I have not been able to ascertain with certainty whether the appearance of the opposite light was instantaneous or not; but if not instantaneous it is extraordinarily rapid—I should call it momentary; and for the present I should compare it to motion which we see in the lightning flash between clouds eight or ten miles distant from us. We see the beginning of this light—I might say its head and source—located at a particular place among the clouds; but it immediately spreads to the surrounding ones, which seems to be an argument that at least some time is required for propagation; for if the illumination were instantaneous and not gradual, we should not be able to distinguish its origin—its center, so to speak—from its outlying portions.[373] *What a sea we are gradually slipping into without knowing it! With vacua and infinities and indivisibles and instantaneous motions, shall we ever be able, even by means of a thousand discussions, to reach dry land?*

SAGR: Really these matters lie far beyond our grasp [...]

[373] Galileo is referring to the delay between the lightning and the flash reflected by the distant clouds.

Lightning flashes

Galilean realism

This is one of the rare cases when Galileo did not have at his disposal either an experimental result, or a *gedanken experiment* (such as the one used in the fall of bodies), or even a valid argument of a mathematical nature. He relied on his staunch, a priori, conviction that no propagatory mechanism can occur instantaneously, irrespective of how immaterial the phenomenon under consideration may be ("the action of light, although very pure"). His attitude seems to be that of a man who is faced with apparently inexplicable events and, not believing in miracles, looks for hidden, not immediately evident, causes. Nevertheless, his need for a practical verification, together with his bent for systematic observation of natural phenomena, spurred him to look for at least indirect confirmation of his assertions. So his thoughts turned to the delay we normally seem to perceive between the moment we see a flash of lightning—the electric discharge that generates light—and the moment the most distant clouds get illuminated, giving rise to the so-called "spread" of light.[374]

[374] This effect may also be referred to as "sheet-lightning".

Galileo, however, was well aware that he was going beyond the limits of the criteria he held to be the basis of scientific precision—experimental evidence together with mathematical and geometrical reasoning ("sensible experience and the necessary demonstrations"). This is why he took care to say "for the present I should compare it" and, shortly after, to add "What a sea we are gradually slipping into without knowing it!". The reader should note the words *for the present*, which indicate Galileo's firm belief that, in the future, scientists would become able to answer the question of the velocity of light in the proper way.

A quick calculation (one Galileo could not make since he did not have a value for the speed of light at his disposal) shows, for example, that the time delay between the "head" and the "spread" of lightning over a distance of 15 km, is 50 millionths of a second. It would thus be completely imperceptible to our senses, which are already unable to distinguish between two flashes of light 20 ms apart (proof of this is provided by neon-tube lights, which seem to us to emit continuous light, whereas in reality they produce a succession of flashes at the network frequency of 50 Hz, corresponding precisely to intervals of 20 ms). Galileo says that to be sure the experiment of the two friends holding the lamps will work, they need to be at a distance of about 15 km from each other. This shows just how far he was from conceiving the extraordinary speed at which light propagates.

γ-movement

What, then, is responsible for the observed effect that there is a delay between the "head" and the "spread" of lightning? The mechanism of electric discharges in the atmosphere is extremely complex, certainly more complex than Galileo could ever have imagined. In fact, the delay observed stems from factors other than the time needed for the light to travel from the point of the electric discharge to the distant clouds. The most likely and most elementary explanation (suggested by Bozzi) attributes the effect to a psychological illusion discovered by Kenkel in 1913, called "γ-movement"—this involves a sensation of movement associated to the sudden appearance of illuminated forms in the dark.

As regards the propagatory mechanism of light, it is worth recalling here that, until the late 1800s, scientists believed that an oscillatory motion of material particles was associated to a light wave, exactly as occurred with sound, which was known to be mechanical-oscillatory in nature. However, light emitted by stars was visible on earth and this light arrived after a journey through interstellar space. Thus, interstellar space had to be pervaded not by a vacuum, but by a substance capable of transmitting the vibrations of light. It was from the need for this substance that the "fairy-tale" of the ether arose, a "fairy-tale" that was still alive for much of the nineteenth century. The ether was supposed to be a material medium so impalpable as to be comparable to a vacuum, yet at the same time present everywhere in the universe.

By 1800 the velocity of light, close to 300,000 km/s, was already well-known. Now, the laws of mechanics predict that a medium, in order to sustain motions with such a high velocity, must be enormously rigid.[375] As a consequence, the ether needed to be at the same time extremely hard, but nonetheless not detectable by our senses. This clear paradox was solved only when Clerk Maxwell, during the second half of the nineteenth century, explained the real nature of light, as an oscillation of the electromagnetic field which can take place and propagate even in a vacuum, since it does not involve matter being transferred. Years later a famous optical experiment, for which the American scientists Michelson was awarded the Nobel Prize, delivered the fatal blow to the ether, demonstrating that it does not exist.

MATHEMATICAL NOTE

Methods for measuring the speed of light

If we ignore the displacement of Jupiter during the course of our year (this is negligible in comparison to the distance covered by the earth along its solar orbit,

[375] It is well known, for instance, that sound travels faster in solid matter than in air.

because the velocity of Jupiter is far lower), the paths of light at the moment when there is an eclipse of satellite Io are, respectively, AB, when the earth is at her closest

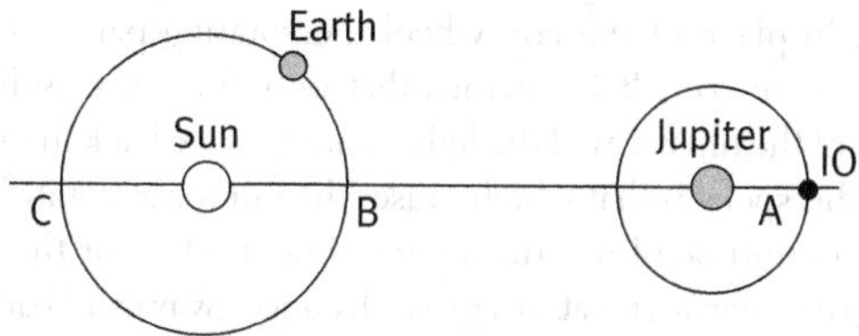

position to Jupiter, and AC when she is at her farthest position. This difference is equal to the diameter of the terrestrial orbit around the sun. The velocity of light is given by the ratio between this diameter and the delay of the eclipse when the earth is at C rather than at B. Using the correct value for the delay between the two extreme eclipses—16.6 min, that is, 996 s—and a diameter of the terrestrial orbit of $2.99 \cdot 10^8$ km, for the velocity of light c one obtains

$$c = \frac{2.99 \cdot 10^8}{996} = 300{,}200 \text{ km/s}$$

The experiment using a cog wheel by Armand Fizeau, dated 1849, is illustrated in the diagram below. A collimated beam of light from a source placed on a hill strikes a mirror located on another hill, 8.6 km away, and is reflected back along its original path. In front of the light source a rotating cog wheel is placed. When the wheel turns slowly, the light beam leaving the source and passing through a gap between two cogs, proceeds with its outward journey; however, upon completing its return journey between the two hills the beam is intercepted by a cog and so does not reach the observer (who should see it by means of a semitransparent mirror placed between the cog wheel and the light source). Now, if the rotation velocity of the wheel is gradually increased, at a given instant the light becomes visible, because on its return journey the beam will have encountered the next gap in the cog wheel. The speed of light can be calculated on the basis of the smallest angular velocity of the wheel that permits an image of the source to be observed.

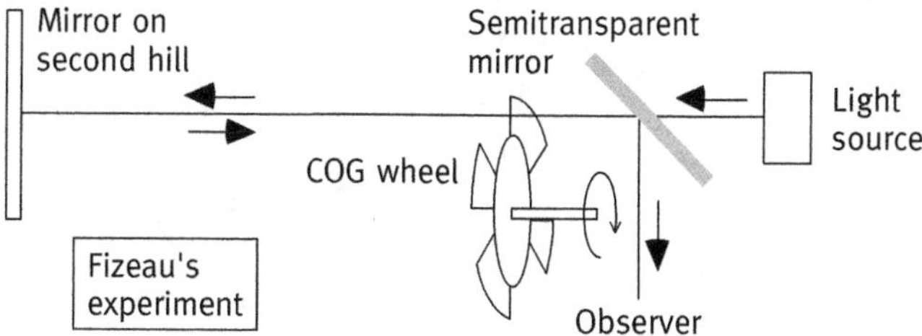

The principle underlying Foucault's variant on the above experiment, where mirrors are used instead of the cog wheel—in the modern version conceived by Michelson at the beginning of the twentieth century—is illustrated in the next diagram. In place of the cog wheel the rotating part is now an eight-faceted prismatic mirror. It is obvious that only for a precisely determined rotation speed of the mirror will the light beam bounce back in the direction of the observer. This speed is that which causes the mirror to rotate by one-eighth of a turn—or a multiple of it—during the time it takes for the light beam to travel to the fixed mirror (located a great distance away) and back again. With this system Michelson performed the first high-precision measurement by sending a light beam from the summit of Mount Wilson to the summit of Mount S. Antonio in California.

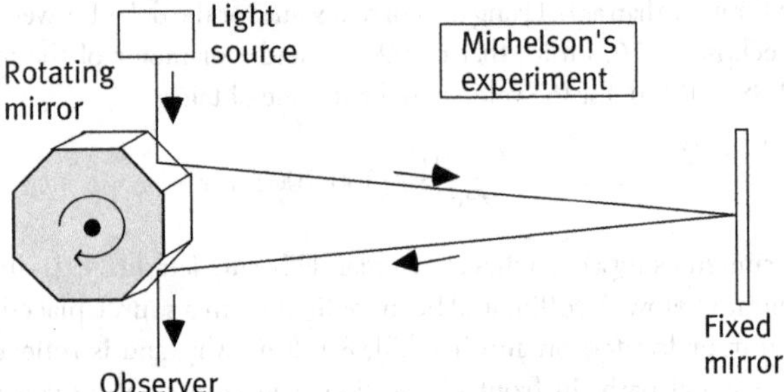

PART IX

ALL THAT GLITTERS . . .

20

Quarrels Among Scientists*

*ànno cercato spogliarmi di quella gloria ch'era pur mia, e . . .
farsi primieri inventori di meraviglie così stupende*

some . . . attempted to rob me of that glory which
was mine by . . . trying to make themselves the
original discoverers of such impressive marvels

This chapter looks at another excerpt from *The Assayer*—one, however, which does not deal with comets, the main topic of the book. The reader will find, instead, an impassioned attack on ethically incorrect scientists, those who, according to Galileo, had usurped his ideas or attacked him out of malice. The way he makes a fool of Father Grassi, his main opponent of the moment, is delightful.

Confidence in the new Pope

So, let us take another look at *The Assayer*. The book opens with a dedication by the Lincean Academicians to Urban VIII, the former Cardinal Maffeo Barberini, newly elected to the pontifical throne. This is packed with flatteries and expresses trust that the Pope will promote and protect science and culture. Indeed, Urban VIII, unlike his predecessor Paul V, appeared to be an intellectual. He had

* From *The Assayer.*

appointed several members of the Lincean Academy to posts in the Vatican and Giovanni Ciampoli, a close friend of Galileo, as his personal secretary. Later he even went so far as to declare that if it had been up to him, the 1616 edict against Copernican books would never have been issued.

Galileo makes the most of the Pope's benevolence, beginning his book with a passionate, stinging, invective against his defamers, on the one hand, and his usurpers, on the other; that is, those scientists whom he held to be either mediocre or in bad faith, and who were a constant nuisance to him. The discourse then goes on to focus upon the emblematic case of Simon Mayr, a German, who, after having published (along with his disciple Baldessar Capra) a pirated Latin translation of the instruction manual for the geometrical compass (written by Galileo), fled back to Germany leaving his pupil "in the lurch", that is, to face the full blast of Galileo's wrath.[376] Galileo's fury against this unfortunate student—whom he took to court and wanted to see annihilated in the eyes of the scientific community—was so powerful it left the judges, and even some of his friends, rather perplexed.

We have already come across, in Chapter 5, Galileo's attack on his devoted pupil Bonaventura Cavalieri—guilty, without malice, of having publicized his teacher's idea on the exact shape of the trajectory of projectiles, before it had officially appeared in print. In fact, complaints against presumed usurpers of his works are commonplace in Galileo's writings. Now, if there is anyone among scientists who availed himself of the ideas of others in order to develop his own theses, it was Galileo. This, though, is how science advances—step by step, contribution by contribution, in an admirable combined effort of the entire scientific community.

[376] The geometric compass, a kind of primitive slide rule, was improved by Galileo, who described its use in a manual entitled *Le operazioni del compasso geometrico et militare* (Edizione Nazionale, Vol. II, p. 363 and following). This text is dated July 1606—that of Baldessar Capra, March 1607. Capra was a medical student, so naive he did not even take the precaution of publishing his plagiarized work outside the Venetian Republic. Galileo brought legal action against him, won it, and obtained the destruction of the pirated work.

A weak point

Less acceptable, however, at least according to the ethical canons of modern science, is to omit, as Galileo did, to cite one's sources of inspiration. It is a plain fact that he failed, systematically, to acknowledge the merits of others. In his dialogues, apart from references to Aristotle (often made, though, in order to confute him) and to "divine Archimedes", time and again Salviati and Sagredo refer to the Author, the Academic, the Philosopher—who is none other than Galileo himself. Whereas no mention at all is made of Giovanni Battista Benedetti, of Guidubaldo Del Monte, of Giordano Bruno, of Girolamo Fracastoro, nor of many others. We have seen, for example, how the admirable *reductio ad absurdum* reasoning about the free fall of bodies was Benedetti's work (Chapter 3); how, to illustrate the principles of inertia and relativity, Galileo borrowed the famous example of the moving ship from Giordano Bruno (Chapter 4); how he drew from Fracastoro and Benedetti ideas concerning the mechanisms of sound (Chapter 6); and how his conviction that no substances exist that are "light in absolute" was taken, once again, from Benedetti (Chapter 14).

Among a variety of ways it may be possible to classify scientists from the moral point of view, one certainly regards the degree of generosity expressed towards colleagues. On the one hand, we have scientists who are happy to attribute to others all the credit they deserve, and at the same time find it exciting that their own ideas should be adopted and developed by others. What matters to scientists such as these is the progress of knowledge *per se*, not to whom it is ascribed. On the other hand, we have scientists who have their personal promotion at heart more than that of science. It seems natural, therefore, to ask: to which of these two "camps" did Galileo belong?

The answer is: more to the first, notwithstanding his all-too-human weaknesses. Making one's findings public and letting others know about them in a free and prolific exchange of opinions is an intrinsic feature of the Galilean method. Testimony to this is not only the extraordinary school he was able to create around him at all moments of his career, right up to the very last days of his internment in Arcetri, but also the great affection for, and devotion to, him on the part of all his pupils. As for the fact that he did not acknowledge his sources of inspiration, the reader should bear in mind that the view Galileo had of science and the world was novel and global. When contextualized within this view, even the ideas that he drew from others became more significant and, in some sense, original. The very fact that these were singled out as valid, among the plethora of absurd opinions and groundless hypotheses in circulation at the time, rendered them, to all effects, Galilean. As Sharratt[377] quite rightly points out, Galileo, more than any other scientist, organized concepts and facts, even if they were already known, in such a way as to create the foundation for physics as we know it today.

In this connection, physicist Giorgio Salvini wonders:[378] "what is left of Galileo if one takes away from him the Leaning Tower of Pisa, the invention, all by himself, of the telescope, plus the first pointing of it at the sky, and the experimental method as it is commonly understood today?". He answers:

> if one takes away the Leaning Tower of Pisa and his first pendulum, there remains the entire foundation of a new science, i.e. dynamics. And this is already one whole lot more. If one takes away from Galileo the first telescope, he still retains the merit of having demonstrated how scientific discovery is directly linked to the resolving power of instruments and the critical analysis of results. If one takes away from him the experimental method . . . , he still holds the merit of having taught freedom from bias in analyzing results; of having reduced a phenomenon to its essentials when conducting experiments; of having liberated physics from undemonstrated and a priori principles, as well as from non-measurable quantities. And this is what his epoch was in need of, especially when one considers that even the great Kepler, though duly starting from experience, was for ever trying to make his observations fit magical geometries or pantheistic virtues.

[377] M. Sharratt, *Galileo—Decisive Innovator*, Cambridge University Press, Cambridge, 1996, p. 10.

[378] G. Salvini, "Mitologia e verità nella figurazione di Galileo. Il Galileo di Bertolt Brecht", *Saggi su Galileo Galilei*, Barbèra Editore, Florence, 1967, p. 10.

Here, now, is Galileo's philippic against Simon Mayr:[379]

The detractors

I have never been able to understand, your Excellency,[380] how it comes about that every one of my studies which, in order to please or to be of service to others, I have seen fit to place before the public has occasioned in many a certain animus to detract, steal, or deprecate that modicum of esteem to which I had thought I was entitled, if not for the work, at least for my intention. In my Starry Messenger *were revealed many new and marvelous discoveries in the sky that ought to have pleased all lovers of true science. Yet it had scarcely been printed when men arose on all sides who envied the praises due to the discoveries thus made, and those were not lacking who merely to contradict what I said did not scruple to cast doubt upon things they had seen at will again and again with their own eyes. My Lord the Most Serene Grand Duke Cosimo II, of glorious memory, once ordered me to write my opinions on the causes of things floating or sinking in water, and in order to comply with this command I set down upon paper everything beyond the teachings of Archimedes that occurred to me, which perhaps is as much as may be truly said about the facts of this matter. And behold! immediately the whole press was filled with attacks upon my* Discourse.[381] *My opinions were contradicted without the least regard for the fact that what I had set forth was supported and proved by geometrical demonstrations, and such is the strength of men's passions that they failed to notice that the contradiction of geometry is the bald denial of truth.[382] How many men, and under what disguises, combatted my* Letters on the Solar

[379] S. Drake and O'Malley, *The Assayer*, in *Controversy on the Comets of 1618*, Philadelphia, PA, 1960 (out of print, available in photostatic form), p. 163. In Italian: *Il Saggiatore*, Edizione Nazionale, Vol. VI, p. 213.

[380] This refers to Father Virginio Cesarini, Lincean Academician, appointed Master of the Chambers by Pope Urban VIII. *The Assayer* is addressed to Cesarini in the form of a letter.

[381] *Discorso intorno alle cose che stanno in su l'acqua o che in quella si muovono* (Bodies that Stay Atop Water, or Move in It), Edizione Nazionale, Vol. VI.

[382] See the Galilean concept (discussed in Chapter 18) that the Grand Book of Nature is written in mathematical characters: this concept attributes to reason alone the power to ascertain truth.

Spots![383] *The material contained therein, which should have opened to the mind's eye so much room for admirable speculation, was completely scorned and derided by many who either disbelieved it or little appreciated it; others, not wanting to agree with my conceptions, advanced ridiculous and impossible opinions against me; and some, conquered and convinced by my reasons, attempted to rob me of that glory which was mine by pretending not to have seen my writings and subsequently trying to make themselves the original discoverers of such impressive marvels. I say nothing of some of my private discussions, proofs, and propositions (many of them not published by me) having been seriously impugned or deprecated as worthless; yet even these have sometimes been chanced upon by others who have then exerted themselves with admirable adroitness to appropriate these honors as inventions of their own ingenuity.*

The Capra–Mayr case

Of such usurpers I might name not a few, but I shall pass them over now in silence, as it is customary for first offenses to receive less severe punishment than subsequent ones. But I shall not remain silent any longer about a second offender who has too audaciously tried to do to me the very same thing which he did many years ago by appropriating the invention of my geometric compass despite the fact that I had many years previously shown it and discussed it before a large number of gentlemen and had eventually publicized it in print. May I be pardoned this time if, against my nature, my habit, and my present intentions, I show resentment and cry out, perhaps with too much bitterness, about a thing which I have kept to myself these many years. I speak of Simon Mayr of Guntzenhausen;[384] *he it was in Padua, where I resided at the time, who set forth in Latin the use of the said compass of mine and, appropriating it to himself, had one of his pupils print this under his name. Forthwith, perhaps to escape punishment, he departed immediately for his native*

[383] Reference is being made here to Jesuit Father Christopher Scheiner, who claimed he had been the first to observe solar spots.

[384] Or to be more precise, Simon Mayr, of Guntzenhausen in Franconia, who had been a disciple of Tycho Brahe and of Kepler in Prague.

land, leaving his pupil in the lurch as the saying goes; and against the latter, in the absence of Simon Mayr, I was obliged to proceed in the manner which is set forth in the Defense *which I then wrote and published. Four years after the publication of my* Starry Messenger, *this same fellow, desiring as usual to ornament himself with the labors of others, did not blush to make himself author of the things I had discovered and printed in that work. Publishing under the title of* The World of Jupiter, *he had the temerity to claim that he had observed the Medicean planets which revolve about Jupiter before I had done so.*[385] *But because it rarely happens that truth allows herself to be suppressed by falsehood, you may see how he himself, through his carelessness and lack of understanding, gives me in that very work of his the means of convicting him by irrefutable testimony and revealing unmistakably his error, showing not only that he did not observe the said stars before me but even that he did not certainly see them until two years afterward; and I say moreover that it may be affirmed very probably that he never observed them at all.*

HISTORICAL NOTE

The dispute with Father Grassi

In 1616, following the condemnation of the heliocentric theory by the Holy Office, Cardinal Bellarmine admonished Galileo to abandon the "censored thesis" (we examined this matter in Chapter 11). In *The Assayer*, published in 1623, in contrast to what he was to do straight afterwards in his letter to Ingoli (see Chapter 4) and later in his *Dialogue*, Galileo took all possible care to appear that he was complying with the injunction he had received. Whenever there is a reference to the Copernican model we find prudent phrases of the kind: "Then as to the Copernican hypothesis, had not the most

[385] Simon Mayr called the planets, rather unimaginatively, *Sidera Brandeburgica* (*Brandeburg Stars*). The book in which he describes them—*Mundus Iovialis* (The World of Jupiter)—was dated 1614. It should be said, however, that some modern historians have tended to give credit to the possibility that Mayr's observations were original.

sovereign wisdom removed us Catholics from error . . .",[386] or "if the movement attributed to the earth (which I, as a pious and Catholic person consider to be most false and vane) . . .".[387] These expressions, clearly hypocritical, were meant to blunt the weapons of potential accusers. They show, nevertheless, that Galileo in no way intended to renounce the opportunity to give exposure to his ideas.

Written in the form of a letter addressed to Father Virginio Cesarini, the book confutes, using a brilliant, biting, polemic, the theory on the nature of comets advanced by Father Orazio Grassi, a Jesuit. Grassi was a reputable professor of mathematics at the Roman College[388] and he had published his theory in a book entitled *Libra astronomica ac philosophica*, under the pseudonym Lotario Sarsi. Even if, as regards the specific theme of the controversy, Grassi was more in the right than Galileo (who defended the hypothesis that comets are optical effects due to the reflection of solar light on vapors surrounding the earth) the book constitutes a lesson on the scientific method which is without equal. It is also a merciless attack on the scholastic philosophy of the Jesuits and upon Tycho Brahe's hypotheses, which they had adopted.

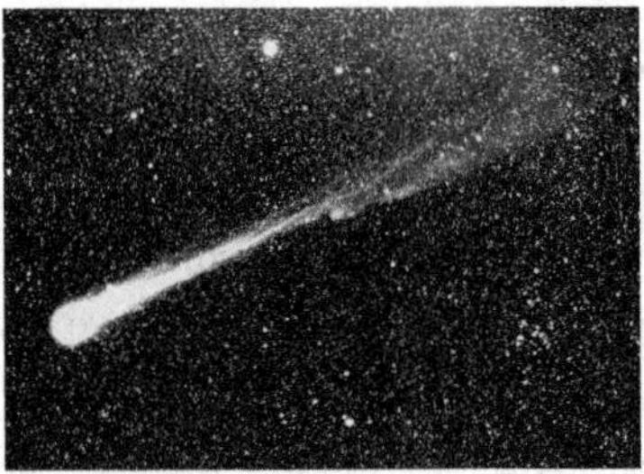

The two large comets Hale–Bopp (1997) and Hyakutake (1996)

[386] S. Drake and O'Malley, *The Assayer*, in *Controversy on the Comets of 1618*, Philadelphia, PA, 1960 (out of print, available in photostatic form), p. 184. In Italian: *Il Saggiatore*, Edizione Nazionale, Vol. VI, p. 233).

[387] S. Drake and O'Malley, *The Assayer*, in *Controversy on the Comets of 1618*, Philadelphia, PA, 1960 (out of print, available in photostatic form), p. 269. In Italian: *Il Saggiatore*, Edizione Nazionale, Vol. VI, p. 311.

[388] Better known, today, as the architect of genius who built S. Ignazio Church in Rome.

Galileo pretended he did not know that the author of the *Libra* was Grassi, so that he could be free to strike home—in his words, to throw "small apples and apple cores" (an insult)—without having to confront the powerful Jesuit order directly. The controversial tone is evident from the title itself, where Galileo is opposing to the *Libra* (the common weighing-scale of the time) the little assay balance used by goldsmiths, to indicate that the themes in question must be weighed with much greater care than Sarsi did.[389] And he goes on to suggest that Sarsi's book would have been better named *The Astronomical and Philosophical Scorpion*, because the comet observed in 1618 had appeared not in the constellation of Libra, but in that

> which our sovereign poet Dante called the figure of that chilly animal which pricks and stings the people with its tail; and truly there is no lack of stings here for me. Much worse ones too than those of scorpions; for the latter, as friends of mankind, do not injure unless first offended and provoked, whereas this fellow would bite me, who never so much has thought of offending him. But as luck will have it, I know the antidote and speedy remedy for such stings, and I shall crush the scorpion and rub him on the wounds where the venom will be reabsorbed by its proper body and leave me free and sound.[390]

These words leave no room for doubt, the polemic was going to be pungent. In fact, despite his token attempts at prudence, Galileo threw himself into it with gusto and irony—and the outcome was to be an increase in his unpopularity among the Jesuits. As further illustration of his attitude here are a few more examples of the sarcasm with which he taunts Sarsi, while teaching him a methodological lesson.

[389] It may be of interest, as proof of the sympathy Galileo enjoyed in ecclesiastical circles, to read the concluding words of the *imprimatur* granted by Father Nicolò Riccardi, qualificator in the Tribunal of the Holy Office, in passing the book: "I consider myself lucky to have been born in a time when the gold of truth is no longer weighed roughly, with a steelyard, but with such delicate assayers". Words which leave no doubt about whose side he was on in the dispute between Galileo and the Jesuits of the Roman College.

[390] S. Drake and O'Malley, *The Assayer*, in *Controversy on the Comets of 1618*, Philadelphia, PA, 1960 (out of print, available in photostatic form), p. 172. In Italian: *Il Saggiatore*, Edizione Nazionale, Vol. VI, p. 221.

Small apples and apple cores

Nor is there any such occurrence as that which Sarsi wishes to split hairs about and to review with all the rigor of geometry. . . . his gain will not equal that of a man who has carefully gone about inquiring which gate of the city one should use in order to reach India most quickly.[391]

*

I desire no more from you than silence, by which you will put an end to this dispute . . . For if you should ever happen to see this essay of mine . . . you will see what a man ought to do when he undertakes the task of attempting to examine the remarks of others; that is, he should pass over nothing without giving it consideration and should not go about, as you have done, like a blind chicken poking its beak into the ground now here and now there, hoping to find some grain of wheat to bite and peck at.[392]

*

Now here you see a great expenditure of words by both Sarsi and me to determine whether the solid hollow of the lunar orb (which does not exist in nature), moving round (which it has never done), sweeps along with it the element of fire (which we do not know to exist), and thereby the exhalations, which accordingly kindle and set fire to the material of comets (of which we do not know whether it is there situated, and are certain that it is not combustible).[393] *Here Sarsi reminds me of the saying of that most witty poet:*

By Orlando's sword, which they have not,
And perhaps which they never shall have,
These blows of the blind have been given.[394]

*

[391] *Ibid.*, p. 220. In Italian: *ibid.*, p. 266.

[392] *Ibid.*, p. 331. In Italian: *ibid.*, p. 368. Here Galileo is addressing Sarsi directly.

[393] Underscoring by the authors.

[394] *Ibid.*, p. 290. In Italian: *ibid.*, p. 329. These lines are by the great Renaissance poet Ludovico Ariosto, particularly loved by Galileo, who knew most of his poem *Orlando Furioso* (*The Frenzy of Orlando*) by heart.

If reasoning were like carrying burdens, where several horses will carry more sacks of grain than one alone, I should agree that several reasoners would avail more than a single one; but reasoning is like running and not like carrying, and one Arab steed will outrun a hundred pack horses.[395]

*

If Sarsi wants me to believe . . . that the Babylonians cooked eggs by whirling them rapidly in slings, I shall do so; but I must say that the cause of this effect is very far from that which he attributes to it. To discover the truth I shall reason thus:

If we do not achieve an effect which others formerly achieved, it must be that in our operations we lack something which was the cause of this effect succeeding, and if we lack but one single thing, then this alone can be the cause. Now we do not lack eggs, or slings, or sturdy fellows to whirl them; and still they do not cook, but rather they cool down faster if hot. And since nothing is lacking to us except being Babylonians, then being Babylonians is the cause of the eggs hardening.

And this is what I wished to determine.[396]

Poor science

In this last passage, the ferocious sarcasm of the concluding phrases is directed against poor science in general. A very similar story—one belonging, however, to the present day—is the following (Please note, it is not fiction; unfortunately in modern science, money and political maneuvering are sometimes just as important as ideas, and the result may be that research projects like the one in this tale get financed).

A scientist receives a considerable amount of money to study the sense of hearing in grasshoppers. After lengthy research, he is invited to present his results before a panel of experts. He takes out of his

[395] *Ibid.*, p. 300. In Italian: *ibid.*, p. 340.
[396] *Ibid.*, p. 301. In Italian: *ibid.*, p. 340.

pocket a little box, opens it, gets a grasshopper out of it, and places it on the floor. "I have made an extraordinary discovery—he says—and now I'm going to give you a demonstration of it". He turns to the grasshopper and orders: "Jump!". The little creature makes a jump forward. The scientist then gets a pair of scissors, cuts off one of the insect's forelegs, and again orders: 'Jump!'. The grasshopper jumps a second time, with some difficulty. Another cut, and off goes the other foreleg. When ordered to jump once more, the grasshopper makes it again, somehow. Only after the scientist has cut off all of its legs does the insect remain motionless. "You see—he exclaims triumphantly—my discovery is that in grasshoppers the organ of hearing is located in the hind legs!".

21

The Scientist and the Cicada*

si ridusse a tanta diffidenza del suo sapere, che domandato
come si generavano i suoni, generosamente rispondeva
di sapere alcuni modi, ma che teneva per fermo
potervene essere cento altri incogniti ed inopinabili

his knowledge was reduced to such diffidence that when
asked how sounds are generated he used to reply
tolerantly that although he knew some of the ways,
he was certain that many more existed
which were unknown and unimaginable.

Again from *The Assayer*, here is the tale of the scientist and the cicada. This is an allegory about learning to "know that one does not know", about discovering that knowledge extends without end, and also about how satisfying curiosity merely serves to stimulate it the more. And, perhaps, about other, more disturbing, truths.

The tale of the man (the ideal scientist: "endowed by Nature with extraordinary curiosity and a very penetrating mind") who keeps on learning of different means by which sound is generated, is, in its apparent simplicity (linguistic, too), constructed with sapient skill. The brief introduction anticipates the moral which concludes the tale, propounding it as a universal value. From the very beginning there is a sweet, fairy tale-like feel, accentuated by the use of

* From *The Assayer*.

diminutives and pet names. The story unfolds in a series of short scenes. These serve to make up the various stages of a journey, in which the progressive broadening of knowledge turns out to be inversely proportional to the certainties acquired. In other words, to the endless widening of the field of investigation (the object) corresponds an ever more acute perception on the part of the subject that he is marginal and non-influential (a sensation Galileo must have felt keenly as, using his telescope, he had been the first to open up immense spaces previously "unknown and unimaginable"). After having learned much, the man "is reduced to such diffidence of his knowledge" as to come to admit that he knows almost nothing.[397]

The object under investigation is sound, the generation of which can occur in innumerable ways. This choice was emblematic of Galileo, who was both a music lover—as recounted in Chapter 6 on harmony—and a modern scientist, always looking for unifying mechanisms capable of explaining different phenomena. Sound is a symbol for transmission of energy, for communication and, in a sense, for life. There is no phenomenon in the world that does not involve energy being transferred to the surrounding environment in the form of waves that stimulate our senses. Thus, the man of the tale, through the various sources of sound he comes upon, experiences sound due to the oscillation of air columns in the human voice and in wind instruments, sound due to the friction of mechanical parts, such as the strings of a violin, the hinges of a door, or the wings of an insect, sound in the open air caused by the wings of a bird, a vibrating glass and the reeds of an organ pipe, and finally sound—unpleasant and foreboding ill—due to the grating of the membranes of the cicada.

[397] A variation on this theme would be found years later in the *Dialogue*: "Such a foolish pretension to understand everything can only originate from not ever having understood anything; for, if someone had tried only once to understand a single thing perfectly and had truly tasted how knowledge works, he would know that he does not understand an infinity of other conclusions". (*Galileo on the World Systems*, translated by M.A. Finocchiaro, University of California Press, Berkeley, CA, and Los Angeles, 1997, p. 111. In Italian: *Dialogo sopra i due massimi sistemi del mondo*, Edizione Nazionale, Vol. VII, p. 127.

The stimulus for research is pleasure. Learning takes place initially by chance and "close to his house": at this stage, the merit of the man is that he admits he has learned from a simple shepherd boy. At the next stage, the man assumes a more active role and decides "to travel far from his home", retaining the humility that is indispensable for learning from others. At the end he finds himself acting in the first person, yet still by chance (when a cicada turns up in his hand), and he discovers that he is unprepared. He has not yet reached the final stage in learning: experimenting in person according to pre-established criteria. So the experiment is bound to fail.

The phrase with which Galileo concludes—"when asked how sounds are generated he used to reply tolerantly that although he knew some of the ways, he was certain that many more existed which were unknown and unimaginable"—may explain why Urban VIII was so fond of the tale that he had it read to him over and over again. His modest scientific training may have led him to believe that these words echoed his "argument", according to which God has an infinite number of ways of making a natural phenomenon occur, which are often inscrutable to man (see Chapter 9). It is clear that Galileo's intended meaning, instead, is quite different. For him the generation of sound, even if it may be manifested in multiple ways, occurs according to a well-defined mechanism—the only plausible one—that is, as a perturbation impressed by a vibrating body upon the air in contact with it.

So let us now read the tale.[398]

Long experience has taught me that with regard to intellectual matters, this is the status of mankind: The less people know and understand about such matters, the more positively they attempt to reason about them, and on the other hand the number of things known and understood renders them more cautious in passing judgment about anything new. There once lived, in a very solitary place, a man endowed by nature with extraordinary curiosity and a very penetrating mind. He raised

[398] S. Drake and O'Malley, *The Assayer*, in *Controversy on the Comets of 1618*, Philadelphia, PA, 1960 (out of print, available in photostatic form), p. 234 and following. In Italian: *Il Saggiatore*, Edizione Nazionale, Vol. VI, p. 279 and following.

many birds as a hobby, much enjoying their songs, and he used to observe with great admiration the happy contrivance by which they would transform at will the very air they breathed into a variety of sweet songs. Close to his house one evening, he chanced to hear a delicate sound, and, being unable to imagine what it could be except some small bird, he set out to capture it. Arriving at the road, he found a shepherd boy who was blowing into a kind of hollow stick and moving his fingers about on the wood, thus drawing from it a variety of notes similar to those of a bird, though by quite a different method. Puzzled, and led on by his natural curiosity, he gave the boy a calf in exchange for his recorder and retired to solitude. Realizing that if he had not chanced to meet the boy he would never have learned of the existence of two methods for forming musical notes and very sweet songs, he tried traveling far from his home in the hope of meeting with some new adventure. The very next day he happened to pass near a small hut, and, hearing a similar tone within, he went inside to find out whether it was a recorder or a blackbird. There he found a boy holding a bow in his right hand and sawing upon some fibres stretched upon a concave piece of wood. The fingers of the left hand (which supported the instrument) were moving, and without blowing the boy was drawing from it various notes, and most sweet ones too. Now, you who are participating in this man's mind and sharing in his curiosity, judge his astonishment! Finding himself to have two unexpected new ways of forming tones and melodies, he began to believe that still others might exist in nature. His wonder increased when upon entering a certain temple he glanced behind the gates to learn what it was that had sounded, and perceived that the noise had emanated from the hinges and fastenings as he had opened the gate. Again, impelled by curiosity, he entered an inn expecting to see someone lightly bowing the strings of a violin, and instead saw a man rubbing the tip of his finger round the rim of a goblet and drawing forth from it a very sweet sound. And later he observed that wasps, mosquitoes, and flies did not form separate notes from their breaths, as did his original birds, but made steady tones by the swift beating of their wings.

In proportion as his amazement grew, his belief diminished that he knew how sounds were created; nor could all his previous experience have sufficed to make him understand or even believe that crickets, which do not fly, could draw their sweet and sonorous shrilling not from breath but from a scraping of wings. And when he had almost come to believe that there could be no further ways of forming notes—after having observed in addition to what has been recounted numerous organs, trumpets, fifes, stringed instruments of various sorts, and even that little iron tongue which when placed between the teeth makes strange use of the buccal cavity as a sounding box and of the breath as a vehicle of sound[399]*—when, I say, he believed that he had seen everything, he found himself more than ever wrapped in ignorance and bafflement upon capturing in his hand a cicada, for neither by closing its mouth nor by stopping its wings could he diminish its strident sound, and yet he could not see it move either its scales or any other parts. At length, lifting up the armor of its chest and seeing beneath this some thin, hard ligaments, he believed that the sound was coming from a shaking of these, and he resolved to break them in order to silence it. But everything failed until, driving the needle too deep, he transfixed the creature and took away its life with its voice, so that even then he could not make sure whether the song had originated in those ligaments. Thereupon his knowledge was reduced to such diffidence that when asked how sounds are generated he used to reply tolerantly that although he knew some of the ways, he was certain that many more existed which were unknown and unimaginable.*

Yet, in this tale of sounds, an apparently false note will certainly not have escaped the reader's attention. The initial "variety of sweet songs" gives way, in the finale, to the "strident sound" of the unfortunate cicada, to whose untimely end several lines of crude realism are devoted. Doubt arises, then, that the moral of the tale is not the one explicitly declared and taken for granted since Socrates' time,[400] or at least that this is not the only moral. Doubt arises that the disturbing conclusion may hint at something else.

Could this "something else" be that aspect of science which is playful but, at the same time, cruel—like the child who, out of curiosity,

[399] This is a reference to the small musical instrument from Sicily called a *scacciapensieri* or *maranzano*. In English this instrument is known as a *jew's harp*.

[400] That is, the more you learn, the more you know that you do not know.

destroys the toy he loves most? Or could it be the clumsiness of some experimentalists who, despite holding the key to a problem in their hands, seem incapable of using it in the proper way? Or could it even be the impossibility of crossing beyond certain limits to knowledge, under penalty of destroying Nature (one's thoughts turn aghast to the instruments of death that humanity devises for itself)? Nothing, however, could be more wrong, or anti-Galilean, than to ascribe prophetic intentions or visionary tendencies to Galileo.[401] It is enough here to observe that, in general, his scientific writings were always guided by his wish to be clear and unequivocal in meaning, but that this particular page not only hints at something other than what it is saying explicitly, but is ambiguous and lends itself to different interpretations (readers will doubtless be able to find some of their own).

We asked some authoritative experts for their interpretations of the tale and we now report their replies. Science historian Enrico Bellone put forward this possibility:

> The context in which the argument of the cicada is placed helps to clarify Galileo's view on the relationship between human knowledge and objective nature. The former is always limited, while the latter is always richer than our ideas and produces "effects by means which we would never think of". The macabre aspect of the killing of the cicada is not, in and of itself, an anomaly. It is, instead, a literary device which Galileo also used elsewhere in *The Assayer*, for example when he describes the ideal observer as one who has been stripped of his biological sensors (eyes, tongue, nose, and skin), so as to be capable of recognizing those qualities which Locke was later to call "primary": motion, number and geometry of bodies.[402]

[401] Concerning Galileo's habit of resorting to parables, his contemporary and biographer Niccolò Gherardini wrote:

> To liberate himself of particular questions that many people asked him out of impolitic curiosity, he diverted the discourse and straight away applied it to another topic, so ably as to make it appear—no matter how far removed it might seem to be—properly related to the matter concerned, to the satisfaction of those who had raised the question. And this he did by telling some small parables, real events or idle tales, of which he knew quite a variety. (N. Gherardini, *Vita di Galileo*, 1654, Edizione Nazionale, Vol. XIX, p. 644 and following).

[402] In regard to this, see Chapter 18.

Physicist Carlo Bernardini had the following comment to make:

> Galileo discovers the existence of what, nowadays, we call destructive tests. From an epistemological viewpoint, even today the fact that tests exist which are inevitably destructive makes our heads spin somewhat. One cannot rule out the possibility that Galileo felt an intellectual jolt when he reflected upon the implications of his little tale. I doubt, however, that this could have anything to do with apocalyptic presentiment concerning the power of science. Times were not such as to induce this kind of concern, even less feelings of condemnation for possible links between knowledge and instruments of death. Nor was pity towards one's fellow beings so strong as to arouse compassion for a simple cicada: hanging and torture were everyday practices.

Echoing Bernardini's comment, another physicist, Giuliano Toraldo di Francia, said:

> Why not add that today we know for sure that in microphysics, in a sense, all tests are destructive? In fact, any observation changes the observed state of the system *in an unpredictable way*. Certainly, Galileo could not have known this, but often we have to acknowledge that the intuition of the greatest geniuses reaches far beyond what they themselves might have believed!

Francesco Guerra, also a physicist, perceived a direct connection between the tale of the cicada and the main theme of *The Assayer*—that pungent controversy about the nature of comets in which Galileo repudiates the theory of Lotario Sarsi (in reality Jesuit Father Orazio Grassi). Speaking for himself and his pupil Mario Guiducci,[403] Galileo, puts forward a theory which is wrong, but about which, nevertheless, he maintains a healthy dose of skepticism. In order to better understand Guerra's comment, it will be worth taking a look at the passage which immediately follows the tale of the cicada in the text of *The Assayer*.

[403] Author of *Discorso sulle comete*, written under Galileo's direct inspiration (Edizione Nazionale, Vol. VI, p. 37 and following). In English: S. Drake and O'Malley, *Discourse on the Comets*, translated by S. Drake, in *Controversy on the Comets of 1618*, Philadelphia, PA, 1960 (out of print, available in photostatic form), p. 21.

I might by many other examples make clear the bounty of nature in producing her effects by means which we would never think of if our senses and experience did not teach us of them, though even these are sometimes insufficient to remedy our incapacity. Therefore I should not be denied pardon if I cannot determine precisely the manner in which comets are produced, especially as I never boasted that I could, knowing that it may occur in some way far beyond our power to imagine. The difficulty of understanding how the cicada's song is formed even when we have it singing to us right in our hands is more than enough to excuse us for not knowing how a comet is formed at such an immense distance. Let us therefore stop at Sig. Mario's and my primary purpose, which is to set forth those questions which have appeared to us to throw doubt upon the opinions previously held, and to propose some new considerations. Let us examine and consider whether there is anything that can give us light in any way and can pave the road for the discovery of truth [. . .].

Guerra says:

I believe that in order to evaluate the tale of the scientist and the cicada properly, one must view it in the context of the controversy about the nature of comets. I find the following words particularly significant: "The difficulty of understanding how the cicada's song is formed even when we have it singing to us right in our hands is more than enough to excuse us for not knowing how a comet is formed at such an immense distance". I would therefore, on the one side, draw a parallel between the comprehensible motions (of varying complexity) of the stars, sun, moon, planets and the sounds (of varying degrees of harmony) of birds, the recorder, and the violin. And, on the other side, I would draw a parallel between the motion and the incomprehensible nature of the comet and the "strident sound" of the cicada.

The comet and the cicada thus represent the current limits to the investigative capabilities of the individual scientist, "a man endowed by Nature with extraordinary curiosity and a very penetrating mind", who "lived in a very solitary place". In frontier zones this man must, anyway, proceed tentatively using hypotheses and conjectures, so as to "examine and consider whether there is anything that can give us light in any way and can pave the road for the discovery of truth", possibly by future generations. In fact, today we can read in an encyclopaedia: The male cicada produces loud noises by vibrating two elastic membranes located at the base of the abdomen (stridulatory system). Comet: body belonging to the class of small celestial objects which describe elongated orbits around the sun, developing gaseous halos and often long luminous tails.

Viewed in this light, Galileo's message becomes, at least in part, positive. On the one hand he affirms his conviction that there are no eternal mysteries, only mysteries with an "expiry date". And on the other hand, he affirms his great trust in the ability of the human mind to arrive, sooner or later, at their clarification. The notion, so dear to Galileo, was that all natural phenomena obey criteria of logic and sound sense—a language which is also the principal characteristic of thinking beings.

EPILOGUE

22

Conviction and Abjuration

con cuor sincero e fede non finta abiuro, maledico e detesto li sudetti errori et heresie . . . di mia propria mano ho sottoscritta la presente cedola di mia abiuratione e recitatala di parola in parola

with sincere heart and unfeigned faith I abjure, curse, and detest the aforesaid errors and heresies . . . I have with my own hand subscribed the present document of my abjuration and recited it word for word

Threatening shadows

In February 1632, with ecclesiastical authorization, the *Dialogue Concerning the Two Chief World Systems* was published. But as early as August of the same year it was rumored that there existed some "considerations regarding the said book whether to correct or suspend it, or perhaps prohibit it".[404] In August, its circulation was, in fact, forbidden, and the Pope appointed a special commission to examine it. In September, the *Dialogue* was submitted to the Congregation of the Holy Office and in October Galileo, under severe suspicion of heresy, was summoned to Rome to defend himself. What had happened, in this short space of time, to explain such a drastic deterioration in the situation?

Pope Urban VIII, once Galileo's friend and benefactor, had been accused by the Spanish cardinals of being a sympathizer of both

[404] Letter of Filippo Malagotti to Mario Guiducci, August 7 1632, Edizione Nazionale, Vol. XIV, p. 368.

France and, even worse, the heretical Sweden of King Gustav Adolf. Almost isolated, and threatened with deposition by the most intransigent of his enemies, he had to prove he was an effective defender of the values of the Counter-reformation Church.[405] In addition to this, on a personal level, he was "extremely angry"[406] with Galileo, because his argument, the so-called "argument of Urban VIII",[407] had been added on at the end of the *Dialogue*, but it had been put into the mouth of simple-minded Simplicio. These two considerations—together with the never-appeased hostility of the powerful Jesuits[408]—account for Galileo's being summoned before the Tribunal of the Inquisition.[409]

From the very first signs of the impending danger, Galileo's correspondence provides ample documentary evidence of his friends' deep concern. The frantic exchange of letters testifies to the steps taken to ascertain the Pope's mood, as well as to the consultations about what should be done and to the words of comfort and encouragement offered to Galileo. Initially, Galileo seemed to be the least worried of anybody, to the extent that he launched himself into an argument with his devoted pupil Cavalieri,[410] rather than looking

[405] For a general picture of the political atmosphere during those months, see: P. Redondi, *Galileo eretico*, Chapter 8, Einaudi, Turin, 1983 (English translation *Galileo Heretic* by R. Rosenthal, Princeton University Press, Princeton, PA, 1987), and L. Geymonat, *Galileo Galilei*, Chapter 8, Einaudi, Turin, 1969 (English translation by S. Drake, *Galileo Galilei: A Biography and Inquiry into His Philosophy of Science*, MacGraw-Hill Book Co., New York, 1965). Also A. Banfi, *Galileo Galilei*, Ambrosiana, Milan, 1949.

[406] From a letter written by Francesco Niccolini to Andrea Cioli on September 5 1632 (Edizione Nazionale, Vol. XIV, p. 383). Niccolini, the Florentine Ambassador to Rome, attempting to defend Galileo, had a heated conversation with the Pope, who, he reported, "lost his temper", showed "a hostile attitude", and could not have been "more ill-disposed towards our poor *Signor* Galileo".

[407] See Chapter 9.

[408] Redondi, *ibid.*, proposes an alternative hypothesis, which will be reported later on.

[409] In the above quoted letter, Malagotti wrote: "in substance, it must be the Jesuit Fathers who secretly do their best in order that the book be prohibited . . ." and "the Jesuits will persecute him with extreme determination". Malagotti suggested also that, as to the *Dialogue*, Galileo and his friends "should show their readiness to add, to cut and change, since here it is just a matter of keeping up appearances".

[410] See Chapter 5.

for a remedy. It may be, though, that this attack on his friend (Galileo was soon to recognize that he had been in the wrong) was the sign of a general and deeper uneasiness he was experiencing. In fact, he came to believe, in the end, that he was the victim of a conspiracy based on envy, and he felt so disheartened that he contemplated burning his own books. He played for time, in the hope that it might be possible for him to be questioned by the Inquisitor in Florence.

In an attempt to at least put off his journey to Rome, he wrote a letter to Cardinal Francesco Barberini, which closed with these words:

> should my advanced old age, or the many bodily indispositions, or the afflicted mind, or the length of a very hazardous journey due to the present suspicions [of plague], not be considered by this holy and lofty Tribunal sufficient excuses for me to be granted exemption or delay, I shall set out on the journey, putting obedience before life.[411]

However, having been warned that his delaying tactics risked being interpreted as a desire to evade judgment, and that they might, therefore, backfire upon him, on January 20 1633, after making his will, he finally left for Rome. That winter was very cold, and the plague—the same outbreak described by Manzoni in his novel *I promessi sposi*, which had caused so many deaths in Milan—was now spreading in the countryside of central Italy. Rome feared contagion, and travellers were put into quarantine, while goods and mail were fumigated at the town gates. It is not difficult to imagine the anguish that must have assailed Galileo, the conflicting hopes and fears, and his disappointment that his long battle had not altered the anti-Copernican intransigence of the Church in the slightest. And also perhaps, his regret at having left, long before, the Republic of Venice, whence his friend Fulgenzio Micanzio had recently written

[411] Letter of October 13 1632, Edizione Nazionale, Vol. XIV, p. 406. Referring to this letter, the Ambassador for Tuscany, Niccolini, who was more realistic, wrote to Galileo: "I fear that the letter may make them harsher towards you, rather then more benevolent; because the more you hint at the possibility of you defending and proving what you have written, the more the thought will grow in them of damning your work in full". (Letter of October 23 1632, Edizione Nazionale, Vol. XIV, p. 417).

to him: "here . . . for sure, no wrong would be done to you". On February 13, after having spent a few days in quarantine, he finally arrived in Rome.

It was the first Sunday of Lent and Galileo was worn out. Yet he stuck stubbornly to his belief that "the fancies of the ignorant must necessarily yield in the face of reasoning supported by facts"[412] and the thought comforted him that "whatever aims at immortality must not fear the storms of the time".[413] While awaiting the trial, due to begin on April 12, Galileo was allowed to stay at the Florentine ambassador's residence, rather than in a cell in the palace of the Holy Office.

The trial

Before his judges Galileo claimed he had not intended to defend the Copernican system, but had merely presented inconclusive proofs in its favour, alongside other proofs against it. If he had erred, he said, it was not out of evil intent, but out of his human vanity ("that natural self-satisfaction that everybody feels about their own subtleties, and in showing themselves to be sharper than most people in finding ingenious and plausible arguments, even for false propositions"). Galileo begged his judges to consider his "declining old age", his poor health, his mental anguish, but also "his honor and reputation" and the fame he enjoyed.[414] He even offered to add one or two days to the *Dialogue* in order to revisit the suspect arguments and confute them.

An apocryphal document, dated 1616, was produced against him, which contained a solemn injunction by the Commissioner of the Holy Office not to deal with Copernicanism in *any manner whatsoever*, under pain of imprisonment. It was pointed out to him that, in the presence of a notary and witnesses, he had promised to obey this official injunction. To this, Galileo objected: "In February 1616, Cardinal Bellarmine said to me that, since Copernicus's opinion, taken in absolute, was against the Holy Scriptures, it could be neither

[412] From Micanzio's letter, mentioned above.

[413] These words were written to Galileo by his friend Archbishop Ascanio Piccolomini, in a letter to him of September 29 1632, Edizione Nazionale, Vol. XIV, p. 399.

[414] Edizione *Nazionale*, Vol. XIX, p. 342 and following.

held nor defended, but it could be taken and used *ex suppositione*". As for the notarial document, Galileo stated: "I do not remember this injunction having been notified to me other than by mouth, by Cardinal Bellarmine". Galileo was, nevertheless, accused of having obtained the *imprimatur* for his *Dialogue* by keeping the injunction concealed from the authorities.

Found "vehemently suspect of heresy", he was sentenced to prison and his *Dialogue* was banned. He knew only too well the punishments that were meted out in such cases and if he wished to avoid them he was left with no alternative but to abjure. Three of the ten Cardinal Inquisitors did not sign the sentence of conviction. In its opening paragraph, among the proofs of guilt, it should not pass unnoticed that reference was made to his correspondence with some "German mathematicians" (meaning heretics), and to the letter to Father Benedetto Castelli (cited in Chapter 11). No reference was made, of course, to his having offended the Pope (which, in providing yet one more pretext for striking at Galileo, may have played its part in his finally being brought to trial).

Galileo's heresy

According to Redondi,[415] the principal reason for the trial and conviction—never made public by order of the Pope—was the Eucharistic heresy related to the atomistic view expounded in the *Assayer* (discussed in Chapter 18). Redondi writes: "Had it come out that the Pope's official scientist was suspected of heresy against Faith, it would have been a scandal . . . the situation was in many respects the same as that brought about when the *Assayer* had been denounced. The difference was that times had now changed, rendering all the effects more serious . . .". The Pope, Redondi claims, had therefore preferred to reduce the charges to the level of disciplinary violations, thereby avoiding ruinous consequences for his papacy and at the same time saving Galileo from a tragic end. This is an intriguing hypothesis, but the evidence on which it is based is somewhat flimsy.

[415] P. Redondi, op. cit., Chapter 8.

Before the judges

Just try, now, to picture the scene. White-haired Galileo kneeling, his head bent to the chest, before ten solemn Cardinals (some little more than youths) sitting high on finely carved chairs. There are some who are favorable to the sickly old sage, others who are determined to hit him hard, once and for all. The accusations heaped on his head are by no means shameful, in his own judgement, and yet in that absurd context they may lead to his annihilation. Galileo is well aware that he has infringed the injunctions, he knows he has played a game of risk, relying, much too naively, on the Pope's protection. He also realizes that he has gone too far, since, in putting the Pope's argument into the mouth of the ridiculous figure of Simplicio in the *Dialogue*, he has ended up challenging the pride and authority of Urban VIII. Even Popes have someone to whom they must answer for their actions—as he is learning to his own great cost.

Try to imagine, therefore, Galileo's despair if he was aware of the charges of Eucharistic heresy which, Redondi claims, were just beneath the surface. What then had been the use of all his displays of respect for the Church of Rome? What had been the use of having hidden his pragmatism? What had been the use of having exhibited faith in and observance of the official religion? Maybe all his efforts had achieved nothing other than to appease the anger of the Church, nothing other than to avoid more severe punishment.

The accusations cut into his heart and mind like whiplashes. To free himself of them he would be prepared to destroy all his work, he would be ready to live his life differently, like the countless individuals who have been pushed to the point where, for the sake of peace and tranquillity, they simply renounce the truth. He is not a hero, the weight of the dramatic situation he is living through is too heavy. It is a situation that he finds totally unreal, and that right up to the very last moment he has been unable to picture to himself. Abjuration appears to him as liberation, he embraces it eagerly and pronounces it almost with pleasure, the self-destructive pleasure of someone who has been pushed to the lowest level of abjection. This is why the famous sentence "*eppur si muove*" ("and yet it moves"), commonly associated with the closing of the trial, was never, in fact, uttered by Galileo. Indeed, he expressed his "*eppur si muove*" in a different way: by immediately setting off to write his *Discourses*, in which, more than ever, he reaffirmed his ideas, almost as if he felt assured that the punishment already inflicted upon him was sufficient to cover the years to come, too.

Del resto son qua nelle loro mani, faccino quel che gli piace
After all I am entirely in your hands, just do what you please
(Galileo, from the documents of the trial)

Conviction[416]

Rome, June 22, 1633.
Noi, Gasparo del titolo di S. Croce in Gerusalemme Borgia[417]*;*
Fra Felice Centino del titolo di S. Anastasia, detto d'Ascoli;
Guido del titolo di S. Maria del Popolo Bentivoglio;
Fra Desiderio Scaglia del titolo di S. Carlo detto di Cremona;
Fra Antonio Barberino detto di S. Onofrio[418]*;*
Laudivio Zacchia del titolo di S. Pietro in Vincoli detto di S. Sisto[419]*;*
Berlingero del titolo di S. Agostino Gesso;
Fabricio del titolo di S. Lorenzo in Pane e Perna Verospio, chiamati Preti;
Francesco del titolo di S. Lorenzo in Damaso Barberino[420]*; et Martio di S. Maria Nuova Ginetto: Diaconi;*
by the grace of God, Cardinals of the Holy Roman Church, Inquisitors General by the Holy Apostolic See specially deputed

[416] G. de Santillana, *The Crime of Galileo*, University of Chicago Press, Chicago, IL, 1955, p. 306. In Italian: Edizione Nazionale, Vol. XIX, p. 402 and following.

[417] The King of Spain's Ambassador, who did not sign the sentence.

[418] A Capuchin friar, younger brother of Pope Urban VIII, was made a Cardinal against his will.

[419] The Papal Nuncio in Venice, who did not sign the sentence.

[420] Pope Urban VIII's nephew, a scholar, founder of the Barberinian Library, who did not sign the sentence. It was he who refused to grant Galileo a delay in his journey to Rome, and later prohibited a mausoleum from being erected to him in the S. Croce church in Florence.

against heretical pravity throughout the whole Christian Commonwealth;

Whereas you, Galileo, son of the late Vincenzo Galilei, Florentine, aged seventy years, were in the year 1615 denounced to this Holy Office for holding as true the false doctrine taught by some that the Sun is the center of the world and immovable and that the Earth moves, and also with a diurnal motion; for having disciples to whom you taught the same doctrine; for holding correspondence with certain mathematicians of Germany concerning the same; for having printed certain letters, entitled 'On the Sunspots', wherein you developed the same doctrine as true; and for replying to the objections from the Holy Scriptures, which from time to time were urged against it, by glossing the said Scriptures according to your own meaning: and whereas there was thereupon produced the copy of a document in the form of a letter, purporting to be written by you to one formerly your disciple, and in this divers propositions are set forth, following the position of Copernicus, which are contrary to the true sense and authority of Holy Scripture;

This Holy Tribunal being therefore of intention to proceed against the disorder and mischief thence resulting, which went on increasing to the prejudice of the Holy Faith, by command of His Holiness and of the Most Eminent Lords Cardinals of this supreme and universal Inquisition, the two propositions of the stability of the Sun and the motion of the Earth were by the theological Qualifiers qualified as follows:

The proposition that the Sun is the center of the world and does not move from its place is absurd and false philosophically and formally heretical, because it is expressly contrary to the Holy Scripture;

The proposition that the Earth is not the center of the world and immovable but that it moves, and also with a diurnal motion, is equally absurd and false philosophically and theologically considered at least erroneous in faith.

But whereas it was desired at that time to deal leniently with you, it was decreed at the Holy Congregation held before His Holiness on the twenty-fifth of February 1616, that his Eminence the Lord Cardinal Bellarmine should order you to abandon altogether the said false doctrine and, in the event of your refusal, that an injunction should be imposed upon you by the Commissary of the Holy Office to give up the said doctrine and not to teach it to others, not to defend it, nor even discuss it; and failing your acquiescence in this injunction, that you should be imprisoned. And in execution of this decree, on the following day, at the Palace, and in the presence of his Eminence, the said Lord Cardinal Bellarmine, after being gently admonished by the said Lord Cardinal, the command was enjoined upon you by the Father Commissary of the Holy Office of that time, before a notary and witnesses,[421] *that you were altogether to abandon the said false opinion and not in future to hold or defend or teach it in any way whatsoever, neither verbally nor in writing; and, upon your promising to obey, you were dismissed.*

And, in order that a doctrine so pernicious might be wholly rooted out and not insinuate itself further to the grave prejudice of Catholic truth, a decree was issued by the Holy Congregation of the Index prohibiting the books which treat of this doctrine and declaring the doctrine itself to be false and wholly contrary to the sacred and divine Scripture.

And whereas a book appeared here recently, printed last year at Florence, the title of which shows that you were the author, this title being: "Dialogue of Galileo Galilei on the Great World Systems"; and whereas the Holy Congregation was afterward informed that through the publication of the said book the false opinion of the motion of the Earth and the stability of the Sun was daily gaining ground, the said book was taken into careful consideration, and in it there was discovered a patent violation of the aforesaid injunction that had been imposed upon you, for in this book you have defended the said opinion previously condemned and to your face declared to be so, although in the said book you strive by various devices to produce the impression that you leave it undecided, and in express terms as probable: which, however, is a most grievous error, as an opinion can in no wise be probable which has been declared and defined to be contrary to divine Scripture.

Therefore by our order you were cited before this Holy Office, where, being examined upon your oath, you acknowledged the book to be written and published by you. You confessed that you began to write the said book about ten or twelve years ago, after the command had been imposed upon you as above; that you requested license to print it without,

421 There is agreement among historians that this claim was untrue.

however, intimating to those who granted you this license that you had been commanded not to hold, defend, or teach the doctrine in question in any way whatever.

You likewise confessed that the writing of the said book is in many places drawn up in such a form that the reader might fancy that the arguments brought forward on the false side are calculated by their cogency to compel conviction rather than to be easy of refutation, excusing yourself for having fallen into an error, as you alleged, so foreign to your intention, by the fact that you had written in dialogue and by the natural complacency that every man feels in regard to his own subtleties and in showing himself more clever than the generality of men in devising, even on behalf of false propositions, ingenious and plausible arguments.

And, a suitable term having been assigned to you to prepare your defence, you produced a certificate in the handwriting of his Eminence the Lord Cardinal Bellarmine, procured by you, as you asserted, in order to defend yourself against the calumnies of your enemies, who charged that you had abjured and had been punished by the Holy Office, in which certificate it is declared that you had not abjured and had not been punished but only that the declaration made by His Holiness and published by the Holy Congregation of the Index had been announced to you, wherein it is declared that the doctrine of the motion of the Earth and the stability of the Sun is contrary to the Holy Scriptures and therefore cannot be defended or held. And, as in this certificate there is no mention of the two articles of the injunction, namely, the order not "to teach" and "in any way", you represented that we ought to believe that in the course of fourteen or sixteen years you had lost all memory of them and that this was why you said nothing of the injunction when you requested permission to print your book. And all this you urged not by way of excuse for your error but that it might be set down to a vainglorious ambition rather than to malice. But this certificate produced by you in your defense has only aggravated your delinquency, since, although it is there stated that said opinion is contrary to Holy Scripture, you have nevertheless dared to discuss and defend it and to argue its probability; nor does the license artfully and cunningly extorted by you avail you anything, since you did not notify the command imposed upon you.

And whereas it appeared to us that you had not stated the full truth with regard to your intention, we thought it necessary to subject you to a rigorous examination[422] *at which (without prejudice, however, to the matters confessed by you and set forth as above with regard to your said intention) you answered like a good Catholic. Therefore, having seen and maturely considered the merits of this your cause, together with your confessions and excuses above-mentioned, and all that ought justly to be seen and considered, we hare arrived at the underwritten final sentence against you.*

Invoking, therefore, the most holy name of our Lord Jesus Christ and of His most glorious Mother, ever Virgin Mary, by this our final sentence, which sitting in judgment, with the counsel and advice of the Reverend Masters of sacred theology and Doctors of both Laws, our assessors, we deliver in these writings, in the cause and causes at present before us between the Magnificent Carlo Sinceri, Doctor of both Laws, Proctor Fiscal of this Holy Office, of the one part, and you Galileo Galilei, the defendant, here present, examined, tried, and confessed as shown above, of the other part:

We say, pronounce, sentence, and declare that you, the said Galileo, by reason of the matters adduced in trial, and by you confessed as above, have rendered yourself in the judgment of this Holy Office vehemently suspected of heresy, namely, of having believed and held the doctrine—which is false and contrary to the sacred and divine Scriptures that the Sun is the center of the world and does not move from east to west and that the Earth moves and is not the center of the world; and that an opinion may be held and defended as probable after it has been declared and defined to be contrary to the Holy Scripture; and that consequently you have incurred all the censures and penalties imposed and promulgated in the sacred canons and other constitutions, general and particular, against such delinquents. From which we are content that you be absolved, provided that, first, with a sincere heart and unfeigned faith, you abjure, curse, and detest before us the aforesaid errors and heresies and every other error and heresy contrary to the Catholic and Apostolic Roman Church in the form to be prescribed by us for you.

And, in order that this your grave and pernicious error and transgression may not remain altogether unpunished and that you may be more

[422] "Rigorous examination" implied torture. It is possible that, in Galileo's case, the inquisitors confined themselves to just showing him the instruments and threatening to use them.

cautious in the future and an example to others that they may abstain from similar delinquencies, we ordain that the book of the "Dialogue of Galileo Galilei" be prohibited by public edict.

We condemn you to the formal prison of this Holy Office during our pleasure, and by way of salutary penance we enjoin that for three years to come you repeat once a week the seven penitential Psalms. Reserving to ourselves liberty to moderate, commute, or take off, in whole or in part, the aforesaid penalties and penance.

And so we say, pronounce, sentence, declare, ordain, and reserve in this and in any other better way and form which we can and may rightfully employ.

Ita pronun.mus nos Cardinales infrascripti
F. Cardinalis de Asculo.
G. Cardinalis Bentivolus.
Fr. D. Cardinalis de Cremona.
Fr. Ant.s Cardinalis S. Honuphrii
B. Cardinalis Gipsius.
F. Cardinalis Verospius.
M. Cardinalis Ginettus.

Il Galileo fu abiurato mercordì mattina nel Convento della Minerva alla presenza di tutti i Cardinali della Cong., e gli abbruciorono in faccia il suo libro, dove tratta del moto della terra . . .

Galileo was made to abjure Wednesday morning in the Minerva Convent in the presence of all the Cardinals of the Congregation, and they burned before his eyes the book where he deals with the motion of the earth . . .

(Antonio Badelli, Rome, June 25, 1633)

Abjuration[423]

I, Galileo, son of the late Vincenzo Galilei, Florentine, aged seventy years, arraigned personally before this tribunal and kneeling before you, Most Eminent and Reverend Lord Cardinals Inquisitors-General

[423] G. de Santillana, *ibid.*, p. 312. In Italian: Edizione Nazionale, Vol. XIX, p. 406 and following.

against heretical pravity throughout the entire Christian commonwealth having before my eyes and touching with my hands the Holy Gospels, swear that I have always believed, do believe, and by God's help will in the future believe all that is had, preached, and taught by the Holy Catholic and Apostolic Church. But, whereas—after an injunction had been judicially intimated to me by this Holy Office to the effect that I must altogether abandon the false opinion that the Sun is the center of the world and immovable and that the Earth is not the center of the world and moves and that I must not hold, defend, or teach in any way whatsoever, verbally or in writing, the said false doctrine, and after it had been notified to me that the said doctrine was contrary to Holy Scripture—I wrote and printed a book in which I discuss this new doctrine already condemned and adduce arguments of great cogency in its favor without presenting any solution of these, I have been pronounced by the Holy Office to be vehemently suspected of heresy, that is to say, of having held and believed that the Sun is the center of the world and immovable and that the Earth is not the center and moves:

Therefore, desiring to remove from the minds of your Eminences, and of all faithful Christians, this vehement suspicion justly conceived against me, with sincere heart and unfeigned faith I abjure, curse, and detest the aforesaid errors and heresies and generally every other error, heresy, and sect whatsoever contrary to the Holy Church, and I swear that in future I will never again say or assert, verbally or in writing, anything that might furnish occasion for a similar suspicion regarding me; but, should I know any heretic or person suspected of heresy, I will denounce him to this Holy Office or to the Inquisitor or Ordinary of the place where I may be.

Further, I swear and promise to fulfil and observe in their integrity all penances that have been, or that shall be, imposed upon me by this Holy Office. And, in the event of my contravening (which God forbid!) any of these my promises and oaths, I submit myself to all the pains and penalties imposed and promulgated in the sacred canons and other constitutions, general and particular, against such delinquents.

So help me God and these His Holy Gospels, which I touch with my hands.

I, the said Galileo Galilei, have abjured, sworn, promised, and bound myself as above; and in witness of the truth thereof I have with my own hand subscribed the present document of my abjuration and recited it word for word at Rome, in the convent of the Minerva, this twenty-second day of June, 1633.

I, Galileo Galilei, have abjured as above with my own hand.

Galileo's "rehabilitation"

Over the last few decades, the Catholic Church appears to have undertaken a small campaign of "rehabilitation" of great thinkers of the past who were anathematized in their own time. It began with Galileo, as seemed natural, given that history and culture have consistently confirmed him as having been in the right. However, seeing that abjuration is a violent imposition, and as such is cause for shame and guilt on the part of those who impose it, it is the Church that should have been rehabilitated, not Galileo. And this rehabilitation of the Church should have consisted not so much of a condemnation of the judges of Galileo's time, as a denouncing of today's Church for committing the same mistakes that led to the prosecution of the scientist: in particular, more than its considering itself the "depositary of truth", its determination to impose its truth upon others.

However, the Church has limited itself to accepting Galileo's strictly scientific propositions, which today can no longer be disputed. It has certainly not accepted his experimental method and the use of reasoning in human life in general, nor the concept of a mutable truth, which is continually updated during the course of history, nor the exclusion of principles that are commonly accepted but have never been proved. All positions, these, which end up clashing with dogmatic creeds of any kind. Only when ways have been found to approach sociology, law, and politics with a scientific mentality, will Galileo's trial, as a rejection of his scientific philosophy, really come to an end.

But there is more. Galileo's rehabilitation by the Church appears to have been a propaganda exercise, as Antonio Beltrán points out. He writes:[424]

The outcome of this propaganda exercise is clearly apparent from the remarkable interest it has created in the media. This is its first "merit". The second consists in its having managed to give a value to the term 'rehabilitation". At such a distance of time, the majority of people certainly believe that Galileo must have done something wrong, and that the Church, showing magnanimity, has decided to pardon him.

And he adds[425]

Obviously, anything the Pope has to say about Galileo's science, or about science in general, is of absolutely no relevance. However, of all the contradictions present in this exercise, I wish to comment on just one, the most grotesque and dangerous of them. The fact that the Pope continues to consider himself an authority capable of saying something relevant about Galileo and his science shows that, on the Pope's side, nothing has changed. He is behaving in exactly the same manner as Galileo's judges, whose mistake he now recognizes.

Nevertheless, this *mea culpa* by the Church was not appreciated even by scientists who have a Catholic background. Among these, physicist Nicola Cabibbo expressed the following opinions: "If we re-examine the trial, we see Galileo was not convicted for what he said, but rather because he tried, somehow, to be a theologian". And even: "Galileo himself used to say, mistakenly: 'Since it is the earth that moves around the sun, we need to change the Holy Scriptures,' implying that when Newton discovered universal gravitation and Einstein relativity, we should have re-written the Bible on each occasion". And elsewhere:

The fact is that in the seventeenth century a series of discoveries began, an innovative process, that was not perceived by either part. Both sides

[424] A. Beltrán Marí, *Diálogo sobre los dos máximos sistemas del mundo*, Introducción, Alianza Editorial, Madrid, 1994, p. LXX.

[425] A. Beltrán Marí, *ibid.*, p. LXXIII.

reasoned within an obsolete logic scheme, which was unable to foresee the gap that was later to open between science and religion. The mistake, though, was made by both parties without malice, and in good faith.

These are groundless statements which, if not strongly rejected, cast a shadow of suspicion over Galileo and provide the Church with an alibi. Ambiguities are possible, and easily exploited. Galileo, therefore, has to be defended once more, today, no less than yesterday: a fate that perhaps will never leave him, or at least not until humankind has learned to appreciate the great beauty and superiority of reason.

So this is Galileo: the first true physicist in history, inventor of the critical analysis of phenomena, of verification by experiment, of the lucid and demystifying use of reason. The Anglo-Saxon world recognized this fact well before the Italian one: John Milton, returning from a visit to the (by then blind) scientist in Arcetri, wrote: "There it was that I found and visited the famous Galileo, grown old, a prisoner to the Inquisition, for thinking in Astronomy otherwise than the Franciscan (*sic*) and Dominican licencers thought".[426] Please note that Milton, who was a poet, not a scientist, says "Astronomy". He does not even hint at religious matters.

Had Galileo actually said or done any of the things he was charged with, it is unlikely he would have escaped being burned at the stake. All his writings show, instead, that he was extremely cautious, and that he went to great lengths to offer the ecclesiastics an interpretation of the Scriptures in historical terms, which would have allowed the two truths, faith and science, to live together. Two truths which, after all, were only such from the Church's point of view. They certainly were not so from Galileo's. The fact that the earth was to all effects immobile for observers located on it, as the authors and readers of the Holy Scriptures were, did not worry him in the least.

He was disturbed however, and in no small measure, that the theologians were in error more as a result of their hurry to make a definitive pronouncement on the matter than of their interpreting the

[426] *Aeropagitica. For the Liberty of Unlicenc'd Printing*, 1644.

Scriptures in an obtuse manner; he believed it would have been preferable for them to assume a position of sage detachment, or at least adopt an approach of "wait and see". In fact, he repeatedly invited them to be prudent, he himself being alarmed at the quicksands into which they were about to sink. In his letter to Father Castelli (Chapter 11), he warned them to be aware that, in choosing to make the immobility of the sun and earth a matter of faith, they ran the risk that one day in the future they might have to condemn as heretics those who declared the earth immobile and the sun in motion, if it should become possible, through physics and logic, to prove that the former is in motion and the latter is immobile. Pope John Paul II, in his "rehabilitation" speech, showed he had picked up on this particular point:

> the new science, with its methods and its freedom of research, obliged theologians to question their criteria for interpreting the Bible . . . The majority of them were unable to do so. Paradoxically Galileo, a sincere believer, proved himself more perspicacious on this issue than his theologian adversaries. The majority of theologians did not perceive the formal distinction that exists between the Holy Scripture in itself and its interpretation, and this led to them unduly transferring to the field of religious doctrine an issue which actually belongs to scientific research.

Returning to the issue which started this discussion, the clash between Galileo and the Church did not stem from the fact that "in the seventeenth century a series of discoveries began, an innovative process, that was not perceived by either part". The exact contrary is the case. Indeed, when the *Dialogue* was published, the Jesuits called the book more pernicious for the Church than the entire Reformation of Luther and Calvin. If the Church was undoubtedly more cautious in dealing with Galileo than with Bruno, Pucci, Campanella, and other philosophers, it was because in his case a system of ideas was involved which could have been proved to be correct (whereas philosophical theories are always open to discussion), a feature that many learned Jesuits did not fail to recognize.

The clash had its origin in the fact that both parties were aware of the magnitude of the change. On the one side, a man who was able to think in new ways and whose only mistake was to believe that the power of reason must, in the end, prevail. On the other side, an establishment, custodian of eternal truths, which in the face of the new could do nothing but systematically hinder its propagation. No clash would have taken place if only, on the Church's side, there had been men as ingenious as Galileo, that is, equally farsighted. The partial tolerance the church has recently shown can be attributed to the evidence, accumulated over the centuries, that faith is protected from the perils of reason: humans are far more easily ensnared by the comforting charm of the irrational, than attracted by the responsibility they must assume through the use of reason.

Persisting Misconceptions—Answers

1. Fall of bodies in a vacuum. Yes it is indeed true, because in the absence of friction forces all bodies fall with the same acceleration, that due to gravity, independently of their mass (and shape).

2. Fall of bodies in a vacuum. The greatest height attained by each of the two balls will be identical, for the same reason as that given in the previous answer: the gradual reduction in the upward speed is always and only due to the downward acceleration of gravity, which is equal for all bodies.

3. Laws of the pendulum. In the absence of friction, the small-oscillation period of the pendulum is independent of the mass of the attached weight (as well as of the oscillation amplitude). The swing speed of the two balls will therefore be identical.

4. Relationship between force and velocity. At all points in its trajectory the body is always subjected to its weight, including when it becomes stationary at the point where its direction of motion reverses. The acting force is therefore directed downwards.

5. Independence of motions. It will take the same time in each of the two cases, for the reason that the fall to the ground depends only on the acceleration of gravity, and is not influenced by any eventual horizontal displacement caused by the throw.

6. Principle of inertia. The arrow advances due to inertia, because once the throw is completed there is no propelling force acting on it. The motion of the arrow would continue indefinitely at the initial velocity if no braking forces were present, in this specific case friction due to air.

7. Atmospheric pressure. The outside surface of the piston is exposed to the air and is subject to atmospheric pressure, but this is not the case for its surface on the inside of the syringe. The force that needs to be applied is thus 1 kg weight for every square centimeter of the section of the piston.

8. Atmospheric pressure. The precise mechanism is that atmospheric pressure is acting on the liquid outside the straw, but not inside it, where the sucking creates at least a partial vacuum.

9. Atmospheric pressure. The atmosphere exercises a hydrostatic pressure on all surfaces which are in contact with it, regardless of their orientation. It therefore also acts on the underside of the scale's platform, thus canceling out its own contribution to the weight.

10. Hydrostatic pressure. Since here, too, the pressure acting on the small balloon is hydrostatic in character, the shape will remain spherical (but the greater the depth to which it had been taken, the more its radius would have shrunk).

11. Conservation of angular momentum. As a consequence of the process by which the planet was originally formed, the rotational motion of the earth is shared by all its parts, including those which are liquid and gaseous. In order to slow down the atmosphere and prevent it from rotating along with the solid earth, a force capable of reducing the speed of this individual part of the system would be required (in scientific terms: a force capable of changing the angular momentum of the system). This would only be possible if, in place of the vacuum existing beyond the atmosphere, there were some kind of immobile material against which friction would be produced.

12. Principle of relativity. The anchor will touch the sea's surface at the same distance from the boat in each case, because after it has been thrown it retains by inertia the speed it had when it was on board, meaning that the effects of the two throws differ in no way from those that would occur if the boat were stationary.

13. Principle of relativity. The reason is the same as for the case of the boat and anchor: all objects aboard a moving vehicle share its speed, so in the carriage the movement of each object in relation to the others does not differ from that occurring when the train is stationary.

14. Force and acceleration. By Newton's law, in the presence of acceleration (or deceleration) all bodies are subject to forces. If no real force is applied, one talks about "fictitious" or "apparent" forces. The forward push in the case of abrupt braking is of such a kind

(as is the centrifugal force which one experiences when undergoing circular motion).

15. Viscous motion. The weight of the parachutist is opposed by the resistance exercised by air against the parachute. The former of these two forces is constant, while the latter increases with increasing velocity. When the two forces become equal, the net resultant force is zero and by Newton's law the body stops accelerating and maintains the achieved speed (called terminal velocity).

16. Buoyancy. In a dense fluid a body falls at a constant speed, called the terminal velocity (the initial stage of acceleration is very short due to the high viscosity of the medium). The terminal velocity is proportional to the difference between the weight of the body and force of buoyancy: since the weight of lead is greater than that of marble, whereas buoyancy is the same in the two cases, lead sinks at a higher speed.

17. Diffuse reflection. If the moon were a perfect mirror it would reflect light only in one specific direction, thus appearing extremely bright from one well-defined observation angle, but dark from all other angles. Its rough surface, instead, produces diffuse reflection, that is, spreading in all directions. This also explains why all parts of the moon's disc appear to observers as equally bright.

Main Texts Quoted (updated for the English edition)

Baldo Ceolin, M., Editor, *Galileo e la scienza sperimentale*, publication of the Physics Department "Galileo Galilei" of the University of Padua.

—— Editor, *Galileo Scientist*, IV Galilean Centenary, Venice, 1992.

Banfi, A., *Galileo Galilei*, Ambrosiana, Milan, 1949.

Battistini, A., *Sidereus Nuncius, Introduzione*, Marsilio, Venice, 1993.

Bellone, E., "Galileo", *I grandi della scienza*, No. 1, *Le Scienze*, Milan, 1998.

Beltrán Marí, A., *Diálogo sobre los dos máximos sistemas del mundo*, Introducción, Alianza Editorial, Madrid, 1994.

Brecht, B., *Life of Galileo*.

Carugo, A. and Geymonat, L., *Discorsi e dimostrazioni matematiche intorno a due nuove scienze, Note*, Boringhieri, Turin, 1958.

Castellani, G., *Galileo ritrovato, Prometheus 14*, Franco Angeli, Milan, 1993.

Caverni, R., *Storia del metodo sperimentale in Italia*, Vol. 2, Florence, 1895.

Drake, S. *Discoveries and Opinions of Galileo*, Doubleday & Co., New York, 1957.

—— *Galileo: Pioneer Scientist*, University of Toronto Press, Toronto, 1990.

Duhem, P., *Études sur Léonard de Vinci*, Paris, 1913.

Favaro, A., Editor, *Pensieri, sentenze e motti di Galileo Galilei*, F.lli Fusi, Pavia, 1907.

—— Editor, *Edizione Nazionale delle opere di Galileo Voll. I–XX*, Barbèra, Florence, 1890–1907; reprinted in 1964–66.

Finocchiaro, M.A., *The Galileo Affair, A Documentary History*, University of California Press, Berkeley, CA, 1989.

Flora, F., *Antologia leonardesca*, Istituto Editoriale Cisalpino, Milan, 1947.

—— *Galileo e gli scienziati del Seicento*, Ricciardi Editore, Milan-Naples, 1953.

Fracastoro, G., *De sympathia et antipathia rerum*, 1584.

Frova, A., *Perché accade ciò che accade*, BUR Supersaggi, Milan, 1995.

Galati, D., *Galileo*, Pagoda Editrice, Rome, 1991.

Galilei, G., *Discorsi e dimostrazioni matematiche intorno a due nuove scienze*, Edizione Nazionale, Vol. VIII. English translation by H. Crew and A. de Salvio as *Two New Sciences*, Dover Publ, New York, 1954.

—— *Il Saggiatore*, Edizione Nazionale, Vol. VI. English translation by S. Drake and O'Malley as *The Assayer*, in *Controversy on the Comets of 1618*, Philadelphia, PA, 1960 (available only in photostatic form).

—— *Sidereus Nuncius*, Edizione Nazionale, Vol. III. English translation by A. Van Helden as *The Sidereal Messenger*, University of Chicago Press, Chicago, IL, 1989.

—— *Dialogo sopra i due massimi sistemi del mondo*, Edizione Nazionale, Vol. VII. English translation by M.A. Finocchiaro as *Galileo on the World Systems*, University of California Press, Berkeley, CA and Los Angeles, 1997.

—— *Dialogo sopra i due massimi sistemi del mondo*, Edizione Nazionale, Vol. VII. English translation by S. Drake as *Dialogue Concerning the Two Chief World Systems*, University of California Press, Berkeley, CA, 1967. Latest reprint by The Modern Library, New York, 2001.

Geymonat, L., *Galileo Galilei*, Einaudi, Turin, 1969. English translation by S. Drake as *Galileo Galilei: A Biography and Inquiry into His Philosophy of Science*, McGraw-Hill Book Co., New York, 1965.

Gliozzi, M., *Origine e sviluppi dell'esperienza torricelliana*, Turin, 1931.

Koyré, A., *Etudes Galiléennes*, Hermann, Paris, 1940. English translation by J. Mepham as *Galileo Studies*, Humanities Press, Atlantic Highlands, NJ, 1978.

Loria, G., *L'infinito e l'infinitesimo secondo i matematici moderni anteriori al secolo XVIII, Scientia 8*, 1916.

Marcolongo, R., *La meccanica di Galileo*, Milan, 1942.

Naylor, R., "Galileo and the problem of free fall", *British Journal of History of Science*, No. 7, 1974, pp. 105–34.

Redondi, P., *Galileo eretico*, Einaudi, Turin, 1983. English translation by R. Rosenthal as *Galileo Heretic*, Princeton University Press, Princeton, NJ, 1987.

Russo, L., *Segmenti e bastocini*, Feltrinelli, Milan, 1998.

Rufini, E., *Il "metodo" di Archimede e le origini dell'analisi infinitesimale nell' antichità*, Rome, 1926.

Salvini, G., "Mitologia e verità nella figurazione di Galileo. Il Galileo di Bertolt Brecht", *Saggi su Galileo Galilei*, Barbèra Editore, Florence, 1967.

Santillana de, G., *The Crime of Galileo*, The University of Chicago Press, Chicago, IL, 1955.

Settle, T.B., "An experiment in the history of science", *Science*, Vol. 133, 1961, p. 19.

Sharratt, M., *Galileo—Decisive Innovator*, Cambridge University Press, Cambridge, 1996.

Tannery, P., *Mémoires scientifiques*, Paris, 1926.

Various authors, papers in *Quaderni di Storia della Fisica*, n. 1, Editrice Compositori, Bologna, 1997.

Zentner, M.R. and Kagan, J., "Perception of music by infants", *Nature*, Vol. 383, 1996.

Additional Reading (updated for the English edition)

Barsanti, G., *Il nuovo universo e la riforma del sapere: antologia / Galileo Galilei*, Le Monnier, Florence, 1982.

Bellone, E., *Caos e armonia. Storia della fisica moderna e contemporanea*, Utet, Turin, 1990.

—— *Spazio e tempo nella nuova scienza*, La Nuova Italia Scientifica, Rome, 1994.

Beltrán Marí, A., *Galileo, El autor y su obra*, Ed. Barcanova, Barcelona, 1983.

Blackwell, R.J., *Galileo, Bellarmine and the Bible*, University of Notre Dame Press, Notre Dame, 1991.

Brunetti, F. and Geymonat, L., *Sensate esperienze e certe dimostrazioni*, Galilean anthology, Laterza, Bari, 1966.

Burstyn, H.L., *Galileo's Attempt to Prove that the Earth Moves*, Isis 53, 1962.

Carugo, A., *Galileo*, SEDI, Milan, 1978.

Camerota, M., *Galileo Galilei*, Salerno Editrice, Rome, 2004.

Crombie, A.C., *Augustine to Galileo*, Falcon Press, London, 1969.

del Lungo, I. and Favaro, A., *La prosa di Galileo per saggi criticamentre disposti*, new presentation by C. Luporini, Sansoni, Florence, 1958.

Drake, S., *Galileo* in *English Literature of the Seventeenth Century*, in: "Galileo, Man of Science", Editor E. McMullin, Basic Books, New York, 1968.

—— *Galileo at Work: His Scientific Biography*, The University of Chicago Press, Chicago, IL, 1978.

—— *Galileo*, Oxford University Press, Oxford, 1980.

—— and Drabkin, I.E., *Mechanics in Sixteenth Century Italy*, The University of Wisconsin Press, Madison, WI, 1969.

Fantoli, A., *Galileo per il Copernicanesimo e per la Chiesa*, Libreria Editrice Vaticana, Rome, 1992. English translation by G.V. Coyne as *Galileo for Copernicanism and for the Church*, Libreria Editrice Vaticana, Rome, 1994.

—— *Il caso Galileo*, RCS-Superbur Saggi, Milan, 2003 (English translation in progress).

Favaro, A., Editor, *Galileo Galilei: pensieri, motti e sentenze tratti dalla edizione nazionale delle opere*, G. Barbèra, Florence, 1949.

Field, J.V., *Cosmology in the Work of Galileo and Kepler*, Annali Istituto e Museo di Storia della Scienza, Editor P. Galluzzi, Barbèra, Florence, 1984.

Finocchiaro, M.A., *Galileo and the Art of Reasoning*, D. Reidel Publishing Co., Dordrecht-Boston, 1980.

—— *The Methodological Background to Galileo's Trial*, in: 'Reinterpreting Galileo', Editor W.A. Wallace, The Catholic University of America Press, Washington, 1986, pp. 241–72.

Galilei, G., *Discorsi e dimostrazioni matematiche intorno a due nuove scienze*, Edizione Nazionale, Vol. VIII. English translation by S. Drake as *Two New Sciences*, University of Wisconsin Press, Madison, WI, 1974.

Langford, J.J., *Galileo, Science and the Church*, University of Michigan Press, Ann Arbor, MI, 1966.

Lattis, J.M., *Between Copernicus and Galileo*, The University of Chicago Press, Chicago, 1994.

Maccagni, G., *Antologia galileiana*, G. Barbèra Editore, Florence, 1967.

Maccagni, C., Editor, *Saggi su Galileo Galilei*, Celebration on occasion of the fourth birth centenary, G. Barbèra Editore, Florence, 1972.

McMullin, E., Editor, *Galileo, Man of Science*, Basic Books, New York, 1968.

Montinari, M., *Sulla libertà della scienza e l'autorità delle Scritture/Galileo Galilei*, Theoria, Rome, 1983.

Pagano, S.M., Editor, *I documenti del processo di Galileo Galilei*, Pontificia Accademia delle Scienze, Rome, 1984.

Righini, G., "Contributo all'interpretazione scientifica dell'opera astronomica di Galileo", *Suppl. Annali Istituto e Museo di Storia della Scienza*, Florence, 1978.

Ronchi, V., *Il cannocchiale di Galileo e la scienza del Seicento*, Einaudi, Turin, 1958.

Rossi, P., *Il pensiero di Galileo Galilei, una antologia dagli scritti*, Loescher, Turin, 1959.

—— *Aspetti della rivoluzione scientifica*, Morano Editore, Naples, 1971.

—— *La nascita della scienza moderna in Europa*, Laterza, Bari, 1997.

Shea, W.R., *Galileo's Intellectual Revolution: Middle Period*, 1610–1632, Science History Publications, New York, 1972.

—— and Artigas, M., *Galileo in Rome*, Oxford University Press, New York, 2003.

Sobel, D., *Galileo's Daughter*, Penguin Books, New York, 1999.

Wallace, W.A., *Galileo and His Sources*, Princeton University Press, Princeton, NJ, 1984.

—— Editor, *Reinterpreting Galileo*, The Catholic University of America Press, Washington, 1986.

Subject and Name Index